Harvest of Dissent

Harvest of Dissent

Agrarianism in Nineteenth-Century New York

THOMAS SUMMERHILL

University of Illinois Press
URBANA AND CHICAGO

FIRST ILLINOIS PAPERBACK, 2008

Manufactured in the United States of America
∞ This book is printed on acid-free paper.
1 2 3 4 5 C P 5 4 3 2 1

The Library of Congress cataloged the cloth edition
as follows:

Harvest of dissent : agrarianism in nineteenth-century
New York / Thomas Summerhill.
p. cm.
Includes bibliographical references and index.
ISBN 0-250-02976-3 (cloth : alk. paper)
1. Farmers—New York (State)—Political activity—History—
19th century. 2. Agriculture—Political aspects—New York (State)—
History—19th Century. 3. Land reform—Political aspects—
New York (State)—History—19th Century. 4. New York (State)—
Politics and government—1775–1865. 5. New York (State)—
Politics and government—1865–1950. 6. New York (State)—
Economic conditions—19th century.
I. Title.
F123.S86 2005
974.7.03—dc22 2004023663

PAPERBACK ISBN 978-0-252-07547-6

For Sarah Jean, with love

Contents

Acknowledgments

In the process of completing this book, I have accumulated a number of scholarly debts and consider myself fortunate to have received so much support from friends and prestigious institutions. Several scholars at the University of California, San Diego, had a key role in helping me formulate this project at its earliest stages. Above all, I would like to express my gratitude to Steven Hahn. Whether in informal conversations or in his painstaking reading of successive drafts of the manuscript, Steve helped me meet the challenge of analyzing rural radicalism in a place few would expect to find it: the Northeast. His influence on my own work will be readily apparent to anyone familiar with the study of agrarian movements, even though my conclusions are different from his in significant ways. In the process, he has become a cherished friend, and I cannot thank him enough. Michael A. Bernstein taught me not only a great deal about how to practice history but also—and more important—to always ask new questions and never be afraid to pursue them. He, Steve, and I, transplanted New Yorkers in California, also had the pleasure of watching the Yankees build themselves back into a dynasty. Rachel N. Klein, Susan A. Davis, and Amy Bridges helped me come to terms with backcountry disorder, mobbing, and the arcane science of analyzing New York State politics through their own work and their reading of mine.

Welcome support came from several institutions. The Department of History at the University of California, San Diego, provided me with both a Regents Fellowship and a completion fellowship. The Smithsonian Institution early on granted me a graduate fellowship at the National Museum of American History that enabled me to conduct invaluable research in Washington, D.C., while having the singular pleasure of sharing ideas and many cups of coffee with other fellows and the staff. I would particularly like to thank Pete Daniel for taking interest in the project, providing input, and being a wonderful host. As well, I received an Albert J. Beveridge Grant from the American Historical Association, and Drake University and Michigan State University generously provided research funds for the revision of the manuscript. Last, my postdoctoral fellowship at the Yale Program in Agrarian Studies thoroughly reshaped my thinking on agrarian movements thanks to the yeasty weekly seminars, my fellow postdocs, and James C. Scott's gift for making others' relatively mundane insights into great ones. His ideas and his friendship have been truly exceptional. I cannot let slip the opportunity to state that he and I still carry the scars from sheep shearing on his farm in the spring of 1997.

This book relies heavily on archival sources, and I was able to construct it in such a way because of the gracious staffs of the libraries at which I conducted my research. A special thank you goes to Wayne Wright, Eileen O'Brien, and Kathleen Stocking at the New York State Historical Association; they bore the heaviest burden as I plowed through NYSHA's seemingly bottomless local history collection. Other individuals too numerous to mention at NYSHA, the Division of Manuscripts and University Archives at Cornell University Libraries, the New York State Library, the Delaware County Historical Association, the Schoharie County Historical Association, Arents Special Collections at Syracuse University Library, the American Antiquarian Society, and the Baker Library, Harvard Business School, offered their services with the utmost kindness and professionalism. And, the late New York State grange master, William Benson, opened previously unseen membership rolls that allowed me to develop a social profile of those who joined the Grange in the nineteenth century. I appreciate his gesture tremendously. Last, I am grateful to *New York History* for allowing me to use material first presented in that journal.

Of my friends and colleagues who read parts of this book, I would first like to thank Charles Radding and Maureen Flanagan. Chip helped me tear down sections of the book that were too large and, though he may not know it, helped me develop as a writer. And Maureen carefully worked through my revised manuscript line by line. Both helped me sharpen the argument and make the book more lively. I would also like to thank the readers for the University of Illinois Press, Stephanie McCurry and Robert D. Johnston, for their penetrating and helpful comments. My editor, Laurie Matheson, made pleasurable the arduous work of turning a manuscript into a book with her sharp eye and constant encouragement. She's the best. And for favors great and small along the way, I am grateful to Thomas Dublin, Alan Taylor, William Deverell, Susan Sleeper-Smith, Jeffrey Bleimeister, Gary Kulik, Jeannie Whayne, Harold Forsythe, Alan Rogers, Myron Marty, Patrick McConeghy, Susan O'Donovan, Jewel Spangler, Robert Bonner, John Haskell, Thomas Abercrombie, Roger Abrahams, Reeve Huston, William Christian, David Ortiz, Carl Dyke, Katie Boardman, Phil Deloria, Cathy Stock, Iver Bernstein, Nancy Hewitt, John Scott Strickland, Sally G. Kohlstedt, James R. Sharp, William Stinchcombe, Fred Adams, Kay Mansfield, K. Sivaramakrishnan, Larry Lohman, Witoon Permpongsacharoen, Juliette Konig, Cecilia Mendez, and, of course, Fionn MacCumhal. Research assistants Jennie Han, Sarah J. Haviland, and Timothy Boring deserve special thanks for a job well done. Claude MacMillan, David Austin, Robert Guinto, Holly Stevenson, Julie Carney, and Michael Eckman graciously hosted me along the way. Three friends and colleagues—Lewis Siegelbaum, David Bailey, and Mark Kornbluh—honored me with great kindness and understanding in a time of need. To them I am especially indebted.

Above all, I would like to thank my family for providing inspiration through their own intellectual pursuits and by the kind of example they set as women and men. My mother, Margot Summerhill, in particular, deserves special recognition here: she raised my sisters, Eleanor and Robin, and my brother, Michael, who has cerebral palsy, on her own. Each of our successes is very much hers, as well. Last, my wife, Sarah, helped me "find the words," and to her the book is dedicated with love, friendship, and admiration.

Harvest of Dissent

Introduction

My hill land is so cold and lean
It is the last to put on green
In the winter first to put on snow
In the spring the last to let it go.
—Will Christman

The Will Christman farm sat along the old Schoharie Turnpike, in the northeasternmost corner of Schoharie County where the Bozen Kill starts its crooked path to Normans Kill, which empties into the Hudson River twenty miles away. Neither Will nor his father, who passed the farm to him in the late nineteenth century, made much money from the land. But Will managed to raise nine children and see them all marry and have families of their own by the time he reached sixty. Will called the land he worked "Untillable Hills" because of the constant labor required to keep the upland fertile and free of stones. But with careful clearing and cultivation, he still scraped out a living selling beets, potatoes, beans, peas, and other crops in nearby villages in the 1930s, even though he confessed, "I'm not doing any more farming than I have to these days."[1]

When Will Christman wrote poetry, however, his love for farming and living in the countryside was readily apparent, even after sixty years of working the rocky dirt of the Helderberg Mountains. In simple verses, he wrote of the people, crops, and animals that had defined his life:

Hop fields are in clover,
Corn, and pasture grass;
Dancing days are over
For each youth and lass.
But one that was in denim
Wishes he were still
Picking hops in Blenheim
By Schoharie Kill.[2]

Though few New York farm men and women who lived in the nineteenth century could express themselves with the simple poetic elegance of Christman, his words capture well their own commitment to farming—hard and unremunerative as it often was in central New York. Like him, their days were filled with labor, raising children, attending church, helping neighbors, celebrating holidays, visiting family and friends, and politicking. And, like Will Christman, they believed farming provided them with a particularly close relationship with the land that purified them and made them better fathers, mothers, and neighbors. In this humble way, they also believed that tilling the land earned them citizenship; indeed, it embodied the most fundamental values of the Republic. It was this sensibility that led central New York farmers into politics time after time in the nineteenth century.

The narrative that follows brings farmer movements to life in one small corner of the nation—Delaware, Otsego, and Schoharie Counties in central New York. No effort is made here to claim that they are representative. What each shared, however, was a common legacy of agrarian dissent that spanned the period. As well, these counties had a common grounding in New York's unique manor system and underwent similar alterations in crop specialization, labor patterns, and family structure. And, because of the richness of archival sources, these counties also present an opportunity to bring into bold relief two often separate lines of inquiry: changes in the socioeconomic lives of individuals and their consequences in politics. The abundance of sources that brings to life the thoughts, ideas, and concerns of men and women like Will Christman has made it possible to let them tell their own stories. I have tried to let them do so whenever possible in order to reveal the texture and contingency of their experiences.

In particular, this book traces how farmers shifted their political and ideological positions and tactics to meet the challenge to rural society posed by the expansion of a national capitalist economy in the nineteenth century. I argue not only that central New York farmers voiced a consistent ideological challenge to the liberal individualism that drove this economic change but also that theirs was not a static, backward-looking stance. Rather, they constantly adapted and reconfigured a core set of beliefs to meet new circumstances. Nevertheless, they gradually abandoned party politics and espoused more conservative social views as the century progressed. These changes were not inevitable but were the product of farmer victories and defeats in the political, judicial, and economic arenas during the three periods covered in this book.

Studies of agrarian radicalism generally focus on the South, Midwest, or West, where the Populist revolt at the end of the century looms as the most successful—if conflicted—farmer movement in American history. But my purpose is to examine agrarian politics in a place where farmers had to interact earlier and more intensely with capitalist economic and social relations. New York, always at the forefront of the commercial and industrial development in the nation, was a natural place to conduct such an investigation and, no less significantly, offered an opportunity to witness the political interaction of farmers and other groups at both the early and latter stages of this process. The potential historiographic dividend of such an inquiry was high because agrarian radicalism clearly thrived in the heart of this economically and industrially advanced northern state, as voiced in the Anti-Rent movement, in the opposition to the construction of the Albany & Susquehanna Railroad, and in the Grange.[3]

Yet my goal is also to contribute to broader scholarly debates over the significance of local, regional, and national farmer movements to other historical developments—party realignments, economic cycles, social and cultural changes, immigration, industrialization. To accomplish this, I have attempted to reconstruct farmer notions of political economy—that is, the intersection of political and economic ideas—as they evolved during the century. Political economy is a particularly helpful concept to apply to agrarian movements because it offers insight into what people hoped to achieve through their actions rather than simply measures how they adjusted to the resolution of various conflicts. It allows us to appreciate the extent to which the rural men and women studied here historicized their own experiences as they struggled to create an enduring republic in the nineteenth century. Their urgent concern with the health—and fragility—of America's republican political and social order animated the movements studied here, finding voice in their letters, diaries, speeches, newspapers, and everyday decisions.[4]

While this is the primary purpose of the book, the process of writing it has led me to several conclusions that deserve mention. Above all, analyzing central New York farm movements made me appreciate the depth of commitment farmers had to republican government, individual liberty, and political democracy. I now realize the inherent radicalism of such a stance, whether then or now. Whatever less-than-laudable views agrarians evinced, and there were several, their refusal to accept the defeat of this set of principles kept their particular notion of political economy before the nation long after other groups had pushed them to the margins of political power. Second, farmers' devotion to political democracy had important consequences within the party system. While national and state party officials attempted to swing county leaders into line on major issues of the day, often as not local leaders cut their own deals between and among factions to ensure their own success. The interplay between these conflicting imperatives offered space for political insurgencies at the same time that it made them unlikely to succeed, thanks to manipulation, double-dealing, and local jealousies. Tip O'Neill's famous line that all politics are local resonates throughout this book.

As a scholar of the Civil War, I was at first at a loss to explain where that greatest event of the period covered here fitted into my tale. My ultimate conclusion surprised me: that for most central New York farmers, the war was important but relatively detached from the ongoing challenges of making a living from the land, a task that could not be interrupted and required virtually all of one's energies. Certainly many young men went off to war and came home changed, and those farmers who stayed at home benefited from high wartime prices for commodities, either intensifying their own labor or investing in machinery to meet demand. But in the end, they did not witness the massive disruption of farming seen in the South, nor the rapid economic expansion that occurred in the West. Theirs would be an abstract experience, unseen but felt in the centralization of economic and political decisions in Washington and in the impact of a more overtly national economy on crop prices, labor relations, and politics that I outline in the last three chapters.

This book, then, contributes to the literature on nineteenth-century farmer movements in several ways. First, it shows that farmer unrest was not an "episodic" response to immediate and discrete events. Rather, it played an ongoing role in the major historical events of the period. Most previous work on agrarian movements has assumed either that such movements were nostalgic and therefore condemned to failure or that farmers primarily organized to rid themselves of social, political, or economic structures that stood in the way of free competition in the marketplace.[5] Second, it depicts how farmers maneuvered within the broader political, economic, and social realms with a keen sense that they had a separate interest to promote and protect, one that in their minds was superior to alternatives and necessary to the health of the Republic. Third, it reveals that the trend toward nonelectoral farmer political action that emerged by 1900 was based not only on a sense of disillusionment with political parties (and their constituencies) but also on a recognition that democratic insurgency could not succeed in the most crucial arena for farmers—economics—thanks to the constitutional and political changes that occurred in the state and nation between 1845 and 1870. Central New York farmers were skeptical of capitalism after 1870 but, thanks to fears of social disorder, could not bring themselves to make alliances with radical movements such as Populism at the end of the century.[6]

And finally, unlike many earlier studies of agrarian movements, this book does not assume that farmers *ought* to have acted in concert. Rather, it sets out to examine divisions of class, ethnicity, crop specialization, and profit motive that made unified political action unlikely. When all of these are considered, the modest success central New York agrarian movements enjoyed, however fleeting, is all the more remarkable. The tenacious spirit of men like Will Christman had much to do with this and could be encapsulated in his explanation, at seventy-one, of why he walked twenty miles from his home to visit his son Henry in Albany: "I get tired behind a harrow and this is the way I rest—I walk."[7] Such people were not easily defeated. This is their story.

1. Republican Fathers: Capital Takes to the Woods, 1790–1825

> [A]s the Purchaser when he holds the Soil in fee sees a probability of making it his own, he therefore builds better Houses Barns and other Buildings clears his Lands in a better and more effectual manner attends to planting Orchards, and in fact looks up as a Man on record with more ambition than he that is settled on any other plan ever yet practiced.
>
> —William Cooper, 1790

Frontier warfare had driven both white and native inhabitants from central New York during the American Revolution. By 1790, the land had begun to revert to nature, giving the appearance of wilderness to the men and women who came to claim the land. They were but the latest wave of newcomers attracted by the land's potential for development. The first land grant in the region had been issued to British colonial official Nicholas Bayard, who secured a patent for the Schoharie valley in the 1690s. Other large grants followed. The Hardenburgh Patent of 1708 engrossed much of modern Greene, Ulster, Sullivan, and Delaware Counties, totalling either one or two million acres, depending on whether one accepted the east or the west branch of the Delaware as its border. Fine tracts of land along Schoharie Creek and at Cherry Valley fell into the hands of men like Lieutenant-Governor George Clarke, who in the years 1738–41 secured patents totalling 100,000 acres, including the Cherry Valley, Long, and Bradt's Four Tracts grants. Finally, Sir William Johnson negotiated the Treaty of Fort Stanwix in 1768, opening the land east of the Unadilla River and south of the Susquehanna to English settlement. Johnson, his military subordinates, and their Philadelphia business associates patented the remaining lands in the region. Johnson secured the rich Susquehanna bottomlands stretching south from Clarke's patents to the Pennsylvania border. By the outbreak of the Revolution, most land east of the Unadilla had been engrossed by great proprietors.[1]

The Revolution only partly challenged this pattern of landholding. When New Yorkers met to frame a state constitution after the war, democratic reformers argued that all Crown grants ought to be voided. The land would revert to the state, which then could take steps to ensure widespread opportunity for small investors to purchase plots. Under their plan, current occupants of the land—debtor-farmers and tenants—would be offered the right to purchase their homesteads before the land was placed on the market ("preemption"). Reformers had too little political and economic clout to accomplish this populist agenda. Indeed, the state's most prominent revolutionaries ranked among the landed elite and were in no mood to unsettle property rights. In the end, the state legislature voided only a few patents held by Tories, then proceeded to grant massive tracts in western and northern New York. The state therefore entered the early national period with a land system that many found undemocratic.[2]

When the great Yankee migration hit New York in the 1790s, frontier patents nonetheless filled with settlers. The impact on the state was profound. Whereas in the 1780s only three new counties had been formed, in the 1790s fourteen new counties—including Otsego (1791), Schoharie (1795), and Delaware (1797)—were organized, primarily in central New York. Villages along the Schoharie and Cherry Valley Creeks and a few scattered hamlets housed a sparse population in 1790. But by 1800, Otsego boasted 21,636 inhabitants, with another 10,228 in Delaware and 9,808 in Schoharie. Between 1800 and 1810, all three counties doubled in population. Expansion remained robust in the next decade, too: Otsego climbed from 38,802 to 44,856; Delaware from 20,303 to 26,587; and Schoharie from 18,945 to 23,154. Yankees made up the bulk of the arrivals, though Yorkers, Pennsylvanians, Jerseyites, Scots, Irish, Germans, and French also crowded into the region. Abandoned, burnt, and overgrown a generation before, central New York teemed with families by 1820.[3]

The inundation of central New York by pioneers raised the stakes in longstanding disputes over who actually held title to lands in the region. In the three counties, most patents had been granted to Tories. Lands held by Johnson, George Croghan, John Butler, William Franklin, and the Wallace family could have been seized and resold. But the state refused to vacate these patents in toto because the chain of title was uncertain (a problem compounded by a number of overlapping grants and unpatented gores). New York officials had long practiced the custom of granting tracts to paid proxies to circumvent Crown restrictions on the amount of acreage an individual could receive. When Tory owners fled, men who supported the patriots stepped forward to claim ownership of tracts they had signed for as proxies for other men. With

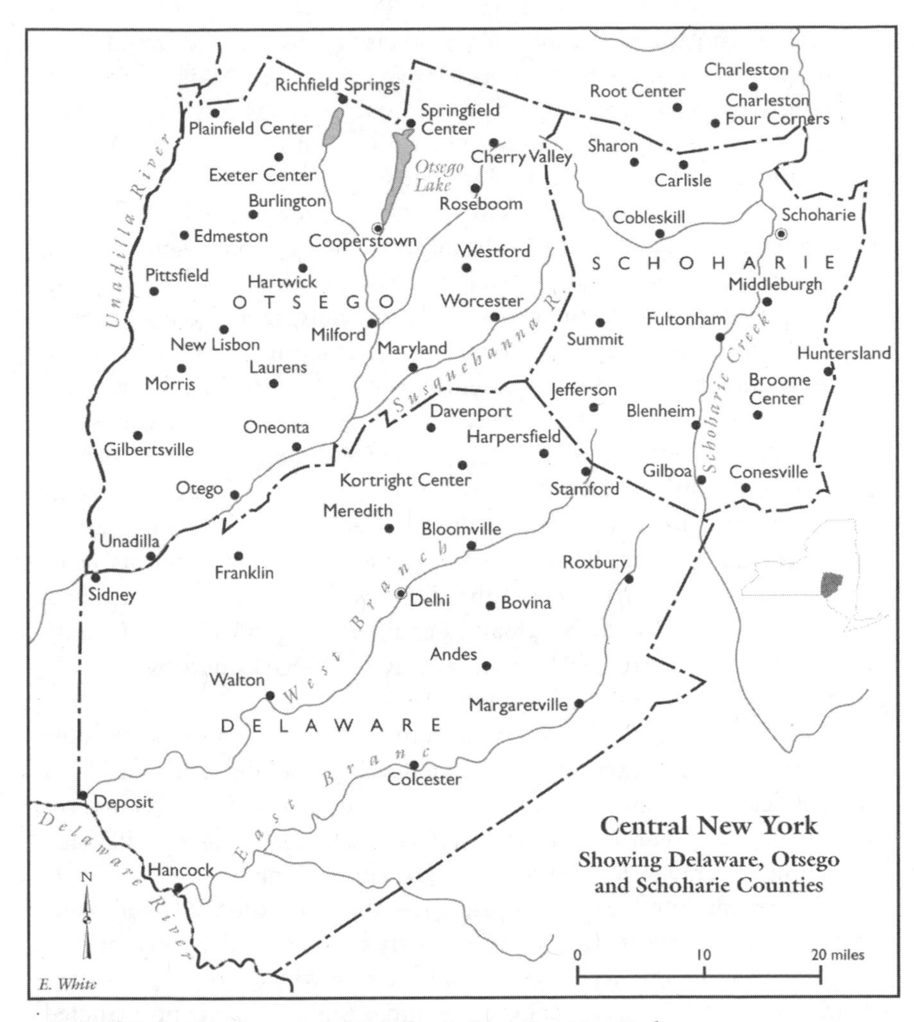

Map of Delaware, Otsego, and Schoharie Counties, New York.

slightly less skulduggery, creditors of the original patentees demanded title to Tory tracts as well. The Wallace, Croghan, and Butler Patents in Otsego and the Hardenburgh and Rapalje Patents in Delaware each became the objects of artful maneuvering by competing creditors. Further, local patriot leaders battled men they deemed "outsiders"—proprietors, proxies, and creditors—arguing that the people in the counties had borne the brunt of fighting against the Iroquois and British and thereby had earned the right to appropriate Tory holdings. Indeed, they had rallied support for the cause by promising settlers Iroquois lands if the Americans won the war.[4]

Creditors consistently outmaneuvered their foes because they could muster the greatest resources. Burlington, New Jersey, merchant William Cooper, for example, purchased a mortgage on Otsego County's 25,000-acre Croghan Patent—also claimed by Benjamin Franklin's exiled son, William—and secured title to the tract in court. More often, patents fell into the hands of creditors who had been members of New York's provincial elite. They lobbied the state legislature to deny calls for the redistribution of land, arguing that the Revolution had been fought to preserve property rights. Preemption or seizure of Tory lands by the state would deny mortgagors equal protection under the law. Goldsbrow Banyar, who as secretary of the province before the Revolution had overseen the patenting of lands and frequently demanded a share in exchange for approval of claims, surreptitiously amassed a 70,000-acre estate to add to his only official patent, the 4,000-acre Banyar tract in Cobleskill, Schoharie County. During the war, he declared himself neutral, keeping his business and political interests intact on both sides of the Atlantic. Afterward, Banyar's associates apprised him of the status of Tory land claims within the legislature. The inside information enabled him to lay claim to the Hardenburgh and Rapalje Patents in Delaware County and the Wallace, Stewart, and Schuyler Patents in Otsego County by purchasing mortgages on them at the right moment.[5]

Banyar also exploited the power vacuum of the immediate post-Revolutionary period. The government had not developed a fully articulated legal and political system, and competing interests within the state had yet to form strong alliances. Many elite families had conflicting land claims with each other and an array of smaller competitors. The ensuing flurry of title suits and countersuits hurt the large proprietors, as the anti-aristocratic sentiments of the people led juries to rule in favor of petty speculators. Banyar countered by negotiating a series of gentlemen's agreements with other proprietors to divide patents, avoid further court cases, and purloin the land unobstructed by the courts. Elites then defended each other's claims against interlopers, patentees, and settlers—with titles unclear, the word of two or more gentlemen proved decisive. So effective was the alliance among landlords in 1789–90 that land agent John Kiersted yearned for similar solidarity against Anti-Rent title challenges in 1848.[6]

Not all lands in central New York, of course, had been held by Tories and therefore did not change hands. Members of New York's great families—the Van Rensselaers, Morrises, Livingstons, and Schuylers—had sided with the patriots and thus retained grants in Delaware, Otsego, and Schoharie. Even when members of leading families had been disloyal, kin who supported the rebellion often kept their real estate from being seized. Jacob Morris, the son of Lewis Morris (who signed the Declaration of Independence), successfully lobbied the legislature to award him an Otsego County patent issued in 1769 to his Tory uncle, Staats Long Morris, as compensation for Bronx property destroyed by British soldiers. In a more controversial move, landlords convinced the legislature to affirm the legality of the leasehold system that dominated the Hudson valley in return for acceding to minor republican reforms: baronial courts, entail, and primogeniture were abolished. Since such practices had long since ceased, the legislature did little more than maintain the status quo. A special problem was posed by patents held by landlords such as George Clarke who claimed to be nonbelligerent Englishmen. Clarke petitioned the legislature in 1789 to confirm the patents issued to his great-uncle, the former lieutenant-governor. By not taking up arms in the rebellion, he pleaded, he had not forfeited his rights to the land. Nor, he asserted, could the legislature seize land owned by foreigners without endangering the property rights of citizens. The state ruled in Clarke's favor.[7]

The dominance of large landowners on the frontier was cemented by the fact that nearly all remaining patents were owned by patriotic merchants, and financiers from Philadelphia, New York, and Albany held some of the choicest land concentrated along the Susquehanna and Unadilla Rivers in Otsego and Delaware Counties. These men made little pretense to the social status of New York's great families, nor did they seek to become country squires like Cooper. Rather, they viewed the property as a financial investment to be disposed of quickly and profitably. The Philadelphians represented the vanguard of the nation's emerging capitalist elite—Thomas Wharton, Richard Smith, and the Gratz brothers. By developing the upper Susquehanna and Delaware valleys, they hoped to siphon trade from Albany and New York. William Walton, who had sent agents to the region before 1770 to identify choice lands, represented a set of New York merchants who hoped to tap the Delaware and Susquehanna trade for that city via new roads or canals that would connect the region to the Hudson. All wished to turn the wilderness into a commercial mart by selling land to freeholders.[8]

The number and size of grants to prominent local men paled in comparison to these large estates. Most were concentrated in Schoharie County, where the Borst, Lawyer, Bouck, Becker, and Richtmyer families—the descendants of Palatine Germans who had ventured into the valley in 1711—received small patents before the Revolution. John Harper, a Cherry Valley resident, garnered the impressive Harper Patent in Delaware County in 1769 in recognition of his military service, but he was unusual. The patents did not rival in value the grants made to Johnson and his friends in 1768, nor were local men cut in for a slice of lands along the Unadilla. The revolution in land ownership in central New York therefore was modest. Sizable areas of the hinterlands changed hands, but the frontier was not democratized to the degree many common folk had hoped. The main concession to land reformers was a preemption law applicable only to Tory estates. For pioneers who had weathered the Revolution and the turbulent economy of the 1780s, the peace dividend that many expected—rights to land—proved elusive.[9]

* * *

Though it is tempting to see pioneers as an undifferentiated swarm, in fact settlers arrived in central New York with widely different prospects. All viewed the land as the patrimony of the Republic to be distributed to its children. Rich and poor farmers came with the same general goal: to buy larger plots of land than they had at home, turn them into working farms, and have enough left over to pass to their children. Craftsmen wanted to practice in new villages with less competition or escape wage labor. The commercial boom of the 1790s made these risky ventures attractive, but wealth, not initiative, determined one's fate. The calculus was simple. Frontiersmen wanted valley land already cleared by the Iroquois or studded with oak, maple, and beech—the mark, in their minds, of good soil. Buying it was another issue. A few brought substantial savings and bought up promising lots, though such farmers usually invested in the richer soils of the Mohawk and Genessee valleys or Ohio. Those farmers who came to central New York with modest savings could purchase good land by mortgage. For poorer migrants, however, freeholding and good land were elusive. They had to mortgage or lease inferior lands, often in hill townships. To observers, it seemed that the better farmers swelled the valleys and dirt farmers roosted in the hills. Whether or not that was true is a moot point; opportunity in central New York came in proportion to a family's wealth.[10]

Virtually all newcomers, regardless of circumstances, faced a common challenge: to negotiate a sale, mortgage, or lease with some of the wealthiest, sharpest, and most successful businessmen in the nation. Adding to the task, each proprietor had a settlement strategy that depended on his or her financial goals, the relative value of the land, and an estimation of the quality of the pioneers themselves. If settlers were fortunate, they dealt with more tractable agents who could arrange terms that fitted their budget. The irregularity of this process created a patchwork of freeholds, leased lands, and unoccupied lots that varied from town to town and county to county.[11]

While proprietors adopted differing approaches, they shared the conviction that commerce was a civilizing force that would transform the continent. Mostly men of means, they considered themselves uniquely qualified to lead that project. Their plans for taming the woods incorporated some version of the following formula: invest capital in land; entice pioneers to clear the land on generous terms; set up credit and marketing structures; build roads, schools, and churches; and lure craftsmen and professionals to villages. At that point, a second wave of settlers would come, willing to pay more for farms. The proprietor made up for initial losses by selling improved land at a premium and providing essential services. Landowners wished to transform an estate in a generation, fearing that if land lay idle longer, pioneers would bypass central New York and take up lands farther west. Enticing worthy farmers to settle locally was the primary goal, for muscle, determination, and financial solvency easily collapsed during the five or six years it took to carve a farm out of the forest.[12]

Selling lots in fee proved the surest method for attracting good farmers, but it was only practical on some of the richest lands in the region. These filled first, mostly the Fort Stanwix grants along the Susquehanna and Unadilla Rivers or Cooper's lands around Otsego Lake. Croghan's creditors—Smith, Wharton, and Gratz—owned much of the choicest. Before the Revolution, they visited, surveyed, and mapped their property. Members of the New Jersey and Pennsylvania merchant elite, they had no inclination to become members of New York's landed aristocracy and wanted to sell quickly. They offered the land in fee simple on generous terms. A standard mortgage ran five years with no down payment. Like Cooper, their strategy rested on the belief that better farmers would arrive to take advantage of generous terms, temper the wild appearance of the land, and stimulate new arrivals. When Yankees poured into the area, these speculators sold the land fast with little added investment of time or money. In cases where a farmer could not meet his payments, land reverted to the mortgagor with a reasonable expectation of immediate resale. But fortune usually smiled on the sellers of prime land and those who could afford to buy it.[13]

Patents dominated by fee simple transfers did not always flourish, however. Where land was poor, proprietors had to be inventive and patient. Sales in fee predominated in the northern section of Schoharie County, much of Otsego's western and northern portions, and the Harper, Franklin, Wharton, and Walton Patents in the western half of Delaware. But hilly lots in these areas filled slowly, even with generous terms. The Wharton brothers, who owned several large tracts of land in Delaware and Otsego Counties, sold fine land in Otsego County's Otego Patent quickly and profitably. But they could not unload hill lots in Meredith, Delaware County, despite the unflagging efforts of their agent, Samuel A. Law, to attract Yankees to the prime grazing land. Law complained that immigrants would not take up lands, squatters ruined the soil and then demanded compensation for the hovels they constructed, timber thieves destroyed forests and drove off potential farmers, and legitimate settlers either refused to grow cash crops or failed in the attempt. They invariably turned to lumbering, often clandestinely making off with the best wood on the Whartons' "commons," as locals called unoccupied plots. As a result, Law leased lots on an annual basis for cash or shares while he awaited the arrival of a better class of farmers.[14]

Despite the benefits of fee simple sales for many owners, a substantial number leased their land. Central New York boasted several types of lease arrangements, depending on the long-term goals of individual proprietors. Several landlords settled patents in the 1790s by offering "durable" leases (perpetual and life), the foundation of manor tenancy in colonial New York. Perpetual leases entitled a tenant and his heirs to occupy the land forever, so long as he paid the landlord an annual rent, usually equivalent to 7 percent of the value of the farm. Perpetual leases had been granted since Dutch settlement and preserved the elevated position of the landed gentry in New York. Such leases existed on the Cherry Valley Patent in Otsego County; the Blenheim Patent in Schoharie County; the Hardenburgh, Kortright, Charlotte, and Goldsborough Patents in Delaware; and several other holdings. The terms of perpetual leases were fairly uniform in New York. Most landlords charged rents in kind—wheat—to ensure that farmers produced for the market, maintained the land's fertility, and raised the value of adjacent plots. Also, wheat rents shielded landlords from declines in real income caused by inflation. Rents ranged from five bushels of merchantable wheat per hundred acres on hill farms to twenty-five in the valleys, with clauses for several fowl and labor service with a team. Perpetual leases generally required rent payments to be delivered to the manor house. Most landlords waived rents for the first five years. Landlords charged cash rents after inflationary fears subsided in the 1790s or when they lived too far away to receive or store rents in kind. On the Kortright Patent in the townships of Davenport and Kortright, for example, the landlord offered 150-acre farms rent-free for five years, then charged sixpence sterling per acre annually. Tenants paid the taxes.[15]

At law, perpetual leases constituted conveyances in fee with certain reservations by the landlord, a contradiction that confounded the greatest legal minds of the nineteenth century, not to mention tenants. On the one hand, tenants owned title to the land in fee. State law granted them the status of freeholder because they paid highway taxes and served in the militia. In a civil sense, then, tenants enjoyed the same rights and performed the same duties as any other owner of property in fee simple. On the other hand, the contracts denied tenants full control of the property. Landlords retained all mineral, timber (except for fences and fuel), and milling rights, demanded a quarter of the proceeds of each sale of the lease, and could nullify a transfer by either rejecting an offer outright or agreeing to purchase the land for the same price. Contracts also stipulated that improvements be made to the land and buildings. Finally, the landlord reserved the right of distraint for the nonpayment of rent or violations of collateral clauses. This effectively nullified any claim that title had been transferred to tenants.[16]

It is revealing that both landlord and tenant considered the transfer of fee accomplished by the contracts as essentially fictitious. Rather than transferring land permanently, the parties established by contract a set of social relationships that placed the landlord in an advantaged position and gave him broad latitude in dictating the management of farms. In the decades to come, Anti-Rent tenants justifiably considered perpetual leases to be "feudal," especially clauses that required labor service. Their character as labor contracts was enhanced by the fact that the landlord usually agreed to pay a tenant for the value of improvements made to the farm at termination. From another perspective, annual rents could be construed as similar to interest payments, with tenant families permanent mortgagees working off a debt to the landlord, which, if not feudal, resembled peonage. Whether leases were designed to be land or labor contracts, the tenants' fictive (but legally defensible) titles provided incentive for continued good behavior and practices by farmers. One thing is certain: landlords who granted perpetual leases intended to become rentiers. They sought a fixed income from the land and designed a set of legal remedies to ensure an uninterrupted flow of cash into their hands. For tenants, leasehold property came at bargain prices, ensured occupancy, and could be sold. As the state assembly reported in 1840, rents in kind shielded tenants from market fluctuations and initially were desirable. Thus, land filled in the 1790s despite the tenuousness of tenant title to the land in perpetuity.[17]

Life leases, for their part, allowed landlords greater power over the development of estates. The aged Goldsbrow Banyar ultimately hoped to sell in fee simple his holdings in Otsego, Delaware, and Schoharie Counties. But in his mind, frontier prices rested below the value of the land he had liberated from its actual owners. He sold to pioneers reluctantly, believing that they made poor husbandmen who likely would fail financially or abandon the property. But they were necessary to clear fields, construct dwellings, cut roads, and provide customers for merchants, millers, and craftsmen and would only do so if they had a long-term interest in the property. In their wake would come better farmers. Hence Banyar fashioned a rough checkerboard system of lease and fee simple lots, renting about 60 percent. He preferred life leases, though he used perpetual and annual leases, too. Banyar's counterparts, the Livingstons and George Clarke, also offered life leases but rarely sold lots in fee. Life leases pervaded Banyar's estate; Livingston lands in the Scott Patent in Schoharie County; Livingston and Hardenburgh holdings on the Hardenburgh Patent and other tracts in Delaware County; Clarke's Long, Bradt's Four Tracts, and Cherry Valley Patents in Otsego, Schoharie, and Delaware Counties; and the Edmeston estate in Otsego.[18]

Life leases specified that a tenant, his heirs, or assigns could occupy a given farm until all of the individuals named in the contract died (one, two, or three people, depending on the custom of the landlord). The Livingston family divided their portion of the Hardenburgh Patent among nine heirs of Chancellor Livingston after the Revolution. They opted for one- or two-life leases and required payments to be made in kind (five bushels of wheat per 100 acres), with the first five years free to encourage settlement (one member of the family required one shilling per acre annually). Life leases resembled perpetual leases in certain aspects. They contained clauses that required the farmer and his family to make specific improvements to the farmstead, thus making up for the landlord's lack of income from rents in the value added by tenant labor. And wheat rents forced farmers to grub stumps out of fields, build fences to keep out livestock, and plow and fertilize the soil. Like perpetual leases, life leases were transferable and required tenants to pay all taxes. Since the land would revert to the owner, however, landlords enforced collateral clauses more diligently to prevent tenants from leaving a farm in dilapidated condition. Having collected rent tax-free and seen the land cleared at no cost, the landlord then sold it at full value.[19]

Shorter-term leases became more common after 1800 as rapid settlement helped increase land values and small holders let excess land to other farmers. One form of tenancy that emerged was share farming. Generally, share leases stipulated that the tenant supply all of the labor and half of the investment in seed and livestock to run the farm for a year, at which point the parties divided the produce, usually in half. Share tenants rented from both large and small holders.[20] Or a farmer could rent for a number of years for cash. Finally, a man might work a farm jointly with a family member or neighbor. Cash and share leases and partnerships were frequent within families, especially when parents retired or died and siblings chose one member to look after the home farm. Family agreements of this nature included clauses for the care of elderly parents. These different forms of land tenure existed side by side, and often farmers held land under two or more arrangements. Between 30 and 40 percent of farmers in the three counties leased their land in 1800.[21]

Some pioneers avoided altogether the formalities of paying for the land. Squatting had its own peculiar traditions. Squatters recognized that they had no claim to the land they occupied, but state law placed the burden of proving title on plaintiffs, and squatters took full advantage of the situation. Some squatters came solely to cut timber surreptitiously and then leave. Generally, however, a pioneer would clear a patch of land, build a shelter, and hope to remain undetected until he had made sufficient improvements to force the owner to negotiate a settlement. When a landlord or his agent arrived to evict the squatter, the pioneer often argued for a generous lease or mortgage—if he and his family had escaped undetected for several years, the farm might be profitable enough to meet the payments. If not, or if the landlord proved intransigent, the squatter usually agreed to accept cash payment for improvements to the farm and moved off. Landlords complained that they paid too much for the shoddy houses and outbuildings, but they preferred mild extortion to having to produce letters patent, deeds, and other documents to prove their claims. Proprietors consoled themselves that the most expensive labor required to start a farm had been performed and the land would be attractive to newcomers.[22]

* * *

The inequalities inherent in this land tenure system ran counter to the values of equality and individualism that had emerged from the Revolution, but that did not stop landlords from declaring it republican. The symbolic head of the New York aristocracy, Stephen Van Rensselaer III, for example, believed that perpetual leases created a partnership between capital and labor that would civilize the land, with the gentry setting an example of enlightened citizenship for farm families to emulate. Society would reflect a single interest, rich and poor would avoid class antagonisms, and republican paternalism would instill virtue in the people. The Revolution's emphasis on natural rights convinced other proprietors—such as William Cooper—to place greater confidence in the individual enterprise of the people. Such men believed that the children of the Revolution would make rational economic choices and move the countryside toward liberal individualism, so long as the gentry used its resources to offer credit, marketing, education, and cultural improvement to the people. The forests and valleys of central New York would foster a vibrant commercial civilization that was self-regulating and progressive. American agriculturalists would outstrip the downtrodden peasantry of England, France, and Holland in agricultural production, trade, and culture. Yet the colonial frontier had furnished evidence that the lower orders, if left to shift for themselves, would not live like civilized folk. Thus, early national landlords continued to monopolize trade, milling sites, and mining rights—even on freehold estates.[23]

The proprietors' sense that even a republican people required guidance translated into politics. Political power rested firmly in their hands in the 1790s, though most New Yorkers could vote. The landed elite supported the Federalist Party and shared its fear that unrestrained democracy would unsettle property rights. Shays' Rebellion in Massachusetts in 1786 made it clear that backcountry farmers were willing to assault contract law, credit, and the courts to promote agrarian interests, and Federalists fretted that the next step might be a call for land redistribution. At their urging, a constitutional convention met in Philadelphia to strengthen the federal government and ensure that neither the states nor the people could interfere with property rights or contract law. Federalists used the momentum of the ratification of the Constitution in 1789 to win control of states like New York. In a state in which deferential politics still held sway, Federalist proprietors and their handpicked candidates dominated elections in the 1790s. And the willingness of proprietors like Cooper to adopt a populist approach helped Federalists neutralize lingering agrarian resentment toward the landed elite.[24]

Many settlers, for their part, tolerated deferential politics but wanted little to do with the sustained effort needed to achieve the proprietors' economic vision. Much of their economic activity was devoted to exploiting abundant natural resources. Better-off pioneers set out to purchase bottomlands in fee simple because they could resell such farms for more than the value of the capital improvements. Prosperous farmers helped draw mechanics and traders to villages and stimulated the establishment of churches and schools to service the population. While even the best frontier farmer rarely lived up to the lofty cultural expectations of the gentry, at least valley townships seemed to thrive. Tenancy reigned in hill country, however, where farmers were either poor or did not believe they could recoup the cost of the labor needed to outfit a farm. Landlords hoped that with prompting, upcountry farmers would embrace the methods and crops of their lowland neighbors and become industrious citizens, but because returns on capital were low, tenants avoided intensive agriculture, seeking instead lands with ample timber to harvest, open forest to graze cattle or hogs, and good enough soil to grow corn. They resisted cropping wheat, making areas occupied by tenants appear scarred and primitive. Landlords blamed such failings on tenants, and unrest erupted when landlords attempted to evict delinquent or lethargic renters, enforce lease clauses, prevent timber theft, or demand rent in wheat.[25]

Though often at odds with proprietors, the pioneers transformed the landscape between 1790 and 1820. During the 1790s, the first settlers, after a season of hunger, sustained themselves by swiftly clearing timber from the land and growing corn. They rafted rough lumber to market or reduced the wood to potash for transport overland. Timber seemed limitless; eastern cities needed it for buildings and fuel, and it brought ready cash. As well, lumbering was a winter activity and did not take time away from growing food crops; indeed, lumber brought enough cash into the household for many settlers to eschew market farming. Merchants accepted lumber in barter transactions and hired young men to glide arks laden with wood, potash, maple sugar, and whiskey down the Delaware and Susquehanna. Many lumbermen braved the spring freshets to drift to Philadelphia or Baltimore to avoid having to sell to local merchants. Other men simply burned the wood where it fell. Great fires consumed the wood, leaving a rich fertilizer for corn, which yielded well in the combined vegetable humus and ashes. Logging bees brought together neighbors with their teams and allowed each family to have its fields cleared before crops needed to be planted. By using oxen, farmers could concentrate the timber in large heaps and easily scoop up the ashes to transport to an ashery. Lazier families simply girdled the trees and planted corn between the graying hulks. A skilled pioneer could clear three to four acres a year and still raise sufficient crops to keep his family, which meant that the profits from forestry could be enjoyed for much of the five-year grace period allowed by landlords.[26]

Forestry fitted well with the preferred agricultural rhythm of pioneers, wherein settlers minimized labor while extracting the bounty of the land. Locals referred to this as "skinning the land": they practiced extensive mixed farming, clearing virgin soil rather than fertilizing old fields, which then "lay to commons" on which they grazed sheep, cattle, or hogs. This was a conservative strategy that reflected pioneer concerns about falling into debt or facing starvation if a single crop failed. They produced commodities that brought high value for the labor invested, were not costly to transport, would not spoil (maple sugar, whiskey, pork, butter, cheese, hides, wool), or, like cattle, could walk to market. Farmers raised corn, rye, buckwheat, and barley (for malt) and planted orchards to make both sweet and hard cider for home use. Before 1810, they hunted game and fished the rivers for trout, shad, and bass to provide protein. Local women retted flax, carded wool, spun fiber into thread, and wove linen and wool into cloth, which was traded to merchants or sold to neighbors. Poorer folk often performed such tasks for wealthier neighbors. Since all of these products could be sold locally, farm families in the three counties achieved a level of community self-sufficiency during the first years of settlement, though few families grew rich from their endeavors.[27]

For proprietors, however, the society that emerged by 1810 hardly lived up to their vision of thriving agricultural communities centered around tidy market villages. Extensive agriculture destroyed rather than increased the value of lands. Proprietors and their agents therefore kept up pressure on farmers to enter cash crop production. Winter wheat was their first choice. It enjoyed ready markets in eastern cities and abroad, maintaining a high enough price to justify growing it in sequestered regions like central New York. Until wheat prices plummeted in 1812, landlord strategies appeared sound. Assured by proprietors that success would follow, farmers planted wheat in the second or third season after stumps had been grubbed laboriously from fields. The first several years of wheat brought bumper harvests with little labor, even on hillsides. Others promoted maple sugaring, sheep husbandry, or dairying.[28]

Proprietors took other steps to promote growth, investing in turnpikes to stimulate trade. Turnpikes required heavy capital expenditures, necessitating the formation of joint stock companies. The gentry, their agents, merchants, and wealthy farmers subscribed to stock, commenced construction, and hoped that enough road could be laid with the initial installment on the stock to fund its completion with toll revenues. Few farmers could afford to purchase stock; even fewer wanted to pay tolls. Thus, the roads tended to serve areas where proprietors held substantial tracts of land or market towns had produced a substantial merchant community. Many were built ahead of demand. By 1810, the three counties were crisscrossed by turnpikes that began in the Finger Lakes region and terminated at Kingston, Hudson, Catskill, and Albany on the Hudson River. Cherry Valley, Cooperstown, Schoharie, and Delhi all had service, and major routes wound through Livingston, Banyar, Clarke, Cooper, Wharton, Hardenburgh, Lansing, Edmeston, and Morris lands. Between investment in crops and internal improvements, land barons committed considerable capital to developing the future prosperity of the region.[29]

The gentry may have considered themselves an inspired, speculative commercial class at the time of the Revolution, yet they found themselves unable to behave according to purely rational economic principles as they confronted the settlers' own set of republican sensibilities. For settlers, the Revolution had promised freedom from subservience to the propertied elite; indeed, some had demanded leases in perpetuity to avoid the kinds of economic exploitation practiced by Hudson valley landlords before the War of Independence. Freedom included, in their estimation, access to the soil and to the value of the improvements on their homesteads.[30]

This notion of freedom was held so widely that few proprietors risked antagonizing tenants or debtor-farmers for fear of general unrest or charges of Toryism. Even their handpicked county sheriffs balked at evicting "honest" farmers. Under the circumstances, the gentry found the subtle pressure of social station, conspicuous display, and political rewards more effective in controlling their underlings than strict adherence to contract law. For their part, while struggling pioneers pursued economic gain, they also lived on the edge of disaster, a situation that made them seek paternalist protection of their economic interests in the land. They came to rely, therefore, on the good graces of the gentry to such an extent that at times they appeared to hold two diametrically opposed views: an urge to shield themselves from the market through paternalism's reciprocal obligations and an unbending pursuit of profit to escape dependency. A shared faith in republican society and the superiority of agricultural over industrial pursuits provided enough common ground to allow the two classes to reconcile most conflicts. Leniency by the landlord could be perceived as a gift from an advantaged republican father to a worthy child, while compliance with the gentry's exactions could appear to be a freely transacted social exchange between democratic partners mutually engaged in bettering the community.[31]

Another potential point of contention arose from the laboring classes' assumptions about republican equality, liberty, and virtue, which centered on the belief that the Revolution established the equality of white male heads of household. Each father represented his family in the public sphere and was responsible for the creation of a republican spirit within the home. By law, men held title to property, enjoyed preferential treatment in inheritance (they usually received land while women received equivalents in goods or cash), voted for the household, and had legal authority to direct their families in their work or bind children to labor outside of the home. Little evidence exists that men conceded authority to their wives, even in artisanal households where women plied crafts with their husbands. During the initial phases of clearing a homestead, men felled the forests, hunted and fished, and planted crops, while women kept a dairy, worked gardens, spun cloth, performed housework, and prepared meals. Ideally, a man provided the wherewithal to survive, even beyond his death, and the family labored at his behest to make sure the family's fortunes remained strong. Socially and religiously, too, patriarchal authority was recognized above all others. A man gained considerable prestige for himself and his family if he conducted himself in a morally upright, Christian manner and worked hard to improve his family's fortunes. So long as the gentry respected household autonomy and the fiction of male equality in the public sphere, relations between the classes remained cordial.[32]

Since resources were not equally distributed, however, the mutual dependency of households became the defining feature of frontier life. Men needed both to give and to receive gifts, credit, labor, and political support in order to navigate the hardships each family faced. An organic society emerged from the circumstances of settlement, with the wealthy occupying positions of authority, but in which the working classes, those who produced wealth, insisted on membership in the Republic. Denied full equality in an economic, social, and political sense by a lack of wealth and education, they instead measured virtue by hard work, humility, and success at providing for one's family and neighbors. Social respect had to be earned within the community, not through the accumulation of wealth, which corrupted men, but by proving one's skill and generosity.[33] Central New Yorkers participated in a rich array of community activities that solidified this notion of republicanism. "Bees" reinforced mutuality through the equitable exchange of labor. Neighbors helped clear land for newcomers; men brought their teams of horses or oxen, piled up the timber, and set it afire to make potash. Neighbors helped during harvests, built houses, washed sheep, and husked corn. At quilting bees, women worked together until after dark, when the young men of the area arrived to dance with the young ladies. Such labor exchanges reinforced the bonds of community. Mutuality found its way into social life as well, with pioneers taking special pains to see that equality was rigidly enforced in everything from schoolhouses to adolescent dances. While central New Yorkers considered personal autonomy and the opportunity to succeed the foundation for freedom, they took it as a given that in a republic, individual aspirations had to subserve to the common good.[34]

In the realm of property rights, pioneers adhered to similar principles because of tradition, hardship, and republican zeal. During the early national period, English common law doctrines still influenced property rights in the state, and these tended to privilege community over individual rights. For example, until 1846 the state constitution left fence law to the counties. In each county, farmers were required to build fences around their fields strong enough to keep out livestock. Unfenced fields, meadows, and woods were considered commons, and cattle, hogs, and sheep roamed freely, with only an earmark or brand identifying who owned them. In short, the open range system was considered a "customary right" that all property holders respected. Indeed, fines could be levied against farmers who did not adequately fence out strays. The open range allowed even the landless to raise a few hogs or cattle for meat, dairy products, or wool. But common law cut both ways. If livestock broke through a fence deemed sufficient by local inspectors, the owner would be fined or lose the animals. And unmarked livestock were considered a nuisance. Any member of the community who found one roaming would be awarded ownership if it went unclaimed for a short period. Owners also were required to place rings in the noses of hogs, the worst offenders, to keep them from burrowing under fences. This system worked well on the frontier, where hardship encouraged individuals to share resources. As well, backcountry notions of republicanism emphasized the "right" of citizens to care for themselves and their families. The open range seemed laudably democratic.[35]

Not protected by law were a set of other "rights" to the "commons" that residents claimed as members of republican society. Most irritating to landlords was stealing timber, or "hooking." The term originated in England where poor or elderly citizens enjoyed the right to pull dead branches—literally with a hook—from the king's forests to fuel their hearths. In early central New York, "hooking" meant harvesting trees on unoccupied lots with abandon, but the continued use of the word suggests that locals believed that custom dictated that they could do so. All strata of society participated in the theft, making it difficult to stop. Landlords rejected the notion that it was their duty to open their valuable timberlands to all comers and were even more alarmed at the way that timber thieves cloaked their actions in the language of republicanism, natural rights, and the labor theory of value. In *The Chainbearer,* novelist James Fenimore Cooper voiced well the people's sensibilities even as he decried them. In one scene, Aaron Thousandacres, a squatter on the Littlepage family's Mooseridge estate, explained that he had a moral right to exploit any unoccupied land: "I begin at the beginnin', when man was first put in possession of the 'arth, to till, to dig, and to cut saw-logs, and to make lumber, jist as it suited his wants and inclinations." These were natural rights "accordin' to the law of God, though not accordin' to the laws of man," that nullified any paper titles landlords might hold.[36] Cooper perceived that the boundary between custom and law was sufficiently blurred on the frontier that people could argue that title by possession and improvement constituted "natural law."[37]

To settlers, customary rights to water, timber, hunting, fishing, and pasturage on commons had to override the statutory claims of landlords if families hoped to survive. Squatting was similar in their minds. All could be passed off as "not hurting anyone," especially if a wealthy absentee landholder were the injured party. Not paying rent was a more complex issue but similar at heart. Farmers believed that landlords had a moral duty to extend leniency to hardworking, honest families. What separated these activities from theft, in their minds, was a reciprocal exchange: landlords gained the value of improvements, political support, and profits from rents, mortgages, and marketing, while farm families earned a competence.[38]

Underlying tension between proprietors and farmers did not lead to the kinds of organized agrarian resistance seen in states like Maine, Pennsylvania, or South Carolina in the 1790s, however, because the wealthy continued their role as republican fathers. William Cooper officiated at weddings, drank heartily with the common folk, and took part in their rough pastimes. He comfortably blended the arts of paternalism with an affinity for the people. Other families cultivated a refined image. Tenants on Stephen Van Rensselaer III's estate, Rensselaerwyck, called him "the Good Patroon" because he built schools, roads, churches, and mills to service the towns of his manor in order to attract and retain worthy settlers. The pioneers demonstrated their acceptance of the ministrations of the elite by relying on them to settle conflicts outside the courts. They might ask a gentleman to evict an individual who proved to be unpleasant, dishonest, or intemperate in order to preserve neighborhood harmony. Such entreaties usually took the form of petitions designed to demonstrate the unanimous feeling of the people. But frontier families also demanded that landlords and their subordinates act honestly and not tread too closely the line between enlightened self-interest and exploitation of the poor. When difficulties between the common folk and agents of landowners arose, local elites had to move quickly to stem controversy, or it could rapidly escalate. If a miller acted in bad faith, for example, the economic welfare of the community could be threatened, and, worse, tenants or mortgagees might withhold payments to punish the landlord for backing a dishonest or incompetent man. A good patron mediated these disputes quickly.[39]

And though landlords periodically attempted to prevent the destruction of forests on unoccupied lands, they generally tolerated such depredations because all members of society needed lumber for buildings and fences, firewood for heating and cooking, and wood to make everything from wagons to hand tools. As maddening as timber theft was for landlords, it constituted a small sacrifice for them while it enabled many others to make ends meet. If landlords failed to respect such claims, the folk resisted by burning timber, failing to pay rent, switching political allegiances, or, from time to time, threatening the landlord and his agents with harm.[40]

Nor were central New York frontiersmen politically passive. As in backcountry Maine, Pennsylvania, and South Carolina in the 1790s, agrarians feared that Federalism posed an imminent threat to democracy and a republic of opportunity. Federalist sympathy for monarchical Britain in its war against republican France, the Adams administration's buildup of the navy, the XYZ affair, the suppression of the Whiskey Rebellion, and the Alien and Sedition Acts convinced many voters already chafing at the stranglehold of landlords over political appointments at both the local and state levels that the Federalists meant to quash political dissent and impose the wishes of the wealthy few on the mass of citizens. Feeding on this discontent and preaching the virtues of an agrarian republic, the Jeffersonian Republicans waged a successful campaign in New York in 1800, with victories throughout the state that crippled the Federalist Party. They called for smaller government, decentralized power, greater democracy in politics, and public education, all issues that played well in central New York. Fittingly, Daniel Shays, whose rebellion in Massachusetts prompted calls for the constitutional convention of 1786, alighted in the mountainous township of Broome, Schoharie County, at this time to live his final days among the outliers.[41]

* * *

But disaster lurked behind the rapid influx of farm families, deceptively large harvests, and high wheat prices occasioned by the Napoleonic wars. The process of clearing the land created an environmental crisis that threatened the financial fortunes of both the gentry and common folk. The denuding of the forests exposed hillsides to erosion that was compounded by overplanting of wheat, rye, and oats. Wolves, bears, foxes, cougars, and deer were slaughtered at will, as were grouse, pigeons, ducks, turkeys, and geese. Even unpalatable crows, which multiplied thanks to the extensive planting of corn and grain, became the target of watchful farmers who collected town bounties for each felled bird. Fish numbers dwindled. Shad runs on the Susquehanna ended before 1820, victims of milling south of Cooperstown and the labyrinth of weirs and nets constructed all along the waterway from Maryland to central New York. Trout all but disappeared from local streams (killed by silting and mill waste), and the Otsego Lake delicacy, whitefish, had to be regulated to preserve a breeding population. Finally, the majestic stands of oak, elm, maple, and beech had been reduced to a fraction of acreage in the counties. After 1820, the state legislature took steps to end game hunting. Such acts were made in the breach. Game, fish, and forest had been nearly wiped out in central New York.[42]

Extensive agriculture masked the effects of continued cropping and deforestation because new fields could be cleared for several years to take full advantage of the nutrients in the forest humus and ashes. But ominous signs appeared by 1820. First, farmers could not climb out of debt, despite the region's skyrocketing output of grains, maple sugar, butter, wool, and timber. Proprietors and merchants reasoned that profligacy among pioneers lay at the heart of the crisis and responded in a variety of ways. William Cooper tried persuasion, only to end up in bankruptcy. Those who retained their estates had to work doubly hard to make a living from them. George Clarke tried to force farmers to practice sounder agricultural methods by inserting lease clauses that required that all manure be used on fields. Others advocated raising sheep to reverse the environmental disruption and stimulate home manufacturing in the wake of Jefferson's embargo, which began in 1807 in response to British impressment of American sailors on the high seas. Some landlords, merchants, and millers opened carding and fulling mills to further encourage wool production and household manufacturing, while entrepreneurs began making spinning wheels, looms, and other equipment needed to process wool and weave cloth. Others concluded that the constant cutting of lumber and its seasonal rhythms of drinking and traveling undercut the morality and diligence of farmers. These landlords stepped up patrols of their land to make sure timber was not being sold off leased property or stolen from unoccupied lands.[43]

Settlers for their part reemphasized safety-first agriculture, which placed them again at odds with proprietors. Farmers responded to the decline of wheat by turning to dairying or sheep grazing, which enabled them to avoid intensive agriculture. But they did not take steps to improve the quality of their livestock, nor did they concern themselves with the strong flavor of butter and cheese made with milk from cows that browsed on everything from wild leeks to tree branches. They also turned to hardier, less marketable grains than wheat (rye, buckwheat, oats, corn) as they cleared more and more acreage to make ends meet. But such crops brought modest returns, leaving freeholders in debt and tenants in arrears.[44]

The frustration felt by both proprietors and farmers found its way into politics. During the disastrous embargo in 1810, state Jeffersonian leader DeWitt Clinton launched an assault on the "unrepublican" and "Tory" manor system in an effort to discredit the Federalists. He singled out George Clarke, whose British birth made him an inviting target; indeed, the invective leveled at Clarke was so vicious that Clarke challenged Clinton to a duel. At Clinton's urging, the state legislature investigated Clarke's title in 1811 and recommended that it be vacated. Though no law to that effect was passed, it was clear that populist politicians had few qualms about tapping agrarian discontent to win elections.[45]

A sustained attack on the manor system was not forthcoming, however, because attention shifted to the debate over the construction of Clinton's pet project, the Erie Canal. The proposed canal stirred controversy from the start. Politicians in places like central New York that would be bypassed by the waterway feared that their counties would suffer economic, social, and political decline as the strip along the canal prospered. Also, New York farmers would be exposed to western competition, a troubling prospect given the perceptible decrease in local soil fertility and crop yields. Some backcountry politicians voted in favor of canal construction after Clinton promised to support internal improvements elsewhere in the state, but many voters feared that these would come too late to save places like central New York from eclipse.[46]

The Erie Canal also reopened the issue of the proper structure and role of republican government. While Clinton adhered to Jeffersonian doctrines and had done more than any other politician to defeat Federalism in the state, his advocacy of the canal created concern within his party about his commitment to limited government. By 1815, when the canal issue was before the people, Federalism was confined to New York City merchants and financiers and the landed aristocracy. These men supported the canal. The state would fund the canal and create a separate canal board to run its affairs, which to many Jeffersonians raised the specter of an enlargement of the government, higher taxes, and few checks on the power of the canal commissioners. Their constituents in backcountry areas were restive, believing their representatives had sold them out for political emoluments. And many were anxious that the already dominant Clintonian faction would use the enormous patronage of the canal to institute single-party rule.[47]

There was considerable legitimacy to this view, for Clinton exercised enormous influence over the workings of state government. Clinton had risen to power by controlling New York's unique Council of Appointments—consisting of the governor and four senators—which appointed civil and military officials down to the county level. Clinton used these "spoils" to vanquish the Federalists and build a loyal cohort of Clintonian politicians across the state. Thus in 1817, Martin Van Buren, who had joined in Clinton's attack on the manor system, led an insurgency within the Jeffersonian party to oust Clinton as the gubernatorial candidate. Labeling themselves "Bucktail Republicans," they contended that the Clintonians had monopolized political office in the state and that loyalty to Clinton, not the people, determined which politicians received patronage. They called on nominating conventions to support "measures and not men" to allow all citizens, not just those in favored counties, to have a voice in government. Clinton won the election and initiated construction on the canal, and many jilted politicians, farmers, and working men who lived away from the canal rallied around Bucktail calls for democratic reforms. They demanded a constitutional convention in 1821 that would end the political "monopoly" of the privileged few over the many.[48]

The new state constitution met Bucktail demands by widening suffrage to include all white men over the age of twenty-one, eliminating the Council of Appointments, and weakening the power of the governor and the state legislature by making county and militia offices elective. While scholars of the period generally argue that Van Buren advocated these reforms as preconditions for creating a mass political party—the Jacksonian Democrats—they have not discussed two important issues that animated the constitutional debates that are relevant to this study: localism and agrarianism. Bucktail delegates argued that the centralization of political power that had occurred under the 1777 state constitution was suitable for an undeveloped agricultural state but that the concentrated financial wealth that came with the commercial revolution left institutions like the Council of Appointments, or frankly the legislature, vulnerable to usurpation by wealthy or influential men. The Bucktail solution was to locate political power in county government while the state government would act "negatively" to balance competing local interests. In this commonwealth system, democratic mobilization at the township level would counterbalance the influence of the powerful. Widened suffrage would strengthen the people's ability to forestall the centralization of governmental institutions. Localism therefore was linked to democracy in the minds of Van Buren's followers.[49]

Equally significant, the convention delegates vied to seize agrarianism as a political force. The debate revealed two separate notions of the place of farmers in the political economy. The Clintonians argued that restricted suffrage advanced the interests of farmers: if freeholders alone could vote, then farmers would be able to fulfill their traditional role as counterweights to the designs of would-be aristocrats. This orthodox Jeffersonianism was skillfully countered by Erastus Root of Delaware County, who contended that the agricultural interest needed an active voice in government. He warned that under the Council of Appointments, mercantile interests in the cities could control appointments in rural districts. For Root, universal suffrage and the expansion of the number of elective offices would encourage voters in townships and counties to band together to promote their interests. Only then could rural districts battle canal towns for political power. Root even called for black suffrage and workingmen's rights to ensure that the rights of the people would be preserved against the growing moneyed interest.[50]

This political transformation coincided with uncertain economic times in central New York from 1819 to 1825. Farmers could not compete with those farmers who had access to the canal, which caused them to abandon wheat production—what had been the only high-value crop grown in the area. And with most of the good land in Delaware, Otsego, and Schoharie Counties already taken and limited prospects for making a living off remaining upland lots, migration to the area declined precipitously. Consequently, proprietors who had borrowed against the value of their land to finance development in the region now could not count on income to make good their debts. As they had done periodically before, many in the mid-1820s sought to collect on arrears, evicting or foreclosing on those most deeply in debt to frighten others to settle accounts. At the same time, the retainers of the old aristocracy—attorneys, merchants, and millers who had accumulated enough property to rent land of their own—endeavored to take the place of the passing elites but hoped to place estates on a more rational footing by charging cash rents and tightening constraints on "customary rights." Settlers for their part demanded leniency, bitterly rejecting landlord demands that rents be paid in wheat and begging to have payments commuted to cash. Others sought to sell their farms but found few to take them at prices high enough to pay off debts. Some attempted to sell leases but balked when landlords refused to pay them for the value of improvements to their holdings. On the Hardenburgh Patent, Clarke's patents in Otsego County, and elsewhere, tenants resisted landlord efforts to evict delinquents, institute greater labor discipline, enforce rent collections, or prevent timber cutting or milling without permission. By the 1820s, a chagrined James Fenimore Cooper would complain that the people no longer respected the great families. However true this might be, the two classes were stuck with each other; the countryside was frozen in debt, and the only way out was to work together.

Thus in 1825, Delaware, Otsego, and Schoharie Counties stood at a crossroads. Proprietors and farmers, in separate ways, believed that the situation could be salvaged. The gentry hoped to reinvigorate the agricultural economy by improving methods and the efficiency of farmers, while farmers sought to pare away the cost of land rents, credit, and government to increase returns. The latter would flock to the Jacksonian banner in 1828 in hope of shoring up the mixed agricultural economy with democracy, localism, and a commitment to small government. Yet events elsewhere in the state would soon challenge Jacksonian dominance and rupture, finally, the social order that had been born on New York's frontier. A political struggle ensued that challenged Jacksonian rule and dashed hopes of a representative government that would not side with elite economic interests. Yet the nature of the political struggle was more complex and more rooted in the particular land system that existed in central New York than even excellent recent studies have appreciated, as the next two chapters will reveal.

2. Democratic Children: Farmers and the Jacksonians, 1825–40

Now to the West we will repair
And wash our face in sweat and tears,
With savages to take our fare,
With catamount, hedgehog and bear,
More human far than Banyar.
—Otsego County anti-proprietor poem, 1828

After 1821, central New York farmers evinced optimism that an agrarian republic was within their grasp, despite the lingering influence of the great proprietors over their lives and evidence that the Erie Canal was shifting economic, social, and political strength within the state. They sensed that the balance of power between themselves and the gentry had switched in their favor. Widened suffrage gave average citizens the ability to effect reforms, and proprietors had to pay heed lest agrarianism become politicized. But farmers did not pursue such solutions because, in many ways, they benefited from the current system—paternalism shielded them from competition or displacement in the rapidly expanding commercial economy of the Jacksonian years. Thus in 1828, an anonymous debtor-farmer on Goldsbrow Banyar's land in Otego, Otsego County, implored the proprietor to grant farmers leniency in paying their mortgages out of a sense of Christian humanity and republican duty. He did not demand land reform but instead sought a return to the paternalist status quo.[1] Delaware, Otsego, and Schoharie County farmers found the economic transformation underway in the rest of the state and the nation more problematic but put faith in the Jacksonian Democrats to check a monopoly of wealth and power among the commercial and industrial elite. This chapter will examine the political economy of the region during the Jacksonian years and how its disintegration under the stresses of the panic of 1837 opened the way for the Anti-Rent movement.

The farm population of the three counties in 1825 had weathered the difficulties of clearing the forest, and though many had failed or moved on, a substantial number had brought their farms to maturity. They had survived through a mixture of skill, diligence, parsimony, and good fortune, creating a world of which they and others were proud. In his 1824 gazetteer of the state, Horatio G. Spafford celebrated the "frugal and industrious" farm people of Otsego County who lived within their means and were "peculiarly plain in their dress, almost wholly the product of the household wheel and loom."[2] Spafford praised the residents of the unexceptional township of Westford because they bred good stock, used gypsum on their land, and made their own sugar and cloth. Their labors revealed their commitment to an agrarian republic: "In no County of this State are there better farmers, nor more honorable examples of this kind of independence. Such men as these are the real back-bone of vital republicanism; their habits, the nerves and sinews; and the surplus, produced by their labors by their frugality and economy from family consumption, forms and fills the arteries and the veins of the community and the Republic." Spafford lauded these farmers for balancing agriculture as a livelihood and as an expression of republican virtue. Waxing philosophic, he cautioned other farmers to follow Otsego's example because, "if they indulge in expensive habits, involve their interests, eat and wear out their farms, they are not the Farmers to whom the Genius of Liberty looks for the perpetuity of our civil institutions." A well-cultivated spiritual tie between farmer and nation, not the profits gained by selling produce, provided the only sure way to link generations of citizens in the cause of freedom.[3]

But it was not enough to economize at home; rather, Spafford cautioned, farmers must recognize that only active measures could protect the agrarian political economy that the people had worked to put in place. The republican cause required increased vigilance because the nation had only partially defeated its old enemy—the aristocracy—while an even more dangerous foe, the entrepreneurial elite, was rising to prominence. He noted that the township of Davenport, Delaware County, was dominated by perpetual leases, "the remnants of feudalism, yet seen largely in this State."[4] Because of this, the dormant landed aristocracy "yet may hatch" and "the only safeguards for the continuance of public liberty and equality, will be found to exist in the multiplication of freeholds."[5] Defeating financial wealth would entail another approach, one that required farmers to pursue long-term strategies. For example, he cautioned farmers to limit lumbering because profits went to merchants, not farmers or landlords, and led progressively to soil erosion, poor harvests, and debts to monied men. He counseled farmers to turn to dairy farming rather than invest the capital needed to grow wheat, which would have to be borrowed and repaid at interest, again enriching merchants at the expense of farmers. Both profits and spending had to be curbed to ensure the survival of the Republic and to safeguard liberties from the grasping hands of monied men.[6]

There were in fact indications that poverty, dependence, and disorder would result from the commercial revolution. Coinciding with the opening of the Erie Canal, both Schoharie and Otsego opened poorhouses, a recognition that poverty was a permanent feature of rural life.[7] British traveler I. Finch remarked on the contrast between poor and rich after a frigid ride through Schoharie and Delaware Counties in February 1825. His driver removed the sleigh bells to avoid being heard by timber thieves "who, with axes as their sole companion, penetrate into these forests to cut down the pine timber. He said, that, being in want of every necessary of life, they sometimes stopped travellers on the road." A more welcome scene greeted him south of Delhi, where he stopped at "the neatest house I have seen in America; every thing was arranged with the most scrupulous exactness. The proprietor had a farm of several hundred acres of woodland and cultivated ground, and the house commanded a fine view. The ladies were engaged in spinning flax. The father was representative for his county, and was at Albany, attending the meeting of the Legislature. . . . Altogether, it presented a charming rural scene."[8] The challenge for agrarians was to contrive a way to foster this kind of prosperity without increasing poverty.

In the absence of a specific crisis that might stir a more positive strategy, farmers fell back on paternalism to protect their world. Tenants resisted landlord efforts to enforce lease provisions by pleading hardship, dragging their feet, and individually negotiating to avoid eviction. Many landlords accepted the chronic indebtedness of tenants, conceded to farmers the right to the value of improvements to the holdings, and turned a blind eye to depredations on forests. In exchange for protection of their security of tenure, renters participated in the humiliating rituals of rent day, performed labor service for the landlord, paid respects to the landlord's family at weddings and funerals, doffed their caps and looked down when addressed by the landlord, and voted for candidates he supported. Even tenants who despised George Clarke for his brusque English manners and quickness to eject delinquents halted efforts to vacate his title when he agreed to pay them for improvements to the land.[9]

The farmers' charitable attitude toward the gentry had much to do with their recognition that the gentry had a stake in the prosperity of the region, in contrast to the commercial elite of the state. Despite his ardent opposition to the manor system, Spafford nonetheless flattered Stephen Van Rensselaer III as a patriot and a friend of the weak. He noted that the landlord earned $100,000 per year in rents but cast him in a positive light because "his thousands of tenants transact their business, pay their annual rents and fealties, and meet with as much favor and kindness as the weak ever experience from the strong, the poor from the rich, the tenant from the landlord. He is a man of most ample resources . . . of singular beneficence and kindness, humane and charitable to the poor, liberal in liberal schemes for the public good, a Patron indeed."[10] Spafford believed capitalism was far more dangerous than paternalism and that it would be curbed only by an alliance of the rich and poor sons of the soil.[11] Still at odds with the gentry over the proper economic strategy for the backcountry and suspicious of their political designs, farmers nonetheless found cause to maintain the social order.

For their part, proprietors not only accepted their continued role as patrons of the community but also sought to reconstruct their image. The disparities of wealth caused by the passing of the frontier and the commercial revolution did not escape their notice. They attributed poverty to a breakdown of the social order that had been accelerated by capitalism. Their solution was to reinvigorate paternalism because it had a human face that the market did not. Proprietors did not promise equality—Van Rensselaer had been the leading opponent of universal suffrage at the 1821 constitutional convention—but they did argue that the poor would not be cast to the winds even during difficult times. Each working in his or her station, a united agrarian order could stave off the disruptive effects of the market economy.[12]

The gentry went further, arguing that they also had been the victims of a predatory commercial elite. Writing a decade later, James Fenimore Cooper compared his class to George Washington's, claiming that the landed elite's ownership of the soil distinguished them from entrepreneurs who gained wealth from wage labor and speculation, exploiting rather than uplifting the laboring classes. When New York and Albany creditors had called in loans during the panic of 1819, creating devastation in the backcountry, the gentry accused the middle class of causing the economic downturn and impoverishing the people by drying up credit in rural areas already starved for specie. The gentry, on the other hand, protected the laboring classes by forestalling debt collections, dealing with hardship on an individual basis, and providing security of tenure. Rural America had produced natural aristocrats—themselves—who had risen to power based on their talents, education, and care for the less fortunate. The social order of the countryside now needed to be protected from the middle-class parvenus seeking to drain the resources of honest farmers.[13]

While the gentry, freeholders, debtor-farmers, and tenants could agree that they had a mutual interest in preserving an agrarian social order in the face of a modernizing commercial culture, proprietors offered a solution all their own. They believed that agriculture had to be retooled to compete in the national marketplace after the completion of the Erie Canal and had endeavored to put in motion a wholesale transformation of the countryside. In 1817, wealthy Otsego farmers, including Clarke, invited agricultural reformer Elkanah Watson to Cooperstown to help launch the Otsego County Agricultural Society. The society identified sheep, dairy cattle, and hops as excellent crops for the sequestered, hilly region and threw itself into agricultural improvement schemes, offering bounties at county fairs, touting fertilizers, and promoting specialty crops. The society spoke of soil exhaustion and conservation and the need to maintain forests and woodlots and to preserve fish populations.[14]

But agricultural development could not take place without a fundamental reorganization of production, which entailed recasting social relations. Initially, the goal of landlords had been the clearing and improving of the land. Now, creating profitable farms and a skilled labor force to ensure the quality and quantity of the crops became the focus of their efforts. The challenge was to do so without inciting unrest. John A. King adopted a hands-off approach, commuting wheat to cash rents on the Blenheim Patent in Schoharie County in the 1820s to allow farmers to innovate on their own. But other landlords wished to direct farm management. George Clarke, for example, winnowed the number of oral contracts he had with tenants and replaced them with written leases with stringent clauses regarding improvements, manuring, and keeping of stock. He also attempted to substitute life leases on his estate with hops and dairy share contracts.[15] Share contracts marked a significant expansion of landlord control over production. As one landlord wrote in 1838, share farming would be more profitable than life leases and "will relieve . . . much care and trouble [by] reserving to us the directing [of] what fields are to be plowed up and how to be cultivated," fertilized, and harvested.[16]

Such new contracts heightened tenant complaints in two ways. First, the process of clearing life tenants from the land hurt the community credit network because life leases were transferable, often changing hands several times, and were offered as collateral for loans based on the value of improvements made by tenants. Neighbors thus would not tell Clarke where original settlers had moved nor reveal whether the individuals named in the leases had died. Creditors who held lease contracts as security likewise kept mum. Second, locals considered share leases exploitative. Leases specified that tenants grow several acres of hops while keeping dairy cattle to produce butter and cheese. Clarke demanded the entire hops crop and a share of the dairy products of each farm as rent, required farmers to breed their cattle with his bulls, stiffened fence requirements, reserved the right to oversee the preparation of hops for market, and stipulated that all manure must remain on the farm. He and his agents rode out periodically to check on the progress of crops. Few would take up farms on shares because they recognized that Clarke meant to force greater market production and increase the time and cost spent on farming. By the time of his death in 1835, Clarke succeeded in letting very few of his farms on shares. The others remained under the original leases with rents collected indifferently by Clarke's attorney, Richard Cooper. Eight years later, George Clarke Jr. faced an indebted tenantry that felt secure in its right to occupy the land by custom.[17]

Debtor-farmers joined the call for forbearance. When Goldsbrow Banyar sent agents into the countryside in 1828 to clear off debt-ridden farmers and sell the lots at prices that reflected the land's value for sheep, dairy, and hops production, farmers refused to vacate unless he compensated them for improvements to the farms. They condemned Banyar for reneging on his paternal obligations to the common people who inhabited his estate. The two sides fought to a draw, a victory of sorts for farmers. And, as with tenant farmers, debtor-farmers enjoyed community support against the proprietor because they too had borrowed from local merchants and respectable farmers who stood to lose if evictions proceeded.[18]

In the end, despite persistent tensions, agrarian discontent never blossomed into outright conflict because both classes still found their interests served by paternalism, especially once the national economic boom between 1828 to 1837 brought better times. In Delaware County, the Livingstons wished to clear life lease properties in the 1820s but discovered that tenant farmers had little interest in creating economies of scale or making way for those who might. Instead, Delaware farmers dabbled in wool and dairying and continued to live off buckwheat, corn, and lumbering. Pastoralism met the agrarian ideal of rural New Yorkers, and both wool and dairy products could be sold locally, to a degree insulating farmers from national market forces. Expending minimal energy on their cash crops allowed them time for leisure, let them set their own work rhythms, and enabled them to still feed their families. And exchanging labor for harvesting and threshing helped neighbors minimize cash outlays. Paternalism provided leniency during lean years, a chance to avoid lawsuits over debt that could lead to the loss of farms, and access to the commons, which became increasingly important as pressure to graze more livestock grew after the construction of the Erie Canal. With at least some income percolating upward from the Delaware lands, the Livingstons gave up their quest to clear the land of life tenants. Paternalism seemed to be a workable compromise for those emerging from the frontier economy.[19]

* * *

In the political sphere, Delaware, Otsego, and Schoharie Counties provided fertile ground for the Jacksonian coalition that began to gather strength in 1826, when Andrew Jackson, Martin Van Buren, John C. Calhoun, and Thomas H. Benton joined together to promote Old Hickory as the people's candidate in the 1828 presidential election. In New York, the alliance that would become the Jacksonian Democratic Party had its origins in the Bucktail insurgency that had helped bring about constitutional reform in 1821. The Jacksonians echoed themes that had animated that movement—decentralization, anti-monopoly, localism, and agrarianism. The Jacksonians refined this message by posing as the defenders of "producers" against "nonproducers," a simple but effective rhetorical device that could unite yeoman farmers, artisans, shopkeepers, and southern planters against a shadowy "monied interest." Charging that dark forces had stolen the 1824 election from Jackson to reestablish Federalism under John Quincy Adams, the party swept into office in 1828.[20]

The vagueness of the Jacksonian message has left scholars grasping to identify its ideological foundations. Interpretations run the gamut, arguing that party ideology spoke for the working class, middle-class men on the make, or agrarians. Resolving such contradictory findings is too cumbersome to be taken on here.[21] However, the course of events in central New York during the period suggests a different Jacksonian agenda, one that attempted to be inclusive rather than exclusive by promoting a localistic, democratic political economy, not the particular interests of one class or segment of society. The Jacksonians radiated an almost blind faith that democracy would balance out competing interests and arrive at the "common will," that is, so long as tyrants were prevented from subverting the political process. Their obsession with maintaining equal opportunity for all groups usually led them to avoid legislative action aimed at aiding one group at the expense of another. And, in fact, they compiled a poor record at advancing the interests of their most ardent supporters, working men and farmers. Moreover, in central New York, the Jacksonians' success at putting in place a political culture that was localistic in structure—and expansive in its praise for the common man but short on concrete actions on their behalf—led to their undoing after the panic of 1837.

Until that crisis, New York Democrats held potential divisions in check by subordinating policy to party and preaching a philosophy of limited government. They believed that a strong central government at either the state or federal level endangered the liberties of the people by concentrating power in the hands of the powerful few at the expense of the many. Democrats were especially fearful that wealthy "nonproducers"—landlords, merchants, factory owners—would monopolize government, which they deemed a natural progression in human history. Related to this, Democrats demanded that government protect individual liberties at all costs, fearing that a strong central government would pass legislation that slowly eroded civil rights or privileged one group over another in the marketplace. To counterbalance the power of the "aristocracy," they proposed to diffuse power among counties and townships, which they believed were more democratic. The legislature then would function as the arena in which representatives of competing local interests forged compromises that best served the people as a whole. The Democratic Party considered its role in this process to be critical. It would gauge and promote the will of the people. And the party would distribute state and federal patronage on a rotating basis to ensure that all localities shared in governance. So long as Democrats won elections, there would be patronage to dispense among loyal politicians and the people they served. Both the party and the negative state would function only to balance the interests of the whole.[22]

This message brought New York Democrats victory in most state races between 1828 and 1836, and backcountry areas responded eagerly to the Jacksonians. Democrats dominated political life in Delaware, Schoharie, and Otsego Counties. They scored a decisive victory in the presidential election in 1832, with Jackson receiving 55 percent of the vote in Otsego, 60 percent in Delaware, and 62 percent in Schoharie. Democratic majorities continued to rise, buoyed by strong political machines in each county. In the gubernatorial race in 1834, Democrat William Marcy thumped Whig William H. Seward in all three counties, polling 58 percent of the vote in Otsego and 67 percent in both Delaware and Schoharie. In 1836, the Whigs were so demoralized that they didn't run candidates in several races, and Democrats achieved a similarly decisive victory, including the election of Van Buren to the presidency. This fit a national pattern in which Democrats did well in rural areas that had passed the frontier stage but were nonetheless isolated from markets, credit, and transportation routes.[23]

Anti-Jackson parties could not generate support in the counties for several reasons. For one thing, the National Republican Party, which ran Henry Clay for president in 1828, was associated both with the landed and financial elite. His promotion of the American System called for the very sort of governmental involvement in economic development that central New Yorkers despised. The Anti-Masons, who reordered political alliances in western New York from 1826 to 1832, also did not gather momentum in any of the three counties, despite preaching an anti-monopoly message. Their primary constituency was the evangelized middle class of canal cities. Their connection to Protestant reform foretold of governmental interference in the lives of individuals, which sat poorly with farmers who wanted limited government and the protection of civil liberties. The Anti-Masons also failed to appeal to the landed elite, whom they blamed for the corruption of politics. Proprietors were glad to see Anti-Masonry ebb with little more damage than the closing of local Masonic lodges. When the Whigs organized in 1832 from both the National Republican and Anti-Mason ranks, they ran afoul of the same difficulties. The party succeeded in industrial and commercial centers, or in rural areas closely linked to market agriculture. Whigs were unified in their support for a "positive state" that promoted economic development, which appealed to businessmen, monied men, and the landed elite. And many middle-class reformers believed that a positive state could be used to promote social reforms like temperance, Sabbatarian laws, and abolition. Thus the Whigs competed well in the Erie Canal corridor, while they labored to get votes just to the south in Delaware, Otsego, and Schoharie, where neither the middle class nor social reform flourished. Democratic rule in the three counties was strengthened by the fact that local Whigs were of the conservative sort, either landlords or agents, millers, or merchants with close ties to proprietors. Rather than a party, the local Whigs were little more than a collection of anti-Jacksonians of various stripes.[24]

The fact that people of the three counties seemed to spurn the political revolution underway along the Erie Canal, much as they did the economic and social changes occurring there, limited the scope of the Jacksonian revolution. Backcountry politicians inherited a political culture in which Clintonians had monopolized patronage, using public office to attain economic advancement. Even after the new state constitution went into effect, political leaders distributed cash and patronage among supporters in an effort to consolidate power at the county level—control of which was more important than ever. William C. Bouck of Fulton, Schoharie County, for example, became sheriff in 1812 thanks to the aid of local political boss Peter Vroman, whose Revolutionary service earned him clout in Albany. Bouck built a political empire by gaining the postmastership of Fulton, then serving in the state assembly and senate between 1814 and 1822. Afterward, powerful friends appointed him surveyor of the Erie Canal and later canal commissioner, where he shared a seat on the board with Stephen Van Rensselaer and other great men. Bouck built alliances downward as well. He used the money and influence he accumulated in office to establish an impressive series of debt relationships with his political allies. In 1835, Bouck's most loyal political supporters owed him nearly $33,000 in credit, mortgages, and other debts. Bouck had no difficulty squaring these activities with his democratic impulses and led his county's leading faction into the Jacksonian camp in 1828. The people of Schoharie, the beneficiaries of Bouck's largesse, viewed him as a benefactor and readily supported his handpicked candidates. He was able to pass leadership to his son-in-law, Lyman Sanford, in 1842 with Democratic hegemony intact.[25]

But Democratic electoral success in the counties masked the instability of the political system in which both parties operated. For one thing, the 1821 constitution ushered in a period of what one scholar has called "distributive politics," in which the state government controlled economic development by chartering all corporations and granting aid to those considered most beneficial to the people. In practice, this led to constant politicking in Albany to charter even the most narrowly local projects. County representatives rarely acted altruistically, shamelessly trading votes to advance pet schemes. Not surprisingly, this spread resentment among spurned localities that festered at both the state and local level. At the county level, the parties were collections of township factions that were based on personal, neighborhood, family, ethnic, religious, or economic ties. This met the needs of the people, since average New Yorkers lived in a world in which labor was shared among neighbors, the margin between success and disaster was small, and the church and family stood at the center of social and economic life. While factionalism was not conducive to tackling the abstractions of public policy, it did help rural New Yorkers fight for the "equality" of their communities and "democracy" among townships and counties. Yet any number of real or perceived slights could lead factions to bolt tickets, switch parties, or simply stay at home, making politics volatile indeed.[26]

To manage the tangled web of fiefdoms that this culture fostered in each county, Democrats relied on a "central junto"—as critics called it—to mediate contending township demands, promote suitable candidates, and get out the vote. While Bouck managed Schoharie, the "Delhi Clique" was led by land agent Charles Hathaway and included a host of friends, family, and business associates at the Delaware County seat. As agent for several landlords, Hathaway developed political connections statewide that helped him secure vital patronage. At home, he and his coterie had a virtual monopoly on the freight trade, with interests in turnpikes, stage lines, and inns to add to income from rent collections. The Whigs, led by former Jacksonian Erastus Root and such men as land agent and entrepreneur Samuel A. Law of Meredith, had difficulty overcoming Democratic influence. Law, for example, consistently found his efforts to build a new turnpike from Meredith north to the Erie Canal blocked by Hathaway and his cronies at home and in Albany. It may appear ironic that Root, who had made a career as a defender of democracy, and Law, who sold land in fee simple, could not generate popular support for the Whigs in a county run by the agents of landlords. But this, in fact, was indicative of what people wanted from their politicians: a slice of the economic and political pie for their communities. The Democrats could deliver.[27] Likewise, Otsego politics revolved around a Democratic central junto. Democrats shared power more than their counterparts, with several prominent citizens running its affairs, including Cooperstown *Freeman's Journal* editor J. H. Prentiss, George A. Starkweather, and Schuyler Crippen. Otsego County had the strongest Whig organization of the three, which followed the wisdom of Cooperstown attorney, landlord, and industrialist William H. Averell, a clutch of Cherry Valley Scots led by the Campbell family, and landlord Jacob Livingston. The economic clout of these men made them difficult for Democrats to ignore completely, but their ties to proprietors curried few votes.[28]

Despite Jacksonian rhetoric, party leaders on both sides found the illusion of democracy far more desirable than actually sharing power with rivals or conducting open elections. As a gubernatorial candidate in 1842, Bouck wanted the Democratic Party to maintain the appearance that nominations sprang from "democratic" township conventions, stating, "We can only return ourselves, by being as wise as serpents and [appearing] as harmless as doves." Bouck insisted that Schoharie Democrats exchange offices to placate localities that otherwise might not be able to advance a candidate successfully but warned that such maneuvers must be kept secret to avoid disputes.[29] If negotiation failed, political factions simply spent money. New Lisbon Whig D. M. Hard asked Otsego party leader William Averell to help raise funds from "gentlemen" prior to the 1838 county convention to ensure victory for their friends.[30] No less important to party success was to maintain the perception that the system was open to men of talent. In 1840, Otsego Democrat Sumner Ely of Cooperstown suggested that he and Joseph Peck of New Lisbon promote a "new" candidate, Cooperstown attorney George Starkweather, for delegate to the national convention in Baltimore. Ely felt it was time to offer a new face to the voters to reaffirm the idea that one of the "people" could rise within the ranks. Appearance was everything in this case, for while Starkweather had not stood before the people at the polls, he had been lavished with numerous military and political posts since 1823. At all levels, then, party leaders orchestrated electoral politics, and, as Ely's words demonstrate, offices were reserved for select individuals.[31]

Living in close proximity and sharing the associations built in politics, the church, and business, the Jacksonian political elite grew difficult to distinguish from the landed elite over time. Jacob Livingston, for example, married the daughter of one of Cherry Valley's leading citizens, Dr. Joseph White. Attorney William Averell, the son of William Cooper's millwright, made a fortune purchasing rental properties from the Cooper estate, opened a cotton factory, and began banking in Cooperstown in 1830. The lands around Otsego Lake became a veritable haven for men jockeying to become certified members of the leisure class. Holt-Averell, Otsego Hall, Hyde Hall, Swanswick, Apple Hill, and other great homes dotted the landscape. As the two classes blended slowly together, a culture emerged that celebrated the princely lifestyle of the gentry and made it seem attainable to political leaders who were, for lack of a better term, the backcountry's middle class. Levi Beardsley, who had risen to political prominence under the tutelage of the brilliant Jeffersonian politician Jabez Hammond of Cherry Valley, fondly recalled the social gatherings of Otsego County's great families. Though he had begun his career as a populist, Beardsley took pains to laud the hospitality of George Clarke, who frequently invited guests to Hyde Hall to enjoy teas, lawn parties, and sumptuous fare. Beardsley, viewing his invitation as a token of his entrance into polite society, defended Clarke against any who might doubt his commitment to republicanism.[32]

As in other parts of the nation, identifying the electoral constituencies of the Jacksonians and their opponents in central New York is an inexact science. Townships that fell into the Whig camp tended to be dominated by Scots (Cherry Valley or Roxbury, Delaware County) or had a political junto that proved strong enough to resist a Jacksonian insurgency (Meredith, Delaware County). Since the three county seats had sizable commercial or industrial activities, all had strong Whig factions. But Democrats held the reins of power in most agricultural townships, whether Yankee or German in ethnic background, suggesting that the agrarian, anti-monopoly message of the Jacksonians, along with their strong distrust of a powerful central government, appealed to central New York farmers far more than the Whig Party's advocacy of an interventionist state.[33]

As the Jacksonian message matured by 1832, it drew an intensely loyal following in central New York. Jackson's Bank War, his unwillingness to use federal funds for internal improvements, and his consistent opposition to the enlargement of government played well among citizens who recognized that they had limited economic opportunity and did not wish to pay for the prosperity of others. Otsego County, for example, was ranked fourth in the state in population in 1825 but fell to twenty-second by 1860, with a large drop between 1830 and 1835. When the Jacksonians promoted commerce by promising no special advantages to any class or locality, Otsego residents could take heart that the decline caused by the Erie Canal would at least not be compounded by new projects. Even a few large landowners found solace in Jacksonian promises to leave to the states the protection of rights of property. Much like their southern brethren, they believed they could better control state legislatures and maintain their prerogatives in the face of mounting pressure from the middle class to use the central government to promote social reform. Tenant and freehold farmers saw the Jacksonian support of the Working Men—an urban artisans' party that advocated free soil—as evidence that the agrarian republic that they had worked so hard to build would remain protected from capital, especially banks and corporations. Farmers who operated on the margin every year or who were in debt in the three counties also warmly greeted promises of smaller government expenditures. Whether politics became more democratic was less important for these groups than protecting a social system that provided security in an advancing economy.[34]

Because Jacksonian rule was so lopsided, Democrats had more to fear from factions within their own party than any external threat. For example, William Bouck considered Anti-Masons and Whigs an annoyance, not a menace. When several families broke away from his clique to join the Anti-Masons or, later, the Whigs, Bouck viewed the moves as acts of personal disloyalty, not party principle. After punishing the dissenters, he pulled all but a few back into his coalition with gifts of small offices. Even when a powerful leader defected, Democratic machines proved resilient enough to repair the damage. In Delaware, Erastus Root wavered in the years 1828 through 1832 between support of the Working Men of New York City and the Anti-Masons and finally broke with the Democracy. Into the breach stepped Charles Hathaway, who controlled the county for the Democrats right through the Anti-Rent years. In each county, single-party rule survived the state and national transformation that established a two-party system in 1832.[35]

Factionalism, however, prevented the parties from developing completely reliable coalitions in the three counties between 1828 and 1836. Leaders complained of the difficulty of convincing township leaders to act according to the party platform. Robert Eldredge, a Democratic assemblyman from Sharon, Schoharie County, viewed the Bank War as a defining moment in American history. "We have the same great power to dread in coming time that the patriots of the revolution did dread . . . 'the power of the purse,'" he wrote, but "men who do not look very deep into public measures" failed to realize the importance of sustaining the president's policy. Despite arguing that the Democrats "must reason men right upon this question," Eldredge forlornly noted that most men acted according to "present interest," abandoning the party's agenda when offered political advancement by "enemies." Eldredge did not fear a defection of voters to the Whigs; rather, he worried that a rival Democratic faction might exploit the issue to unseat Bouck's junto.[36] At the height of the Jacksonians' strength in 1836, the party could not hold members in line in Otsego County. Democratic candidates for Congress and state senator ran nearly 800 votes behind Van Buren's tally for president as Anti-Bank Locofocos scooped up the dissenters—who made up 9 percent of the votes cast. This minor revolt, which began among New York artisans, would develop into a major rift after 1837.[37]

* * *

Though partly insulated from the dramatic economic and political changes taking place in the rest of the nation during Jackson's presidency, these events began to having a recognizable impact on central New York in 1836. Political upheaval came first. On the Holland Land Company's lands in western New York in February 1836, 300 to 500 debtor-farmers sacked the company's land office in Mayville, Chautauqua County. Decrying the loss of the lands "on which the fire and vigor of their manhood had been expended" to owners who "had grown wealthy by the industry of the settlers, and their agents rolled in fatness," farmers smashed open the safe and burned "half a cord of books and papers" that contained records of their debts. Farmers in eastern New York followed events carefully.[38] Meanwhile, artisans chose 1836 to mount the Locofoco political insurgency in New York City. Locofocos embraced the free soil ideas of Thomas Skidmore and George Henry Evans, who voiced a producer ideology similar to that expressed within the agricultural community. In the 1820s, Skidmore demanded recognition of laborers' rights to the products of their toil and argued that rights to the soil devolved naturally to citizens. Led by men like Evans, Locofocos mounted a political insurgency in New York City that year, demanding a thorough reform of the Democratic Party to eliminate corruption and promote the needs of the producing classes. They met with enough success that many Democrats feared a mass defection of voters from the party that would cost them control of the state in 1838 or the presidency in 1840—a prospect made more alarming by the success, albeit limited, of Locofocos in rural districts.[39]

What linked Chautauqua County debtor-farmers, New York City working men, and the people of Delaware, Otsego, and Schoharie Counties was a sense that despite the economic boom of the Jacksonian years, the party had not delivered on its promise to protect the interests of producers. Rather, party insiders seemed to have monopolized political office and used distributive politics to award economic advantages to themselves and their friends. While the laboring classes languished or backcountry counties fell further behind economically and politically, Jacksonian political leaders seemed more interested in maintaining themselves in office than in addressing the pressing needs of the people. Such potentially explosive political issues as land reform and workingmen's rights were left unattended by Democrats, despite Whig agitation.

In central New York, conditions that could unleash unrest had accumulated over time, masked by prosperity. One factor that ironically would cause instability was the success of many farmers. An example of those who made it was William Clarke of Burlington, Otsego County, who had emigrated from England in the 1820s, took a pork redemption lease on William Cooper's estate, and turned his farm into a bustling enterprise, raising hops, hogs, cheese, butter, corn, and grain. But this took concentrated effort. Unlike many of his less industrious neighbors, Clarke traveled to New York to sell his produce to obtain maximum prices. He also hired men to work on his farm, an indication of the intensity of his agricultural pursuits. His rise to affluence, however, did not rest solely on his own endeavors. His in-laws in Britain extended him inexpensive credit that helped him redeem his lease early. Clarke and his children remained one of the most prosperous families in the township for decades.[40]

Many other farmers, however successful, could not climb this agricultural ladder. Tenants in particular found themselves in a frustrating position. John A. King's tenants in Blenheim, Schoharie County, recalled that they believed that they finally brought their hillside farms to full bloom in the 1830s: "The hardships of the earlier days had passed away. The log houses were snug and comfortable, barns sheltered numerous sheep and cows, and every yeoman owned an ox-team. There was plenty in every household." But they could never achieve the kind of independence enjoyed by Clarke, for they held their land under perpetual lease.[41] Odd as it may sound, King's tenants were the lucky ones. As population levels peaked and land became scarce by 1835, large property owners finally could institute stricter lease terms. Share farming contracts, for example, became more competitive, forcing prospective tenants to prove their skill, discipline, business acumen, and temperament to secure holdings. Sending letters to procure a better farm became a seasonal ritual. Wealthy landlords only accepted "worthy" tenants.

George Potts of Middlefield, Otsego County, chose his words carefully when he implored William B. Campbell to let him a farm in 1836: "I have under stood that Mr. Allison and Mr. Davison were a going from your Farm next Spring if so I would like to take it or rint it as we could agree not that I wish to under mind Mr. Allison and Mr. Davison but if they do quit I would like to get it I have worked Jeremiah Reads farm where John prince lived on Shears I took it for 2 years from last Spring and he has sold it and wants me to leave it next Spring I worked it at halvs and he says what he has got for his part is worth over $200 and if I get yours I will try to do as well by you."[42]

Landowners could dictate levels of production not previously seen. Other proprietors required debtor-farmers to grow certain crops, and merchants demanded specific commodities in payments on debts.[43] Others were not above preying on the misfortune of the poor, such as land agent David Morse of Jefferson, Schoharie County, who hoped to take advantage of the crisis by accepting cows in lieu of rent from struggling tenants in order to start his own dairy business.[44] Evidence suggests that share farming impeded progress up the agricultural ladder for most families. Samuel Conklin of Blenheim, Schoharie County, for example, found leasing on shares in the 1830s and 1840s disastrous. His son Henry recalled that "my father was always taking farms to work on shares or hiring some place of somebody so he was constantly clearing or working someone else's land, never having any of his own. . . . No wonder we were half naked and sometimes hungry and cold and no place to lay our heads that we could call our own. . . . It was a continual struggle to keep body and soul together."[45]

Between tenancy, mortgages, and land contracts, most rural New Yorkers could not claim to be masters of their own domain despite the economic boom and agrarian rhetoric of the Jacksonians. Though it is difficult to accurately gauge tenancy rates in the 1830s and 1840s in Delaware, Otsego, and Schoharie Counties, estimates range between 30 and 40 percent. In 1848, Governor John Young reported that 1.8 million acres of land in the eastern half of the state were held under durable leases and that 300,000 common folk made their living from the lease land situated in the triangle formed by the Hudson, Mohawk, and Susquehanna Rivers. Tenancy was only part of the story, however. Few farmers held their land in fee simple free of debts, even though most claimed freehold status. Since mortgages proved difficult to satisfy with the extensive farming methods employed by frontiersmen, families who technically enjoyed yeoman status actually lived in constant danger of losing their land.[46]

At the same time, the opening of the Erie Canal and soil exhaustion forced changes in the social relations of production on backcountry farms. The declining ability of men to provide for their families through grain production and lumbering, the expansion of wage labor, the diminution of women's outwork, and the elevation of traditionally female agricultural tasks like dairying to a central role in the household economy all contributed to rural men's growing sense of displacement. Occupational statistics are sketchy before 1840, but in that year most males in Schoharie (85 percent), Delaware (80 percent), and Otsego (78 percent) listed agricultural occupations. Whereas the lines between farmer, craftsman, and laborer were blurred before 1830, and many men relied on work outside of the homestead to generate income, pressure to specialize in cash crops increased as tanneries and sawmills shut down between 1835 and 1840. Yet while farmers who switched to sheepherding from 1825 and 1835 to replace grains found they needed less labor, farms in Schoharie, Delaware, and Otsego Counties averaged three to four agricultural workers per 100 acres of improved land, usually family members. This labor redundancy was magnified by the industrialization of cloth manufacturing, which soon eradicated spinning and other tasks that could occupy family members left idle by the pastoral economy. Even before wool prices crashed in 1837, therefore, central New Yorkers had begun to abandon sheep farming. Because the patriarchal household remained the ideal farm organization, local farmers increasingly adopted mixed dairy farming because butter- and cheese-making involved all members of the family and allowed those dislocated by economic changes to contribute to the family income.[47]

Women's economic position within the family also began to change, commencing a long process of redefinition of the farm household and the social relations embedded in it. Mary Conklin of Schoharie County, for example, spun cloth and knitted socks and gloves, earning over half of the family's annual income. She also cooked, cleaned, watched the children, and made butter when they had a cow. Neighborhood women also earned extra cash by knitting, keeping hens, or, in the case of one woman, selling guinea pigs as pets in Albany. Mary Conklin's preadolescent daughters helped with the housework, and young children of both sexes helped prepare the wool and yarn for the loom. When children reached their teens, they either helped one of their parents in one of their gendered tasks or were bound out to neighbors, their income supplementing the family's other endeavors.[48] Much of the work done by women like Conklin was outwork. Farm families received raw wool from a carding mill, spun yarn and weaved cloth, then returned the finished product to a fulling mill for final preparation. Thus, while only seven woolen mills operated in the three counties in 1835, ninety-six carding mills and ninety-three fulling mills served the area. But it became difficult to count on women's production to support the family. As household manufacture of cloth declined steadily in Otsego, Delaware, and Schoharie Counties between 1825 and 1845, cloth production became so deskilled that women spurned it. Women instead increased butter and cheese production, tasks that proved more pleasant and monetarily rewarding than the time-consuming process of spinning yarn and weaving cloth.[49]

Cloth factories not only contributed to the decline in home manufactures but also altered family relations. Located south of Cooperstown, the Union, Hope, and Phoenix factories increased production throughout the antebellum period. The Union Cotton Mill employed seventeen men, sixty-one women, and thirty children in 1832, while the Phoenix Cotton Mill had fifteen men, seven boys, and fifty-four women (twenty-four weavers). All three mills housed workers on-site and had a number of complete families working for them. Workers averaged twelve hours per day and agreed to work all year. The manager of the Phoenix mill noted that he had to offer 20 percent more per day to women because "the confinement of females so many hours in the day, renders them unwilling to engage in this business without more wages than they could make otherwise."[50] Young women composed the bulk of the workforce, laboring under agreements made between their parents and managers. Their families expected a substantial portion of their income. Labor problems occurred—usually during the hop harvest and when friends or family moved to new mills—that indicated the beginnings of the young women's attempt to take control of their lives independent of family obligations. These developments were not welcomed by parents who relied on their daughters' income. As a result of changing relations of production, therefore, protecting the patriarchal household economy had grown difficult for men by 1836.[51] If Jacksonian democracy had meant anything to central New Yorkers, it was support for the equality of men in the political and economic sphere and protection of the autonomy of patriarchal households. These were no longer certain.

* * *

The panic of 1837 hit in the midst of such economic and social dislocations. Farmers, the gentry, and merchants all suffered in a depression that persisted for the next decade and forced central New Yorkers to reevaluate the political system that governed their lives. Class politics, elusive during the Jacksonian years, suddenly became a tenable alternative to the single-interest society that rich and poor believed was the advantage that the backcountry enjoyed over cities. Jeffersonian agrarianism, which accepted the existence of a natural aristocracy, suddenly appeared anachronistic. In this milieu, the emerging farmer's movement played a central role in redefining the political landscape of New York.

Of all central New Yorkers, farmers seemed best equipped to weather the depression, perhaps because they did not have far to fall or could shift production. Many farm families sold off sheep, avoided investment in expensive crops like hops, and focused on mixed dairy farming, which offered moderate profits that were insulated from price declines because of continued demand in cities. In the midst of the panic, farmers proudly touted their accomplishments. They boasted of their lush harvests as evidence that their communities had fulfilled the primary goal of the agrarians: successful family farms. James M. Wilson wrote in 1840, "We raised last seson a bought one hundred bushels of wheat and I think it done well for Delawar." Wilson's bumper harvest would likely bring him little profit because of the economic depression; rather, he expressed personal and community pride.[52] Attorney Levi C. Turner of Cooperstown voiced a sense that farmers had escaped the depression unscathed when he wrote in 1839, "Hard times in the City but plenty of money among farmers—Farmers are the most Independent."[53]

But many suffered nonetheless. Traveling through Otsego and Delaware Counties in 1837, Ithaca merchant Samuel J. Parker noted that the "large and grass-growing region . . . was in a very doleful state" with "woods with openings cut and half cultivated . . . acres of stumps and blackberry fields and the plainest log huts, with a husband farmer in tow-cloth clothes, a wife in homespun, and several children barefooted, ragged in half clothes, torn straw hats, and in the fall weeds and a sense of defeated farming attempted on a larger scale than the inhabitants could possibly carry out."[54] The lack of currency made cash payments difficult and increased the possibility of losing all one had accumulated. Peter Shaver, of Shavertown, Delaware County, had to seek financial help from his brother-in-law and from his son in Ohio to save his property when his mortgage came due in 1840.[55] Tenants such as Samuel Conklin fared worse. The Conklin family arrived in Blenheim, Schoharie County, in 1838. For the next seven years, Conklin moved annually, subletting land from in-laws and binding his teenage children to them to make ends meet. He worked the fields, growing buckwheat, oats, and corn, and had an ever-changing, motley collection of individual sheep, cattle, and hogs. Samuel and his sons supplemented their small harvests with work as carpenters, teamsters, laborers, and woodsmen. His wife, Mary, her daughters, and the small children spun and wove cloth, knitted, cooked, cleaned, and worked in the fields when necessary. They barely survived.[56]

While Samuel and Mary Conklin avoided wage labor and dependency by drawing on a robust kin network, others were not so fortunate. When the panic struck, surviving evidence shows, skilled labor commanded between seventy-five cents and a dollar per day. However, wealthy landowners and entrepreneurs took advantage of the hard times to institute, insofar as they could, lower wages and stricter controls on labor. Samuel Law hired men to cut timber, process the lumber at his sawmills, raft it down the Delaware or Susquehanna for sale, and return with goods for his store. He also hired harvesters on his farms, sometimes contracting for only a few days' work. Law kept careful track of the number of days men worked, how much credit he extended them, which goods they received at his store, and whether they took ill, visited families, or ran off. He tried his best to avoid paying men in cash, preferring to pay them in goods. The contract he gave to John Crayton in 1840 typified his arrangements with common laborers. He agreed to pay Crayton $40 for three months' work ($12 for June, $14 for July and August), but Crayton had to accept payment in clothing from Law's store. Law could well have paid cash; indeed, he routinely paid businessmen in specie. But cash would grant Crayton freedom to spend the money as he wished—in effect making him independent of Law.[57]

Wage earners negotiated with such determination because their sense of self was intimately connected to a belief that, first, labor was dignified and thus should be compensated justly and that, second, as independent men, their autonomy should be respected and preserved. Ornan Crane's negotiations with Law in 1840 and 1841 demonstrated this well. Crane agreed to work at haying and harvest in 1840: he would work in August for $16.50 in specie, in September for $20 in clothing, and, if he chose, in July for $14 in clothing. Evidently more skilled than Crayton, Crane bargained for and received a better wage and more freedom to spend it as he chose. The option to work in July allowed Crane to seek better pay, a substantial concession from Law. But Law was not content with this arrangement, for Crane acted more like an independent contractor than an employee. The next year, Law and Crane emerged with a contract that granted Crane $14.50 per month for haying and harvest with the proviso that Crane would take partial payment in clothes as often as possible and that any days off would be deducted. Crane had gained freedom to ignore Law's calls to take payment in kind but at the expense of the value of his work. Law, for his part, clamped down on Crane's penchant for taking days off at full pay by inserting provisions that docked his worker for such holidays. Thus did men like Law chip away at the prerogatives of skilled agricultural laborers. Though the process was not complete on the eve of the Anti-Rent wars—Law felt it was his duty to care for Crane with room and board when he injured his shoulder—the panic pushed central New Yorkers closer to wage labor.[58]

Great proprietors, for their part, faced a graver crisis and did not have the luxury of making do while waiting for the depression to lift. The Jacksonian boom encouraged them to invest in western lands, state bonds, and stocks with an expectation of windfall returns. Most coupled their zeal for these investments with a strategy of nonconfrontation with tenants and mortgagees at home. When the panic struck, western land prices plummeted while states and corporations went bankrupt, leaving New York's landed elite unable to meet their own financial obligations. Their only recourse was to collect the unpaid debts of farmers or extract a greater return from undeveloped lands. In response, proprietors took the offensive, forcing collections of arrears, holding sheriff's sales to make up bad debts, or evicting the most recalcitrant debtors. The sudden switch in tactics had much to do with the proprietors' acceptance of the general impression that farmers could in fact pay their rents but chose to withhold payments and live off the good graces of their creditors. They were convinced that the radical free soil ideas of urban artisans had penetrated the countryside and led their erstwhile honest farmers to throw in their lot with land redistributionists. In this context, such practices as timber hooking once again became a source of controversy as proprietors now believed them to be part of a broader, politicized larceny. As Law observed, "Timber-hooking is so Rife—and Reputable" that he rarely could catch thieves.[59]

For his part, James Fenimore Cooper did not blame farmers so much as the middle class. The leveling doctrines of the labor movement were the product, he reasoned in 1838, of the greed of the bourgeoisie. To Cooper, the middle class had forsaken culture, education, dignity, and republicanism to pursue economic and political power, regardless of the cost to society. Wage labor, factories, and poverty were the unhappy results of the bourgeoisie's failure to minister to the needs of the weak. Worse, the bourgeoisie mercilessly assailed property rights, particularly slavery. No matter how regrettable the institution, the nation protected all property rights equally. The laboring classes, Cooper thought, were too shortsighted to perceive that the landed elite was all that stood between themselves and wage slavery. The Jacksonians and their middle-class Whig opponents, according to Cooper, appealed to the greed of the wretched, the poor, and the propertyless to gain votes, perpetuate political machines, and advance their personal interests by directing their anger at the natural aristocracy. Soon, he argued, democracy would lead to the tyranny of the majority against the rights of the minority—land redistribution in particular. The source of this moral decline, according to Cooper, was the Yankee, sly and self-serving in his pursuit of someone else's property. Yankees were cantankerous, illiterate levelers under whose machinations all property had become insecure—from timber to mill seats to contracts to land titles. Cooper, indeed, would have upbraided Connecticut-born Samuel Law for taking advantage of the panic to force the working poor into degrading wage labor. To prevent this from becoming general, the novelist called for the revivification of paternalism in central New York and in the nation as a whole.[60]

Cooper came across as a curmudgeon desperately trying to stuff the commercial revolution genie back into the bottle. But he did throw into bold relief the growing division that it created between the landed elite and the middle class in central New York in the 1830s. Commercial expansion elsewhere had brought merchants wealth and power that seemed to elude their counterparts in secluded rural districts. When backcountry merchants traveled to Syracuse, Utica, Albany, or New York, their disadvantaged position became obvious. They were faced, therefore, with the Jacksonian rhetoric of equal opportunity, the Whig message that the best men should rule, and the unavoidable conclusion that so long as the social order encouraged by large estates held sway in the countryside, men of their station would be denied the fruits of commercial success. Occupying a secondary status in rural society, these merchants helped launch a series of events that weaned most of the middle class away from their loyalty to the Jacksonians.[61]

From 1828 to 1836, the Jacksonian Democrats provided enough support to merchants to prevent them from jumping wholesale into the opposing camp. Most significantly, the Jacksonians identified the old landed and financial elite as the source of merchant troubles, charging that the aristocracy's handpicked merchants, millers, and craftsmen monopolized local trade. Shattering such unnatural restraints on trade would open avenues for new forms of economic activity dictated by political consensus and the invisible hand of the market. The delivery system would be distributive politics, which would enable local merchants, farmers, and artisans to petition the legislature for charters for banks, turnpikes, canals, and railroads. Vetted by the people's representatives, the most worthy would be approved, after which the market would determine which were actually completed.[62] Merchants embraced this political economy with zeal, chartering numerous corporations in the 1830s to reconstruct the fortunes of the region. Henry C. Noble, who clerked in his father's store in Unadilla, Otsego County, voiced the optimism of this class of men when contemplating a projected railroad to link Cooperstown to Catskill on the Hudson River: "When such a project shall be carried into effect, then I think our part of the country shall flourish again, for it is the only thing that will shake off this curse that was put upon us by the construction of the Erie Canal."[63]

Such enterprises joined merchants, proprietors, and wealthy farmers together and helped cultivate good relations between potentially antagonistic groups. Local boosters had chartered dozens of turnpikes between 1830 and 1850. Samuel Law, a prominent Whig, sheep raiser (boasting a herd of over 1,200 in the late 1820s), land agent, attorney, and shopkeeper, had helped charter the Meredith Turnpike to link his township with the Erie Canal via Otsego County in 1839. He and his compatriots lobbied the legislature for a charter, purchased the shares of the company, paid to have it surveyed, and complained when farmers refused to support the project. But men like Law were not put off by the public's lack of enthusiasm for such undertakings and generally assumed the burden of building roads themselves as a "self-tax."[64]

Nonetheless, internal improvement schemes like the Meredith Turnpike often failed in the 1830s and 1840s because of jealousies between townships or political factions over economic development. Intensive local politicking preceded efforts to secure a charter from the legislature and, more often than not, continued once the initial camaraderie between township leaders wore off. Law and his fellow promoters, for example, obtained a charter for their road from the legislature by piecing together an alliance among leading men in Delaware County. Primarily Whigs, Law's partners seemed to achieve a coup when they secured the support of Delhi Democratic boss Charles Hathaway. Delhi was both the county seat and commercial hub of Delaware County and stood to lose a great deal of political and economic clout if the road achieved financial success. Soon after its charter, however, the Delhi men made an unsuccessful attempt to take over the corporation to prevent Meredith from dominating its affairs. Then they spearheaded local and legislative attacks on the road, with Democrats leading the charge.[65]

The undoing of the Meredith Turnpike was the result of more than just political jockeying between townships, though at times it was difficult to separate political from other considerations. In point of fact, such projects had a low success rate statewide, which inhibited investment among a public fearful that such projects would raise either taxes or shipping costs. Company officials found that Meredith Turnpike investors were easily spooked about the prospects of the road, even wealthy merchants and farmers who stood to benefit most from its construction. The few locals who took stock did not intend to invest large amounts of their own money in the enterprise. They assumed that after they paid their first 10 percent installment, their subscriptions would not be called upon again but would be used solely as evidence of public support to solicit aid from the state legislature. That is, they hoped to reap a tidy windfall, expecting the road to be built with the initial 10 percent outlay by subscribers and state grants or loans. Thereafter the stock would accrue dividends that would pay remaining subscription calls. Promoters thus met resistance from stockholders once a road failed to secure state funding to cover the construction costs. Law and his fellow promoters learned this lesson in 1843 when skittish investors abandoned the project. "Some of the Gentlemen say they never expected to pay the amount of their subscriptions,—had no idea that the stock would ever be taken up, and of course they should never be called upon for payment &c.," E. G. Barnes confided to Erastus Root, "They set themselves to work to devise means to arrest the progress of the road, in which they have been too successful."[66] Law complained of "pure mob spirit" against his road and that his enemies "drew in enough of their Kin, with making and adding a few dupes" to seize control of the company and shut it down. Another observer believed that low prices for butter caused locals to cut their losses. But it is more likely that the stockholders considered such projects boondoggles and refused to invest their own capital in the project.[67]

Local leaders chartered a number of railroads, too, but as these required larger capital reserves than turnpikes, few survived the planning stages. Even the promising Canajoharie & Catskill Railroad, chartered in 1836 to link the Erie Canal and the Hudson River via the Schoharie Valley, failed when investors defaulted on their subscriptions after the panic of 1837. The company procured a $300,000 state grant in 1838, but by 1842, the money ran out and stockholders demanded the abandonment of the road rather than pay their premiums.[68] Gabriel Bouck summed up the frustration of local boosters in 1848 when he wrote, "If the people possessed the least degree of ambition, so as to have aided in the construction of the Canajoharie & Catskill Rail-Road—Schoharie would have been one of the first counties in the State—instead of being as it is now, an isolated *Paradise.*"[69]

Merchants' bitterness at the continued failure of the Jacksonians to reinvigorate the local economy was compounded by their sense that elites and farmers could escape the suffering. The gentry could live on inherited wealth and rents, while farmers could revert to mixed farming to weather lean times. Yet capital earned through discipline and austerity was a merchant's only path to prosperity, and the pall cast on the region by the Erie Canal made that a difficult road indeed. They had to continue to buy and sell goods to earn a living. And entrepreneurs who had invested in factories, carding or fulling mills, or tanning faced the prospect of bankruptcy as industries sprang up along the canal or in the Great Lakes region. No help seemed to be forthcoming from Democratic politicians, either, for they appeared content to line their own pockets, using their offices to mete out political and economic favors to friends and family. Merchants therefore were the first to seek new solutions to the economic hardships they faced at the end of the decade.[70]

Entrepreneurs became strident in their denunciations of the gentry and common folk after 1837. For one thing, many merchants had invested in a variety of stocks and bonds and western lands and saw their wealth disappear in the crash. The local economy was not structured in a way that would allow merchants to weather this downturn. In a modified version of barter, farmers brought their produce to local merchants, who extended farmers credit to purchase goods at their store. A man might exchange labor for store credit. When the economy faltered, however, farmers either refused to pay creditors or sought additional credit from merchants to service mounting debts. Disillusioned merchants decried the developing agrarian tendency to treat credit as unrepublican and thereby voidable, despite the fact that merchants had carried farmers' debts for years. At the same time, merchants found the gentry no more cooperative. Merchants thus chafed at their social and political subservience, which survived despite their critical role in maintaining marketing and credit networks.[71]

Merchants made their final break with the rest of the rural community as the protracted economic depression that followed 1837 created a dearth of liquid capital that made credit extremely tight. As James M. Wilson of Andes, Delaware County, observed in 1840, "The times is vary hard heare mony is vary scarse but provisions vary plenty of awl kinds."[72] Businesses either shut down or limited new investment. Despite banking reforms in 1843, business recovery would begin only in 1847. In the meantime, village merchants went bankrupt by the dozen as farmers stopped paying their debts and banks ceased extending credit.[73] Shopkeepers, whose republican sensibilities dictated that a man who failed to pay his debts lacked virtue, felt mortified by their failures and betrayed by the caprices of their neighbors. George W. Reynolds, a tanner from Stamford, Delaware County, tried to weather the depression for six years. Finally, in 1842, despite all of his efforts, his business failed:

> To be, or not to be—to owe money & have nought wherewithal to pay it—*that, that* is the question—For the want of a few paltry dollars I am ground to the very Earth—driven about, harassed and abused. The tender chores of the soul, touched with a ruthless hand oh who can tell its bitterness—I can feel—I can realize just how I ought to treat every one of God's creatures—And just how they should treat me—but selfishness & covetousness—what can they see—what can they feel—nought, *nought* but a brute may see & feel—There is no scarcity of human brutes, human *hogs* at the present age of man. Oh that I were out of reach of their deadly fangs.[74]

A victim of forces beyond his control, Reynolds felt pained that people he knew had abandoned him. The mutuality of his republican world had been shattered. Reynolds, like many of his peers, found solace through salvation, converting to evangelical Methodism at the height of his troubles. But there would be political consequences as well.[75]

* * *

In 1838, President Martin Van Buren was learning the hard way that two years is an eternity in politics. The Little Magician had glided into office as the handpicked successor to Andrew Jackson in 1836, thanks to the Democratic coalition of urban working men, southern planters, and backcountry farmers that he had worked so hard to build in 1828. But the Democrats could not weather the economic depression unleashed by the panic of 1837. Cotton prices sagged, banks suspended specie payments, credit evaporated, states went bankrupt, merchants failed, and thousands of workers lost their jobs. The last proved especially troubling for Van Buren, for in his home state of New York, the labor vote was critical to Democratic electoral success—if the president lost labor's support, he might not be able to control the state's delegation at the party's national convention in 1840. Thus he made a fateful decision: he came out in support of a number of demands made by "Locofoco" Democrats, New York City working men who advocated, among other things, free soil in the West. Southern Democrats, already alarmed by abolitionist agitation of the slavery issue, concluded that Van Buren was sympathetic to antislavery and threatened to bolt his candidacy in 1840. More immediately vexing, conservative Democrats in New York State criticized Van Buren for embracing Locofoco "agrarianism" and alienating southerners; they began backing anti–Van Buren candidates. Thus began the famous split between Van Burenites ("Barnburners") and conservatives ("Hunkers") that would plague the party in the next decade, with Whigs and third parties like Anti-Rent the direct beneficiaries.[76]

The Locofocos, in addition to creating sectional schisms in the national Democratic Party because of free soil, also demanded political reforms that would curb corruption. The current political system encouraged parties to promote special interests, which in turn led to "logrolling" in the legislature (exchanging votes in favor of pet projects or candidates) and the elevation of spoilsmen to high positions in the parties. In either case, Locofocos believed, Democratic leaders had traded away the long-term interests of the "producing classes" for personal emolument. The party was ruled by a tight inner circle, establishing the very sort of monopoly the party was pledged to eradicate in political life. The Locofocos advocated fundamental changes in government that would alleviate these problems. They called for an end to distributive government, wherein the legislature chartered all corporations. The process violated the public interest by granting the right to incorporate only to the well connected or to communities with representatives skilled in the arts of trading votes. The result of this political interference in economic development, Locofocos noted, were thousands of unemployed workers, hundreds of businesses in bankruptcy, and state finances in disarray. Laissez-faire government would be more democratic. At the same time, the Locofocos demanded that the party become more attuned to the needs of its core constituency by supporting specific policies and running candidates who were sound on issues. When Van Buren bent to Locofoco radicalism, local political leaders who opposed these changes were forced to reconsider their loyalty.[77]

Whigs correctly gauged in 1837 that a number of Democratic regulars would bolt if offered sufficient rewards, since many disliked the reformist politics of the Locofocos. D. M. Hard gleefully reported to William Averell that dissatisfaction in the Otsego Democracy's ranks could be exploited, perhaps for the long term. He counseled Averell to approach one of his town's wavering bosses: "If Capt. John Bell could be detached from them it would create a grate revolution in this town." The stakes were high all over the county. Henry Ogden of Otego told Averell that the Whigs stood their best chance ever of securing Otsego County. "This Town will in all probability revolt from the Regency in toto," he predicted, and the rumored Democratic county ticket "will meet with strong and decided opposition from the leaders of the old Jackson party here but this should not be told abroad . . . unless a nomination is made to suit the views of certain men here, there will be as I have said a compleat revolt."[78] Though predictions of a complete collapse of the Democratic Party proved premature, it was clear that factionalism posed a lethal threat to its hegemony.

Whigs stepped up pressure in 1838, taking on the Democratic machine in Schoharie County, which lay in shambles because Van Burenites had managed to seize control of the county convention from William Bouck and put forward candidates sympathetic to the Locofoco agenda. Whigs convinced former Democrat Mitchell Sanford, whose brother had married Bouck's daughter, to run for Congress. The young attorney, like many businessmen and professionals, held the Democrats responsible for the economic depression and was convinced that the leadership of the uncontested party had grown complacent, entrenched incompetent party regulars in office, and denied young, energetic, and talented men an opportunity to hold office. After begging Bouck to join him in a bolt to the Whig Party, Sanford accepted the nomination.[79] Mixing bravado with a keen sense of the widening rifts in the Democratic Party, Sanford "wore out two horses" riding "30 to 40 miles a day" for three weeks to speak at political meetings in Schoharie and Greene Counties. When Sanford staged a rally in Fulton, Bouck's hometown, Democrats dispatched Bouck's son with a mob to stifle the daring Whig: "Personal violence was over & over threatened, my meetings were tried to be broken up, horns were blowed around the school house in Fulton & Joe [J. W. Bouck] at the head of it, untill the blood . . . waxed warm in my veins & I guess I gave some blows as well as I could. . . . I first wrote him [Bouck] a letter to know whether he countenanced such actings & doings contrary to equity & good conscience & he gave me no answer & thinks I silence gives consent. . . . The Borsts & Stanton & Shaver entered the field to prostrate me for I have attacked them all."[80] The attack on Sanford typified the family-based, factional politics of the county. The Democrats' use of the tin dinner horn to instill fear and warn of the impending assault drew on traditional forms of rough justice that foreshadowed the Anti-Rent insurgency. The attack warned factions allied with him not to challenge the status quo, but Sanford's pugnacity pleased the electorate. Although he lost the election by failing to carry Greene County, he carried Schoharie County. Nonetheless, Democrats healed the breach under Bouck's watchful eye, aided by the Whig Party, which degenerated amid a series of family disputes one year later.[81]

While Sanford's campaign seemed to represent a triumph of the popular will against an entrenched Democratic Party, the Whig achievement actually was an adept navigation of factionalism. Jacksonian-era rhetoric identified "monopoly" and "aristocracy" as the greatest danger to the rights of the common man, but in rural New York these terms rarely were used to describe individual relationships. Rather, party outsiders appropriated the language to demand more equal distribution of patronage among outlying towns. "Democracy" primarily applied to contending towns or counties and only secondarily to individuals. Dissatisfaction with the Democratic juntos led leaders from outlying towns to argue that "equality" between towns had broken down and to rally to recoup rights lost to "tyranny." As one Otego Whig put it in 1837, "This Town will no longer submit to Cooperstown & Cherry Valley dictation."[82] Angry at the constant passing over of men from the western sections of Otsego County for nominations, the political managers of seven western townships—Burlington, Exeter, Edmeston, Plainfield, Richfield, Pittsfield, and New Lisbon—banded together to oppose the Cooperstown and Cherry Valley clique. They resolved to present candidates to the county convention on a rotating basis by town and, acting in concert, overcome the recent dearth of patronage. From 1835 to 1845, all three counties underwent movements to redraw township or county lines as citizens chafed under real or imagined political wrongs engendered by the existing boundaries. New townships and counties would increase patronage and enable more men to profit from the perquisites of office.[83]

Pressure to decentralize control of county government manifested itself in protracted struggles over the placement of government buildings, ironically leading to an alliance of Whigs and Democrats in the county seats to block such moves. When the Otsego County courthouse and jail burned in Cooperstown in 1841, politicos from the southern and western townships agitated in the state legislature to have the buildings placed in Hartwick or Oneonta. While opponents cited Cooperstown's distance from southern townships, their constant fight over offices with the central Whig and Democratic juntos at the county seat made it clear that they hoped to wrest from Cooperstown its primary claim to ascendancy—the courthouse.[84] The "Hartwick Plan" and the "Oneonta Division," two proposals before the legislature, failed under pressure from bipartisan attacks. Schuyler Crippen wrote to fellow Cooperstown Democrat Seth Doubleday that the Whigs must send a prominent party member to Albany to dispel the Democrats' belief that the Whigs supported either plan. Crippen, revealing the alarm of the county's central powers, proclaimed, "I shall remain here until I hear from Cooperstown doing all in my power to prevent success of the disturbances of the quiet of our county." Democrats contacted Whig leader William Averell to help block the plans. The willingness of the leadership of both parties to throw aside their differences to defeat these measures underscored the importance villagers placed on maintaining firm control of political affairs in the county.[85]

The party system in rural New York continued to break down over the next decade because family, township, and neighborhood ties—which often coincided—still shaped political culture. Feuds over nominations, perceived affronts, and outright double-crossing left wide rifts in local parties that often cost them elections. Democrat Halsey Spencer of Edmeston, Otsego County, complained bitterly in 1845 that a rival faction headed by Ransom Spafford brought "the lame and the lazy" and most of his relatives to the town meeting to seize Edmeston's delegation and nominate himself for sheriff. Rumors that Cooperstown politicians engineered Spafford's candidacy incensed Spencer: "If Spafford is nominated there is a blow up here, any other Candidate if he be liveing out of the County we will support. It was said in Cooperstown last winter in my hearing the next day after Crippen was nominated for Surrogate in a public bar room that Spafford must be

Sheriff and it could not be prevented. I do hope the sober thinking men of this County will wake up and put these things as they should be." Spencer angrily declared, "You must know that I am opposed to dictation and forestalling public opinion in these matters and if this kind of logorolling is to be carried out I must stop going to Elections."[86]

Because of this disorder, Martin Van Buren and the Albany Regency recognized in 1839 that they were in grave danger of losing the countryside and, with it, the 1840 presidential race. With his party having lost the governorship to Whig William H. Seward the year before, the president took what at the time was an unusual step, traveling to rural New York to secure support for the party that fall. But convincing citizens to rally around abstract party policies, such as the subtreasury plan, proved frustrating. Cooperstown Whig attorney Levi C. Turner wrote that party leaders touted Van Buren's economic policies as the harbinger of a prosperous future in local papers, yet when the president visited the village in September, only 1,200 people came to hear him speak. The dreadful showing indicated just how unpopular the president had become; indeed, a circus that came to town during court week outdrew the president by 1,800 people. A large number of local Democrats who normally could be counted on to drum up money and transportation to send their neighbors to Cooperstown to demonstrate party strength had chosen to stay home. Van Buren's decision to refashion the Democratic Party, however critical to retaining Locofoco votes, had run aground in the countryside as voters and politicians wondered what benefits the change would have for them.[87]

Having spoken for central New York farmers, mechanics, and merchants for over a decade, the Jacksonian Democrats now found themselves badly divided. Partly it was their own fault, for they had, as Sanford charged, developed a rigid party hierarchy that appeared undemocratic. Moreover, party ideology preached limited government and local autonomy, a message well suited to a populace bent on preventing "federalism" from reemerging in America but woefully inadequate for formulating policies that might protect the interests of Democratic supporters. Indeed, the Democrats' refusal to push legislation that might appear to benefit one class over another resulted in few laws that would protect farmers, working men, or merchants once the panic hit in 1837. Instead, localism had bred faction, further hamstringing the party as it tried to deal with the crisis. The way was open for either the Whigs or third parties like Anti-Rent to rise to the fore.

3. The Anti-Renters: Agrarians and the Politics of Faction

CAMP. OF. THE. LEATHERHEADS,
ROCKY. MOUNTAINS & MOON
G. H. EDGERTON. YOU. ARE. REQUESTED NOT TO. MEDLE. PLOT.
NOR. PLAN. WITH. THE. LANLORDS. OR. THEIR. AGENTS. LIKEWISE.
TO. SETTLE UP. ALL DIFFICULTIES. WITH. ALL. TENANTS. OWNING.
OR. OCUPING. SUCH. LANDS. BE ASSURED. THAT THIS WARNING. IS.
NOT. FOR. NOTHING. IT. TAR. AN. FEATHR. WANT. MAKE. YOU. STOP.
YOUR. ROBEERIES. A TOUCH. OF. THE. HAK. OR. A. DOSE. OF. PISTLS.
PILS. WILL. CURE. YOU. OF. YOUR. DISORDER.
TAKE CARE,
THE. PUBLICK. FEEELING, is. AGAINST. YOU.
TAR. and. FETHERD.

The Anti-Rent movement began as the Democratic Party teetered on the brink of disaster in 1839 and would help speed its decline in the 1840s as slavery, free soil, and nativism unsettled the political system. The tenant uprising in eleven eastern New York counties has been interpreted in a variety of ways: as a popular rural expression of Jacksonian democracy, as a byproduct of a failed agricultural system that stymied progress, as a free soil precursor to the emergence of the Republican Party in the 1850s, and as an example of the failure of the Jacksonian era's factionalized political parties to enact meaningful (and desirable) land reform.[1] These interpretations have focused primarily on the politics of land reform, but this chapter examines Anti-Rent as an expression of a range of rural concerns with the political economy of late-Jacksonian New York. The movement took place amid a series of political conflicts that grew out of the social, political, and economic consequences of the state's rapid development, and yeoman farmers, artisans, laborers, and even some merchants found common cause with tenants on issues ranging from political reform to debtor relief to land redistribution. Furthermore, calls for expanded local democracy helped unite Anti-Renters and other disaffected groups. Anti-Rent became the most effective vehicle for voicing this melange of discontents because the rent issue could draw together Whigs and Democrats across local boundaries. Its ultimate guise, reformist rather than radical, emerged only after protracted infighting at the local and state level to channel this unrest into electoral insurgency.

Given its ultimate effect on the transformation of politics, the Anti-Rent movement's origins were humble indeed. The movement was partly a product of the panic of 1837, for landlords found themselves bankrupted when their investments in western lands, bonds, and railroad stocks crashed. They had no choice but to squeeze tenants for back rent to salvage their fortunes. Unrest first surfaced in 1839, when the heirs to the Van Rensselaer manor in Albany and Rensselaer Counties began evictions of tenants in arrears. Albany County tenants beseeched Stephen Van Rensselaer IV to extend leniency to delinquents. Many were willing to purchase their farms, but the landlords refused to bargain. Tenants responded by forming an Anti-Rent association and "calico Indian" tribes—disguised insurgents—to resist the law. Early Anti-Rent activity focused on protecting producers' natural rights to the soil and enforcing the notion of common weal that defined rural communities. Calico Indians employed traditional folk rituals—shivaree, rough humor, tarring and feathering, mummery—to humiliate opponents, enforce community discipline, and intimidate enemies. Tenants formed central committees led by township-level politicians to disseminate information and coordinate activities. Still, these efforts did not fundamentally challenge local power structures, nor did they threaten the political system. If anything, tenants sought a return to the status quo.[2]

Though state law certainly supported the proprietors' actions, tenants made their traditional appeal to landlords to uphold their paternalist duty to ensure that honest tenants would not be evicted from the land. When Van Rensselaer IV refused to negotiate with tenant delegations, he denied them the dignity of making a "free bargain," irresponsibly exercised his economic power over their lives, and reduced them to the status of serfs or slaves.[3] Albany County tenants issued a "Declaration of Independence" in 1839 to plead for protection of the agrarian republican order they had fought so hard to build: "We have counted the cost of such a contest, and we find nothing so dreadful as voluntary slavery. Honor, *justice* and humanity forbid that we should any longer tamely surrender that freedom which we have so freely inherited from our gallant ancestors, and which our innocent posterity have a right to receive or expect from us."[4] But Van Rensselaer's continued intransigence evoked a sharp response. One tenant warned that the "landlord's pleasure" would no longer be "law." Violence broke out when the landlord sent lawmen to serve process on farmers; tenants organized disguised bands of calico Indians to stop sheriffs from serving process on them and to fight off militiamen sent by Governor William H. Seward to uphold the law in December 1839.[5]

Anti-Rent would not be a movement of the destitute; rather, tenants were drawn from the middle strata of rural society, which had important consequences in shaping the movement. Most tenants listed on the rent rolls of three estates in Delaware, Otsego, and Schoharie Counties that produced Anti-Rent activity, for example, owned at least part of the land they farmed in 1850. Ironically, tenant heads of household were more likely than their neighbors to own land. In Delaware, only 11 percent of Margaret Livingston's tenants—her tract stretched through Middletown, Andes, Delhi, and Bovina—reported no real property in 1850, while 15 percent of Delaware County farmers and 33 percent of all county heads of household owned no land. Of John A. King's Schoharie tenants living in Gilboa, Blenheim, Jefferson, and Summit, 23 percent were landless, compared to 14 percent of Schoharie County farmers and 40 percent of all residents. All but 15 percent of George Clarke's Otsego County tenants in Cherry Valley, Milford, and Middlefield owned at least some real estate, while 28 percent of Otsego County farmers and 36 percent of the other heads of household reported no land. No wonder King's tenants recalled in the 1890s that they considered themselves successful farmers in 1839.[6]

However, these tenants did labor under certain disadvantages. They were less likely to have been born in the state, and in Delaware nearly half of the tenants were foreign-born. Thus, they did not have the same kinship networks to ease credit or labor shortages or weather bad years as other farmers. And tenants had larger households, suggesting that they had more children to offset the lack of kin or friendship support.[7]

Table 1. Real Property Ownership and Value among Household Heads in Central New York, 1850

	Delaware		Otsego		Schoharie	
	Tenants	All HH	Tenants	All HH	Tenants	All HH
HH who owned R.E.	89%	67%	85%	64%	77%	60%
Value R.E.						
Ave. all HH	$1,709	$1,196	$2,093	$1,484	$1,482	$1,352
Mean	$1,400	$1,510	$1,500	$650	$1,200	$500
Ave. for Farmers	$1,721	$1,607	$1,973	$2,030	$1,719	$2,147
Mean	$1,400	$1,200	$1,500	$1,250	$1,400	$1,400

Source: Manuscript United States Census, 1850, #432, Schedule I, Delaware County, Rolls 494, 495, Otsego County, Rolls 579, 580, Schoharie County, Rolls 595, 596, National Archives, Washington, D.C.

Table 2. Age, Household Size, Occupation, and Birthplace of Tenants and All Household Heads in Central New York, 1850

	Delaware		Otsego		Schoharie	
	Tenants	All HH	Tenants	All HH	Tenants	All HH
Ave. age	49	44	48	44	46	43
HH size	7	5	6	5	7	6
Occupation						
Farmer	84%	66%	81%	60%	82%	48%
Laborer	1%	12%	0%	6%	6%	19%
Other	8%	13%	14%	26%	8%	25%
No occ.	7%	9%	6%	8%	4%	8%
Birthplace						
New York	57%	75%	61%	69%	78%	87%
Other U.S.	3%	14%	33%	25%	14%	7%
Foreign	40%	11%	6%	6%	8%	6%

Source: Manuscript United States Census, 1850, #432, Schedule I, Delaware County, Rolls 494, 495, Otsego County, Rolls 579, 580, Schoharie County, Rolls 595, 596, National Archives, Washington, D.C.

Note: Figures may not add up exactly to 100%.

But farmers could not necessarily pay their rent or expand their farms to sufficient size to provide for their children after their death. The seventy-five Livingston tenants who owed arrears (62 percent of the 103 traced in the 1850 census) averaged $60 in back rent, a sum difficult to repay in an era in which workers earned about $300 annually. And among King's tenants, the average value of their property lagged behind that of freehold farmers, suggesting that they had poorer holdings and limited prospects for improving them. Last, in an era in which wealth was closely correlated to age, tenants, who were slightly older than other heads of household in their counties, had accumulated holdings of only average value. With less productive years remaining compared to their peers, tenants no doubt felt concerned for the future, despite their accomplishments. Besides offending their republican sensibilities, therefore, paying rent to a "nonproducing" landlord also represented a constant drain on limited resources.[8]

The structure of Anti-Rent associations indicates that tenants embraced the agrarian values discussed in the previous chapters, which explains why some freeholders, artisans, laborers, and shopkeepers also joined the movement. By any measure, the men who organized Anti-Rent associations in the counties lived up to Jacksonian standards of civic virtue and earned leadership positions in the community through hard work, personal integrity, and service. For example, Thomas Peaslee and Benjamin P. Curtis—the leaders of the Blenheim, Schoharie County, Anti-Rent forces—were very respectable citizens. Peaslee was sixty-seven years old in 1850, was married, had three of his children (including his married son, Nathan—an active Anti-Renter) living at home, and listed himself as a farmer with $3,000 worth of property. His neighbor, Curtis, was a fifty-seven-year-old farmer, was married, had six children living at home, and reported $4,000 worth of real property. Both men were Methodist lay preachers. The Mayham family, which included Anti-Rent leaders John and Steven, had been among the first settlers of the township and were prominent in county politics. The "chiefs" who led calico Indian bands were less prosperous but not poor.[9] For their part, rank-and-file calico Indians, according to one study, were younger and not established as farmers, on average thirty years old and married, with a similar quantity of land as freeholding neighbors, though with less improved acreage. Roughly 60 percent of the Indians were in their twenties, 64 percent were farmers, and 18 percent laborers. The Anti-Rent hierarchy thus mirrored the broader patriarchal social order of the countryside.[10]

From the start, Anti-Rent activity evinced a commitment to localized democratic action to achieve tenant goals. Anti-Rent associations acted much like the committees that central New Yorkers convened to nominate candidates, run town meetings, or negotiate with elites. Such bodies claimed to represent the people because they employed ritualized democratic procedures—electing presiding officers, drafting and presenting resolutions. Committees were democratically elected, though the best men generally received the support of the people. These formalities helped reinforce the system of mutuality that held rural society together. Thus, Albany County tenants formed central committees to disseminate information, coordinate activities among different townships, and petition the legislature in 1839 and 1840. But, significantly, they focused on security of tenure and enforcing the notion of common weal that defined rural communities. They had straightforward demands: they did not wish to pay taxes on lease property; they wanted the right to challenge landlord titles in court and be compensated for improvements to the land; and they asked landlords to sell the land at a fair price.[11]

Seemingly at odds with the decorum of citizens' committees was the intricate folk drama of dressing as Indians to protect communities and enforce conformity to the Anti-Rent message. Yet calico Indian activity sprang from the same commitment to the defense of the patriarchal household and the rights of the weak in a paternalist society. In 1839, calico Indians (usually dressed in bright calico disguises, sometimes simply with blackened faces) focused on preventing law officers from ejecting tenants or selling their personal property, presented an inspiring spectacle at Anti-Rent celebrations, and intimidated or punished opponents. Composed of younger men acting with the blessing of respectable leaders of Anti-Rent associations, Indian

Former Anti-Renters posing in disguise in the 1890s. Negative contained in the Henry Christman Collection. Courtesy of the New York State Historical Association, Cooperstown, New York.

bands initially tended to support rather than challenge prevailing social and political structures.[12] It was only after the creation of a state Anti-Rent Association in May 1844 that the Indians became the driving force behind the radicalization of the movement and, for critics, became the symbol of the lawlessness of the movement.[13]

The rituals and tactics the Anti-Rent Indians used and the way they identified their own helplessness against landlords with that of Native Americans against the onslaught of "white progress" have been extensively covered elsewhere. What concerns me here is less the forms of dissent than its ideological basis and political consequences. For as tenant resistance escalated, landlord determination hardened, and the 1840 election saw leaders of both the Whigs and Democrats maneuver for tenant allegiance.

The intermingling of Anti-Rentism with local, state, and national party upheavals came quickly. With the 1840 presidential contest looming, Governor Seward tried to take advantage of the opportunity to woo the traditionally Democratic tenants into the Whig fold. Believing the lease system to be oppressive and anti-republican, he challenged landlords to allow tenants to purchase their land so that they might have the incentive to improve their farms, escape poverty, and become productive members of commercial society. Meanwhile, at the governor's prompting, the Whig legislature passed bills designed to encourage the voluntary sale of manor lands to tenants or face state seizure under eminent domain.[14] Some conservative Whigs attacked Seward's handling of the manor question as Locofocoism and ominously resolved not to support the reform wing of the party. Yet most conservatives got caught up in the thrill of defeating the Democrats. Long-suffering Samuel A. Law of Delaware County, for example, exclaimed six months before the election, "The Harrison ball rolls on. Log cabins, cider & baked beans bid fair to beat Hickory Poles all hollow."[15] In what is widely considered the first mass political campaign in American history, Whig William H. Harrison defeated incumbent Martin Van Buren by mobilizing voters with hard cider, tales of Harrison's exploits in the West, and parades and festivals centered around Whig symbols. Even New York voted for Harrison. Merchants in particular abandoned the Democrats. How tenants voted is not known, but many undoubtedly sided with the Whigs.[16]

The Albany Regency fell victim to a bitterly ironic fate. In an ideological sense, Van Buren and his emerging Barnburner coalition shared many of the sensibilities of the tenantry—indeed, Van Buren had cut his political teeth fighting landlord power in Columbia County. Barnburner support of Locofoco reforms, especially free soil, was based on the notion that producers were the bone and sinew of the Republic and had earned the right to have access to productive resources, much like the Anti-Renters' calls for an end to manor tenancy. But other elements of the Regency program proved politically debilitating. Both tenants and their leaders largely lived in underdeveloped sections of the state where cash was scarce, transportation routes were arduous, and populations were declining. Barnburner calls for reforming the party or ending distributive politics foreshadowed a direct decline in the economic and political fortunes of the backcountry. Though Van Buren supporters probably held a plurality of votes in manor districts, they could not count on winning elections if Whigs and Hunker Democrats made common cause.[17]

The 1840 election demonstrated that this danger was real, for conservative Democrats had been busy subverting Van Buren's candidacy. Schoharie County was the center of activity as William C. Bouck, smarting from rough handling at the hands of Regency men and Whigs in 1838, laid the foundation for the defeat of Barnburnerism in the state. Bouck secured the Democratic gubernatorial nomination by posing as a moderate, but even as he headed the state ticket, he secretly funded the publication of a pro-Harrison newspaper, *Big Paw,* in Schoharie. Bouck's primary motive was to defeat Van Buren and Locofoco radicalism, which he considered only slightly less dangerous than Seward's flirtations with antislavery. Harrison and his running mate, John Tyler, were conservative Whigs and therefore more acceptable to

Bouck and his Hunker friends. Practically, too, the move made sense. With Van Buren's faction defeated at the federal and state level, Barnburners in the counties would get no patronage while Bouck could count on a few favors from a Whig administration in Washington. Bouck had made a very clever calculation. By running on the same ticket as Van Buren, Bouck could demonstrate absolute loyalty to the party. Then he could claim in 1842 that the reason for Democratic defeat was that Van Buren had tried to push too radical an agenda on voters. Better yet, having lent a hand in the election of Harrison, he could ask conservative Whigs to stand aside in a second contest with the Seward wing of the party. The trick would be to ensure that Anti-Rent, which had spread into Schoharie County, did not derail his efforts. But Bouck had that covered, too, for some of his strongest political allies resided in the manor townships in the southeastern half of the county.[18]

The political situation in Washington and Albany proved fortuitous for the Hunkers. President Harrison died soon after taking office, which placed Tyler, an ardent pro-slavery Virginian, in office. Tyler vetoed laws passed by the Whig Congress, believing that the Seward wing of the party meant to attack slavery through tariffs and an array of other laws that favored northern interests. Tyler was so alienated from his own party in New York that he lavished patronage on Hunker Democrats rather than see reform Whigs obtain any more power. Seward had troubles of his own in Albany. Whig spending stretched state finances to the breaking point as the depression continued unabated, inviting Democratic efforts to pass a "Stop and Tax Act" in 1842. For their part, conservative Whigs grumbled that legislation to adjudicate the rent issue and enact debt relief threatened the sanctity of contracts while encouraging tenant resistance to the laws. Meanwhile, tenants peppered the legislature with petitions for relief that the hamstrung Whig party could not act upon.[19]

With the tenant vote up for grabs, Bouck took decisive steps to win over Anti-Renters. First, Bouck deployed his township lieutenants with great tactical skill, encouraging them to mobilize tenants—Hunker, Barnburner, and Whig alike—in favor of the Hunkers. Prominent Schoharie County Anti-Renters like David L. Sternberg of Livingstonville, John Mayham of Blenheim, and I. W. Baird of Summit had long associations with Bouck and had no difficulty squaring support for him with representing the tenants they served. Second, early in 1842 he funded the publication of an Anti-Rent sheet, the *Helderberg Advocate,* edited by William H. Gallup, late editor of *Big Paw.* Though he kept his involvement in these moves secret, Bouck's decision to court Anti-Rent votes gave the rent issue political legitimacy and placed the prestige of a powerful political leader in the tenants' corner.[20]

Bouck benefited from this alliance but soon had cause for concern. Following the lead of their Albany County neighbors, tenants in Schoharie County began organizing Anti-Rent associations on the estate of conservative Whigs Jacob Livingston and John A. King in Broome, Conesville, and Blenheim early in 1842. Calico Indians soon appeared and began causing landlords and their agents trouble. In Albany, Seward admitted defeat on the rent issue and announced that the laws of the state would be enforced. Landlords and their agents, buoyed by the legislature's refusal to take any further action in April 1842, launched attacks on Indian strongholds throughout eastern New York to prevent a bipartisan union of tenants. In Schoharie, Livingston struck decisively. One Barnburner approvingly wrote, "When I arrived at Livingstonville on my way I found the place all in commotion caused by a visit of the Sheriff with a posse of Constables with bench warrants agt. some of the Livingston tenants. The delinquents had made their escape to the mountains & I saw that the alarm had passed round to a considerable extent."[21] Events in Schoharie County also caught the interest of former Irish Chartist, Locofoco, and current National Reformer Thomas A. Devyr, who flooded the *Advocate* with editorials stigmatizing landlords as "nonproducers" and calling for the destruction of the manor system, the opening of western lands to actual settlers, and the creation of an urban-rural working-class party to unseat the elites who led the Democrats and Whigs.[22]

Whatever Bouck had in mind when he funded the *Advocate*, it now was moving too close to Locofocoism for comfort. Gallup, taking cues from Devyr and deploring the high-handed tactics of lawmen, stated in one article that two young men caught in Indian disguise could not receive justice in the courts because proprietors had co-opted them. Another article advocated vigilantism: "It is *high time* that *more energetic measures* should be adopted to ensure to the tenants that *respect* which is justly due to them as citizens of a *free* Government." Gallup supported the Schoharie Indians and their demands for land redistribution: "The *Indians* it is true have been rather troublesome in Schoharie County but it is *a question in the minds of the tenants* whether it would not be *equally proper* to issue a proclamation proclaiming all these *great Landlords a nuisance in Society and banishing them from this land of liberty and Equal Rights.*" A Schoharie County grand jury was convened to consider whether the *Advocate* was seditious. The jurors found an article that supported Irish tenant uprisings and "revolutions," no doubt penned by Devyr, the most dangerous. The grand jury wrote that the *Advocate* "contains sentiments dishonorable to the character of a free & Enlightened people" and that the newspaper "is in a high degree seditious and calculated to unsettle the right of property in the County." The *Advocate,* they declared, promoted a general disrespect for the laws of the state and county. Gallup shut the paper down, but Anti-Rent had gained considerable momentum from its short career.[23]

Above all, the *Advocate* gave radicals like Devyr a platform to urge Anti-Renters to renounce paternalism and the manor system, as well as to recast property relations and remove nonproducers—landlords, bankers, factory owners, entrepreneurs, attorneys—from their dominant position in society and politics. This message found a keen audience after three years of legislative failure to act on behalf of the tenants, who believed the body was hopelessly compromised by the landlords. Most agreed with Devyr, who wrote in 1842 that landlords, "content to forego the dignity of man's nature[,] are willing to live by preying upon the produce of others' toil. . . . To this genus belong Dukes, Earls, aristocrats, and Highwaymen." By appealing to the tenants' labor theory of value, Devyr deftly nullified the notion of a beneficent republican gentry that the agrarians of the 1820s had been willing to concede. He also brought farmers closer to the ideas expressed by Locofoco workingmen; that is, landed and financial capital *both* oppressed labor.[24]

Tenants now concluded that landlords had not willingly upheld the reciprocal relations of paternalistic society but had skimmed the wealth of both land and people. H. G. Munger, of Kortright, Delaware County, recalled that during his childhood, "the tenants became enraged to desperation, they having made all the improvements to the land, built up houses, churches, school houses, etc., while [Goldsbrow Banyar] the supposed owner lived in New York City [and] had not paid a cent of taxes toward the expenses of the County, but had collected about $40,000 a year as rents."[25] Devyr echoed the tenants' assessment of what had gone wrong in the Republic: "Here is a man-formed right of *property,* destroying the God-formed right of *existence.*" Only democracy—a mobilized people—had ever held proprietors in check. Democracy offered the only way to bring moral order back to a Republic stained by the exploitation of the many by the few.[26]

As the movement evolved, tenants also targeted lawyers, capitalists, and merchants for criticism. In the Albany *Anti-Renter,* Devyr charged that "Lawyer Legislation" had created the battle between labor and wealth, and "the practice of legal profession, is inconsistent with a pure morality or strict consciousness" because attorneys accepted money from whoever paid them, regardless of guilt. The human devastation wrought by the amoral practices of those in government, preponderantly attorneys, was evident in the fact that the working classes *"do not respect themselves."* They became "willing dupes" of wealthy elites and allowed political parties to set laborers against each other. Farmers, Devyr argued, had been degraded the most and were "spat upon, or neglected, by . . . men of influence."[27] The resulting social condition was "false, a vicious one. Selfishness—individual aggrandizement is the motive power which sets every man to work." Elites had thus put "other men into an unnatural and dependent position."[28]

The politicization of Anti-Rent—not to mention the fiery rhetoric of Devyr—magnified an already desperate factional struggle in Delaware, Otsego, and Schoharie Counties in 1842. Escalating Barnburner and Hunker squabbles threatened to negate Bouck's efforts to curry favor with tenants. At the Otsego County Democratic convention, secessionists from the western part of the county, led by Laurens Hunker Levi S. Chatfield, stirred controversy by opposing the nomination of Bouck for governor, which the rebels believed was being rammed down their throat by Cooperstown leaders.[29] George A. Starkweather of Cooperstown informed Bouck that he had to take extreme action to bring dissidents into line. He seized one politician in the assembly chamber and warned "that if he desired to preserve his *good* standing in the State & especially the County of Otsego, he must not be among the number to put a rasor to your throat."[30] Otsego Democrats believed that Bouck authorized an attempt by fourteen Schoharie delegates to renege on their agreement to nominate Hunker S. S. Bowne for Congress as part of a rotation between the two counties. This was no fight over ideology; rather, Schoharie Democrats, no doubt prompted by Bouck (though he denied involvement), feared that an Otsego Democrat would be susceptible to Barnburner entreaties.[31] Whigs endured similar squabbling. The Otsego County Whig fight over nominations for sheriff became so ugly in 1842 that the party's state committee angrily instructed William H. Averell to clear it up immediately before it doomed the entire ticket. Most alarmingly, conservatives led by landlord Jacob Livingston of Cherry Valley threatened to leave the party should Seward, Thurlow Weed, and other reformers continue to dominate Whig councils.[32]

The election vindicated Bouck's strategy, and he handily defeated Whig Luther Bradish in November. But privately Bouck feared that "this is a grave matter and may lead me into difficulty, but I must do my duty and take the consequences. The tenants I fear expect more from me, than my duty will allow me to do."[33] His calls on landlords to voluntarily sell farms to tenants fell on deaf ears once Hunkers and Barnburners officially split in 1843. Barnburners gored Bouck on the rent issue. In counties controlled by Van Burenites, such as Columbia, confrontations between lawmen and tenants escalated as Whigs and Barnburners tried to drive a wedge between the governor and tenants. Barnburners correctly believed that Bouck would not publicly accept the Anti-Rent challenge to property rights and would ultimately have to use force against his erstwhile allies. With conservative Whigs and Barnburners condemning Bouck's "soft" stand on Anti-Rent in the rest of the state, Democrats would have to replace Bouck if they hoped to win in 1844.[34]

With land reform stalled once again, frustrated tenants formed their own political organization in 1844, the Anti-Rent Equal Rights Party. Because both reform Whigs and Hunker Democrats failed to achieve any progress on the rent issue, tenants undoubtedly realized that unless they demonstrated electoral solidarity, Barnburner Democrats also would not respond to calls for relief. Thus Anti-Renters created a state organization at Berne, Albany County, in May 1844 and decided to form a new party. The latter was strongly advocated by radicals Dr. Smith A. Boughton of Columbia County—who was known in Indian disguise as the orator "Big Thunder"—and the tireless Thomas Devyr. Boughton was no stranger to agrarian radicalism, having traveled to Canada to fight in McKenzie's Rebellion, which had begun in 1837. On his return he took up the cause of the tenants, lobbying the legislature incessantly for several years. Boughton now agreed with Devyr that tenants had to adopt more strident methods. Both men favored running separate candidates, but Whig, Hunker, and Barnburner delegates refused to do so for fear of helping one or the other faction achieve political ascendancy in the state. Instead, the Anti-Rent Equal Rights Party would endorse candidates from either party who pledged support to the cause. If any moment in the history of the Anti-Rent movement explains its ultimate unraveling, it was this, for it guaranteed that the major party factions would court Anti-Rent votes with abandon. But it also provides a window into the Anti-Renters' understanding of their place in the political system. They believed government had failed them, and they were willing to take increasingly radical actions to achieve the end of the manor system, but they were not willing to renounce factional loyalties or, by implication, the political, personal, kin, and township networks they were based upon.[35]

The nonpartisan approach did not prevent the convention from aggressively promoting the new party. Delegates decided to send representatives into forty-two New York counties to organize new associations, Indian tribes, and an Anti-Rent political party in hopes of influencing the 1844 election. The new associations levied a tax upon members of one and one-half cents per acre to finance activities. New Indian bands were set up to prevent sheriff's sales, and the state association provided lecturers and material aid. Boughton and Devyr spoke at numerous Anti-Rent meetings, exhorting farmers to expand their movement, join with organized labor in throwing off the shackles of unrestrained capital, and forge a free soil movement. Along with National Reformers, Anti-Renters denounced all forms of monopoly and called for free access to the public lands. George Henry Evans lauded the convention in the *Working Man's Advocate,* while other National Reformers, such as Alvan Bovay, stumped the manor counties to receptive audiences. Local associations performed much of the work of spreading the movement into neighboring districts. The Blenheim, Schoharie County, association "colonized" nearby Roxbury, Delaware County, which in turn organized nearby Ulster and Greene Counties.[36]

The coalition of Anti-Rent and National Reform made sense ideologically once the major parties failed to respond to Anti-Rent's moderate early strategies. Many Anti-Renters, especially those aligned with the Locofoco wing of the Democratic Party in the 1830s, agreed with Evans that "in strict justice, the landless ought to be put in immediate possession of their share of the appropriated soil." Democratic tenant leader Dr. Jonathan C. Allaben of Middletown, Delaware County, championed the *Working Man's Advocate* as a friend of the dispossessed: "The principles advocated by the National Reformers are in perfect accordance with my own view, and their sentiments relative to the distribution of the public domain harmonize with the system of Anti-Rent. Inseparably connected, one cannot move without bringing with it the other."[37] Devyr felt so enthralled by the prospects of an alliance between Anti-Rent and National Reform that he founded a new paper, the *National Reformer,* to espouse the cause. A unified urban and rural workingmen's party, he hoped, was about to emerge in New York.[38]

With the radicals in the vanguard, Anti-Rent swept through eleven eastern New York counties in 1844. The calico Indians had much to do with the success. Under the state organization, Indian bands were reorganized into secret cells and received greater support. Although Anti-Rent associations denied affiliation with the calico Indians, the same men led both groups, and the associations supplied money to equip the Indians.[39] While this is well known, less often recognized is the enthusiastic response by local folk—tenant and freeholder alike—to the radical message; in fact, citizens seemed to push the movement toward an attack on all debts: rents, store credit, mortgages. This turn of events sent up a howl of protest from local conservatives, particularly in Delaware County. Samuel A. Law warned his employer of "a pocket, all round us, of tenants,—refusing to pay rent, a next o'kin of mobocracy, is broiling and boiling up confusion. And agitation and discontent edge along,—from tenants to free holders,—from rent payers to Interest payers,—whispering that abomination, Repudiation!"[40] Deputy Sheriff Timothy Corbin Jr. of Middletown wrote Governor Bouck on July 12 in a state of near panic because "tremendous assemblies of men are daily congregating together using the most inflamatory Language and threatings against the civil policy and good government of this State and to the disgrace of our republican institutions." Corbin demanded law and order because "peaceable citizens [are] daily threatened by Lawless individuals *assuming* to be Savages and is our chief Magistrate quiet and at Ease when all about him is tumultuous Mobs and viscious assemblies Threating and Menacing their Fellow citizens with bloodshed and devastation if the laws are Enforced to Enable the Creditor to Collect his honest dues or the Landlord his rents I cannot believe that you would be Silent upon this grave Subject did you behold its gravity is Lynch Law under your administration to be substituted for the Statute."[41] Indians vindicated Corbin's fears soon after by giving him a coat of tar and feathers.[42]

Participation in the bands radicalized the tenantry as they expanded the scope of their activities. Indian bands organized in 1844 proclaimed that the "natives" had exclusive right to the soil. Kortright tenant H. G. Munger recalled that the local association "issued a proclamation to their brethren that as the land did not belong to the landlords they having no title, and that the State would not take any action in the matter saying they had not right to it, it must of course belong to the Indians." Private property was to be abolished and the land redistributed among its original owners—the metaphoric "Indians."[43] Meanwhile, Devyr explained to Anti-Renters that they could incorporate "Indian" rights to the land with more traditional republican ideals. He argued that each citizen deserved an equitable share of land. No individual or power could usurp that right, and no man should possess land he could not cultivate. Devyr's republic would require "hardly any 'law.' . . . None for the collection of debts. None except a Defining Record for the holding or transfer of possessions in land. No occasion for borrowing and therefore no Interest. In a State so constituted, a fixed limit to the possessory farm, say fifty acres now, and if necessary that lessened to forty, thirty, or twenty, more or less, in succeeding years."[44] Devyr envisioned a nation in which the state held all land in fee and could apportion it to individuals without conveying absolute title. A man would settle his family on an unoccupied plot of land, clear it, and farm it. A man's heirs would retain rights to the plot so long as they remained disposed or able to cultivate it. Rents, mortgages, and debt could no longer pose a threat to occupancy of the land.[45]

Though moderates were uncomfortable with National Reform and calico Indian radicalism, they countenanced it in the face of legislative hostility to the cause. In Schoharie County, for example, Hunkers—no friends of Locofocoism—nonetheless used language similar to the radicals when they begged

Bouck for help in August after nearly two years of frustration, "as Citizen of a free and inlightened government think They have a right to ask it. we are opposed to Large Land monopolies as will as all other monopolies, in our free government especially those Lands held by grants from dispotit governments, whose dates are back of our acknowledged independence, and Let or Leased not to Tenants or Surfs might well be said, under the old dispotick Fudal Sistim. we have petitioned the Law makers for relief—& we have been denied that relief at their hands. . . . When we turn to the law for Redress we are their Estoped."[46] But by August, Bouck, like Seward before him, had admitted defeat on the rent issue. Rather than cause further divisions in the state party during a presidential election year, Bouck deferred to the Albany Regency and stepped aside so that a "neutral" candidate, U.S. senator Silas A. Wright, could run for governor. The Anti-Rent Equal Rights Party endorsed Wright, in part because the Whigs ran conservative Millard Fillmore, who had called for a "law and order" suppression of the tenants. Wright won the election and carried the manor counties, providing hope that he would address the manor issue. The Anti-Rent Equal Rights Party had reason for optimism, having won assembly seats in Albany, Rensselaer, Schoharie, and Delaware Counties.[47]

Anti-Renters would soon rue their decision to support Wright, though they had no alternative. Wright had made no public pronouncements on the rent issue in 1844 but made clear his opposition to Indian radicalism before he took office. The governor-elect had grave doubts about the tactics of the Anti-Renters. Then Indian "outrages"—including the fatal shootings of two men in Rensselaer and Columbia Counties and the arrest of Smith Boughton in connection with the killings in December—exacerbated his fears of Anti-Rent radicalism. He persuaded Bouck to take strong action against the Anti-Renters in Rensselaer and Columbia. The show of force increased the already tense situation. By January 1845, state troops occupied Hudson, at least one rescue mission by the Indians had been scuttled by Boughton (who feared it might provoke civil war), and Anti-Renters had threatened to boycott New York merchants if the city sent militia units upstate. Wright assured tenants that he would not negotiate until the Indians ceased their activities. While the manor leases contained a number of obnoxious clauses, he said, the people had a duty to support the laws of the state. He also made it clear that Democratic Anti-Renters would be punished if they resisted law and order under a Barnburner administration. In his first address to the state legislature on January 7, 1845, Wright, under pressure from the landlords' Freeholders Committee for Safety, resolved to end the Anti-Rent ferment by any means necessary and asked for and received a law making it a felony to appear in disguise.[48]

Calculated to drive a wedge between moderates and radicals in the Anti-Rent ranks, Wright's actions had the opposite effect. In Schoharie County, for example, tenants won a majority of seats on the board of supervisors in February 1845 and thus controlled the county's expenditures and patronage. Moderates were not keen to have the Indians continue their activities—there was a profound difference between Indians defending the people from a corrupted government and having vigilantes ride around the countryside enforcing the will of the ruling party. More troubling, Schoharie Indians had ventured into Otsego County to prevent the sheriff from serving papers for a store debt. Moderates did not support this broader assault on other forms of debt and made efforts to curb Indian activities. When the legislature rejected a petition for redress signed by 25,000 tenants in mid-February, however, moderate tenants ceased efforts to curb Indian tribes. A surge of violence followed.[49]

"Law and Order" forces—Barnburners and conservative Whigs—undertook coordinated attacks in Delaware and Schoharie Counties. On February 11, Delaware County undersheriff Osman N. Steele arrested Daniel W. Squires for his part in the tarring and feathering of Timothy Corbin. Anti-Renters did not retaliate, so Steele and Charles Parker went to Andes on March 10 to take more prisoners. The Andes Indians took them hostage to secure Squires's release (and to tar and feather them). The two escaped to a tavern and held the Indians at bay until a Delhi posse came to rescue them. Two days later, Sheriff Green More led a four-hundred-man posse to Kortright, where More had been turned back by seventy-five armed Indians on March 10, to sell the personal property of tenant John McIntosh. The Indians, boasting that it would take a thousand men to force a sale, remained in the woods at the sight of the large force. Later that day, a posse of eighty men under Steele, Erastus Edgerton, and James Howe clashed with an estimated one hundred Indians in Roxbury. In a hand-to-hand fight, an extremely rare occurrence during the Anti-Rent conflict, the posse took ten prisoners and marched them to Delhi.[50]

Delaware County teetered on the brink of civil war. The Indians demonstrated on the outskirts of Delhi and threatened to storm the village and turn it into a "turnip patch" if the prisoners were not released. The Indians sent to Blenheim for reinforcements. Hard-pressed themselves, the Schoharie Indians chose not to help. The Delaware sheriff again began collecting prisoners, bringing in fourteen Roxbury men indicted for the Corbin incident on March 14. Delhi authorities called up three hundred men from "loyal" townships to defend the village on March 15. Four of the twenty prisoners were convicted in May, and three were sentenced to two years in prison.[51]

The Schoharie Indians' decision revealed a fundamental weakness in Anti-Rent—localism. The movement could not succeed if the different associations remained divided by neighborhood, county, or party affiliation. Landlords and lawmen labored under no such liabilities. Realizing that the Delaware and Schoharie tribes had to be neutralized at the same time, Schoharie authorities attempted to round up Anti-Renters. Schoharie County sheriff John S. Brown and his assistant, Tobias Bouck, rode into Blenheim on March 24, 1845, on behalf of John A. King to serve papers on Stephen Mayham and other members of the association. Instead, 150 Indians dragged them out of a tavern in North Blenheim. They clubbed, stamped, beat, choked, and poked the officers, then took them to a small hill called Baldwin's Heights, surrounded them in a hollow square, and burned their papers. Tar and feathers were produced, but the chiefs maintained order and sent the two lawmen home.[52]

Many Schoharie County residents, including some Anti-Renters, feared the Indians had gone too far—especially under the tense circumstances—and sought to curb their activities. Anti-Rent leaders wavered but in the end found the people committed to the radical course. At an open-air meeting during general militia training, Blenheim Anti-Rent leader Thomas Peaslee counseled his listeners "to live within the law." The crowd, however, reacted enthusiastically to calls for more resistance and shouted an anti-Indian speaker from the stage. Two days later, 1,500 Indians rallied before a large crowd in Summit. The tenantry had reached a critical point. They rejected the moderation of neighborhood leaders like Peaslee, the repositories of patriarchal authority. Confident that they were about to usher in a new order, replete with free land and equal rights, renters coalesced behind the insurgents.[53]

Developments at the state level complicated these local divisions. In April, Ira Harris, a Whig Anti-Rent politician from Albany County, and C. F. Bouton, an Anti-Rent Hunker Democrat from Rensselaer County, bankrolled the Albany *Freeholder*, hoping to forge a coalition of Whigs, Anti-Renters, and Hunkers to bring a negotiated settlement to the conflict. They offered Thomas Devyr the editorship, despite his opposition to working with the two major parties. The alliance foundered within months. Not only did the Indians' war with the authorities discredit the movement outside the manor counties and threaten potential alliances with other groups, but also Devyr used the paper to denounce moderates:

> Some of the Anti-Renters are denouncing any violent resistance to injustice, and resolving that they will only resort to peaceable and constitutional means. This is bad policy, and may be the means of producing the evil they aim to

> prevent. In surface matters it might be best to rely on peaceable remedies; but when inalienable rights are concerned, those who are robbed of them are daily losing part of their existence, and should never voluntarily resign the quickest way of recovering their rights. It may be well to suffer where wrongs are sufferable; but reserve, by all means, the right to decide when they are sufferable. If you leave this matter altogether with majorities, majorities may rob you till death. If a man were to preach non resistance to slaves, would he not be deficient in intellectual honesty?[54]

In June 1845, Harris and Bouton removed Devyr from the editorship as Indian violence threatened any hope of a legislative remedy to the manor issue.[55]

The violence of the summer culminated in the death of Delaware County undersheriff Osman Steele at the hands of the Indians in August 1845, which ended radical influence in the movement. Governor Wright declared a state of insurrection in the county and sent state militia to Delhi to help Charles Hathaway and the Barnburner faction put down tenants by force. Initially, government tactics appeared justified and citizens statewide denounced the tenants. As local lawmen and state militia scoured the countryside for Anti-Renters, the Barnburner machine pressured Anti-Rent moderates to disavow the Indians or face prosecution for the murder of Steele. Fearful of losing any chance of reform, moderates scrambled to distance themselves from the Indians and the radical wing of the party. But Law and Order tactics soon swung public opinion in favor of Anti-Renters. Barnburners used the opportunity to force Hunkers, such as David L. Sternberg, postmaster in Livingstonville, Schoharie County, from office. Such partisanship led observers to wonder whether "law and order" was a mask for a Barnburner war against wayward Democrats. And the sheer number of prisoners taken made the administration seem more willing to act on behalf of the landlords than bring an equitable resolution to the crisis.[56] Many agreed with Democrat G. H. Noble of Unadilla, Otsego County, who believed the Barnburners had overplayed a good hand. "Noble believed that the Delaware County trials, which resulted in two death sentences, several life terms, and numerous lesser penalties, had been staged. Land agents Charles Hathaway and N. K. Wheeler, for example, helped prosecute the offenders. As a result, Noble wrote, Anti-Rent would achieve electoral success 'as soon as it is shorn of its savage & unlawful features.' "[57] The Wright administration and its allies in the countryside bungled an opportunity to establish the credibility of the state's legal system, failed to vindicate their measures against the tenants, and damaged the credibility of the Barnburners' advocacy of free soil.

Factionalism stalled attempts by Law and Order partisans as much as it had Anti-Renters when they prepared to field a conservative Whig/Barnburner ticket to defeat the moderates heading the Whig/Hunker Democrat Anti-Rent ticket. Vowing to forgo old party affiliations until Anti-Rent was crushed, a Schoharie County Law and Order convention nominated Democrats Tobias Bouck for sheriff and Thomas Lawyer for assembly, as well as Whig Harvey Watson for assembly. Lawyer, a Hunker, was a surprise candidate, but (as we will see below) there were local reasons for the choice. The convention charged that all Anti-Renters were responsible for Steele's murder, which hardly won friends among moderate tenants. The Democratic convention confirmed the nominations of Bouck and Lawyer, even though the Hunker Schoharie *Republican* warned Democrats to avoid Law and Order meetings.[58]

Next door in Delaware County, Barnburners and Whigs failed to unite, as conservative Whig Matthew Griffin of Middletown confided in his journal, "The old (Locofoco) Democratic party, After the Death of Steel Refused to unite with the Whigs and make a union law and Order Ticket. they therefore made a Nomination of the Old School party—the Antirenters therefore usd Considerable Economy they adopted the Whig Senator Wm. H. Vanschoonhoven—and made a County Nomination for Assembly & Coroner they Nominated one Whig & one Locofoco."[59] The histrionics of conservative Whig Samuel Law hurt the Law and Order cause. He charged that Anti-Renters had been bitter enemies of the Whigs before the crisis and would not follow Whig principles if elected to office. Whigs should avoid becoming embroiled in the Anti-Rent issue "because the legitimate tendency of 'equal rights' or Anti Rentism, is to the perpetration of manifold crime to stain the sanctuary of domestic life with BLOOD unrighteously shed—to render civil and religious liberty a nullity—to make possession of property honestly obtained, uncertain—and finally, to render life dependent upon the caprice of ruffianism." The vitriolic message backfired so badly that it was withdrawn.[60]

In the end, Law and Order candidates seemed unable to fathom that Anti-Renters, and many of their neighbors, were angry not only at the manor system but also at the monopoly of political power held by party leaders. Lawyer gave voice to this blindness on the eve of the election, when he ruminated, "Our Anti-rent friends in every quarter must see our prostration and ruin in a political sense; *in which they are involved*—Demagogues will address and encourage them to answer and subserve their selfish purposes—They will in the end have to throw themselves into the arms of the Democracy for a redress of their grievances; and the sooner they do so, and abandon the Work and forcible resistance to the executioners of the law the better will their case be. . . . Will They not yield to the solemn voice of experience? Wisdom? and Moderation?" Law and Order forces thus failed to realize that only substantive concessions to moderate Anti-Renters would make the insurgency go away.[61]

And it was, in fact, the moderate agenda that the Anti-Rent Equal Rights Party promoted in 1845, a message that appealed to a wide population in manor districts. First, they tried to redirect Indian radicalism away from land redistribution toward "monopoly" in general. Continued violence would discredit Anti-Rent calls for democracy, which would allow a new elite—financiers—to assume power, one whose influence was already felt in government: "When in power they have been engaged in chartering companies—in building up an aristocracy. They have condemned and set aside men of talent, who were too honest to give their sanction to measures which violated every principle of good professed faith. Yea, and they have built rail-roads over the Constitution. That sacred instrument has been trampled underfoot, and its principles violated to gratify stock-jobbers and speculators."[62] Characteristically, their solution to this danger was the cultivation of a decentralized, democratic political system. Charging that the state legislature was too far removed for people to participate in its functions and too large to hear the voice of the people, Andes Anti-Renters demanded political reforms that would place greater power in the hands of county boards of supervisors.[63] They also believed that townships deserved greater representation in county government. The Delaware County Equal Rights Convention, held at Hamden on October 16, declared that "the old federal doctrine of concentrating all the power of the Union in the central government is as odious when applied to the towns of a county as to the states of the Union," and

> that inasmuch as the convention held at Delhi on the 8th inst. violated every principle of democracy by totally disregarding the popular will of the county, [a]nd treading under foot the right of towns and the reputation of respectable citizens—thus affording abundant evidence that it was the work of the Delhi clique of political gamblers, we do not recognize it as an act in accordance with the republican principles of the Democratic party. That as republicans, as men, and as citizens of independent towns in the county, we are bound to oppose with all means in our power, this aristocratic distinction. And we call upon every good citizen of this county to unite with us in giving a signal rebuke to the authors of this odious system, and to assist us in sustaining honest practices and Equal Rights.[64]

Charging that the Wright administration and Charles Hathaway's Delhi Clique posed a threat to democracy, Anti-Rent rhetoric equated the tyranny of the Barnburners with that of the manor system. With the former out of office, the latter finally could be destroyed.[65]

The marriage of Anti-Rent and local demands for a sharing of power was most apparent in Otsego County, which had a relatively weak tenant organization. Associations had been formed in only a few towns on the lands of George Clarke: Milford, Middlefield, and Westford. But tenants found allies in western townships that wished to form a new county. Secessionist Hunker Democrats S. S. Bowne (who defended several Delhi prisoners), Sumner Ely, and Levi S. Chatfield encouraged the two groups to create a union party. In its first campaign, the union ticket proved that Anti-Renters, Whigs, and dissident Democrats could wield considerable strength if they acted in concert—even if they came together largely over local representation. Union candidates outpolled their opponents in races for state senator, sheriff, and assembly, with Barnburners the losers.[66] Delaware returned equally promising results, with the Whig/Anti-Rent Party achieving a gratifying victory on election day, despite the concerted efforts of state and local Barnburner leaders to quash the movement. Anti-Renters won both the state senate and assembly races. But the low voter turnout indicated that many Anti-Rent Democrats and Law and Order Whigs remained at home rather than cast votes for a Whig Anti-Renter.[67]

The Schoharie results clouded Anti-Rent victories elsewhere. Local issues derailed the Anti-Rent campaign, thanks to an unexpected reunification of Hunker and Barnburner Democrats. William Bouck and Lyman Sanford decided that if they hoped to regain Hunker control of the county, they could not lose the office of sheriff (and its patronage) to a Whig. Fortuitously, the courthouse in the village of Schoharie had been destroyed by fire, and three townships—Schoharie, Cobleskill, and Middleburgh—now competed for the political power and lucrative trade that came from being the county seat. Democrats set to work dividing the Whig/Anti-Rent coalition over the placement of a new courthouse. The Whig/Anti-Rent coalition nominated George Badgley, who had been an active supporter of the Anti-Rent cause, for sheriff. Playing on sectional jealousies, Democrats spread rumors in northern townships, which were not affected by the lease system, of Badgley "harboring and encouraging the Indians," while in the south they charged *"that he was no Anti-Renter"* but a tool of other political interests. Democrats from Cobleskill, Schoharie, and Middleburgh all opposed Badgley. He lost to Barnburner Tobias Bouck of Schoharie township, who had rounded up tenants with zeal as deputy sheriff, despite rumors that he had received $5,000 from landlords to ensure his election. Badgley lost the race by thirty-two votes—the precise number he lost in the Anti-Rent townships of Blenheim, Broome, Conesville, Fulton, Jefferson, Middleburgh, and Summit, where voters feared he would join efforts to keep the county seat in the northern section of the county. While Whig/Anti-Rent candidates won the race for state senate and one of the county's assembly seats, Law and Order Democrat Thomas Lawyer defeated Anti-Rent Democrat Adam Mattice, 2,952 to 2,300, for the other. Mattice's strong tenant support could not overcome Lawyer's successful wooing of courthouse hopefuls in the northern section of the county. Though the editor of the Albany *Freeholder* wondered why the courthouse issue should split tenant votes, local representation, jealousies, and entitlements overrode Anti-Rent principles.[68]

Even with the Schoharie results, moderates engineered a remarkable victory in the fall elections. Anti-Rent candidates swept into office from the manor counties. Statewide, a broad coalition of voters, including Whigs, joined Anti-Renters in voting for a constitutional convention to be held in the summer of 1846. Soon after the election, the legislature formed a committee under Democrat Samuel J. Tilden to investigate the manor system and propose a solution to the Anti-Rent troubles. Several Anti-Rent assemblymen served on the committee, which promised to address seriously the tenantry's grievances.[69]

The electioneering preceding the election of delegates to the constitutional convention indicated that many Anti-Rent leaders in the townships believed that it was time to return to their old parties—the rent issue seemed in the process of resolution, and few party men wanted to have lifelong enemies write the new constitution. The process unfolded with remarkable precision in Schoharie County, where Bouck, having used the courthouse controversy to divide the Whig/Anti-Rent alliance in November, now defused it so that his faction could win seats in the convention. He slyly made no effort to influence the February town elections, then afterward persuaded town supervisors to sign a petition asking the legislature to appoint commissioners to locate the courthouse. Demosthenes Lawyer of Cobleskill gathered signatures in the north and west, and Lyman Sanford of Middleburgh collected them in the south and east, to avoid arousing sectional suspicions.[70]

With the courthouse issue dead, the Bouck faction concentrated on outmaneuvering the Law and Order and Anti-Rent nominating conventions. Bouck lieutenant N. T. Rosseter of Blenheim helped persuade Law and Order Democrats not to nominate a ticket while reporting good news indeed: "From all I can learn, I believe there will be a *Split* in the *Anti Rent Convention* & a Democratic ticket nominated by the *seceders*—Now then if our Convention is held first, our Candidates will be before the People & would undoubtedly receive the nomination of the *seceding Anti Renters*—There must of course be *some* disappointed men at the *Anti Rent Convention,* if our nomination is ready for them they will take it up—if not ready they will be obliged to make their own Selection & we should either have to *follow their lead,* or by not regarding it, possibly *alienate* some of our Anti Rent friends." Schoharie Anti-Renters began to move back into the Bouck camp. Rosseter proudly announced, "You see we gave you a thumping vote at our Town Meeting—I intended addressing the people myself, but as they were mostly *Anti Renters,* I thought any remarks would come best from one of themselves, so I got *John Mayham* into it & as you see with good effect."[71] Mayham's motives for returning to the Hunker fold became clear in September, when he held the Blenheim Anti-Rent "meeting breathless while portraying the oppression of our government by denying any but the wealthy men the privilege of holding office that will be of benefit to them."[72] Bouck and Sanford prepared to reap the rewards of giving Anti-Rent townships influence within the Hunker faction. The Anti-Rent convention, as Rosseter predicted, dissolved into a battle between Whigs and Democrats. The convention adopted the Whig nominees, John Gebhard Jr. and Elisha Hammond, over Democratic protests.[73]

The Hunker coalition came together rapidly after the Whig/Anti-Rent party convention. Hunker I. W. Baird assured Bouck that Summit Anti-Renters would support him: "Our Democratic portion of the Anti Renters, as a general thing, regret their course, and will, I think, be willing for a season to come to sustain the democratic cause. They are well aware that nothing is to be made in helping build up the Whig party—The Whigs, for a season, have been quite fortunate in converting the Anti Rent excitement to their own benefit."[74] Baird understated the recommitment of Democratic Anti-Renters to their old party. Baird and Whig Seymour Boughton, both members of the correspondence committee that complained of land agent John Westover's treatment of the tenants in 1844, gladly helped him distribute cash to readjust the minds of Summit voters.[75] Anti-Rent Democrat Elisha Brown of Summit renewed his support for the Hunkers and campaigned for Bouck: "I feel that it is all important that we elect men to fill an office of so much importance that are honest and competen in other respects then the rites of the labouring class will bee secured to them as well as the captolist and incorporated companies that is I think that the producers of wealth should be heard as those that possess it." Brown echoed classic Hunker support for internal improvements while paying tribute to his Anti-Rent belief in the rights of working men to the rewards—in this case, political—of their labor. The Whigs' belief in the centralization of power ran counter to the decentralization he and many Anti-Renters sought, Brown added, and "I am appost to having a federal elected in this County." Brown, like many Schoharie County Anti-Renters, set aside the rent issue in order to prevent a Whig- and Barnburner-dominated constitutional convention. Bouck easily won the election, though Whig/Anti-Rent candidate John Gebhard Jr. took the other seat from Schoharie. Bouck no doubt found him more acceptable than a Barnburner.[76]

Genuine change appeared to be at hand as Whigs and Anti-Renters joined to dismantle distributive government and the gentry's grasp on political power. The opening of the state constitutional convention on June 1, 1846, heralded a new era for the ideological foundation of government in New York State and the resolution of the tenants' greatest concerns. Tenants sent sympathetic delegates from nearly every manor county. Delegates from the three counties—Bouck and Gebhard of Schoharie, D. S. Waterbury and Isaac Burr of Delaware, and Levi Chatfield and Joshua Spencer from Otsego—enjoyed Anti-Rent support. The convention followed suggestions put forward by the Tilden Committee in March. The judiciary would be elected, interest from leases would be taxed, leases would no longer exceed twelve years in duration, and quarter-sales and other reservations in leases that impeded the transfer of title were eliminated. Leases made before the new constitution would stand, but the tenants could take solace from the legislature's readiness to challenge manor titles in the courts. Anti-Renters appeared satisfied.[77]

What few tenants understood clearly, however, were the long-term ramifications of their alliance with the Whigs. The new constitution reflected the sensibilities of the emerging middle class, the core of Whig strength in New York State. In particular, the constitution withdrew the state legislature from direct intervention in economic affairs, a measure both reform Whigs and Barnburners could agree upon. Corporations could be formed without legislative approval, leaving economic policy in private hands. What little governmental oversight existed came through the courts, not representatives of the people. The 1846 constitution emerged as one of the most liberal in the nation and opened the state to a frenzy of railroad construction, business speculation, and the expansion of commercial agriculture in the 1850s. In their effort to undermine the landed gentry, tenants had placed considerable economic power in the hands of merchants and financiers, who proved to have little regard for the plight of tenants.[78]

But in 1846, the achievement of much of the Anti-Rent program within the constitution vindicated the course pursued by the moderate wing of the movement and its alliance with the Whigs. Moderate Whig and Hunker Anti-Renters began a final, successful effort to purge Devyr and the radicals, hoping to consolidate their gains and head off future public antagonism.[79] Already in disrepute, radicals suffered a decisive setback at the Albany County Anti-Rent convention in August 1846. Moderates disliked Devyr's abolitionism, calls for amalgamation with National Reformers, and support for women's and immigrants' rights. Led by Whig Ira Harris, the convention denied Devyr and his delegates seats. Undeterred, Devyr and the National Reformers formed a Free Soil ticket opposed to the Whig/Anti-Rent nominee for governor, John Young.[80] Devyr miscalculated his strength in the manor counties. Local associations ardently supported Harris and greatly preferred Young—who promised to pardon Anti-Rent prisoners—to Barnburner incumbent Silas Wright.[81] Devyr warned Anti-Renters "that should they chime in with the corruption that seeks only to strengthen the Whig party . . . in sackcloth and ashes they will repent it—or should they give Silas Wright & Co. their votes—equally great will be their remorse."[82] The tenants resoundingly voted for Young.

Despite the fact that Anti-Renters held the balance of power between the Whigs and Democrats, divisions continued to weaken the party. Increasingly dominated by Whigs and Hunkers, local associations read Devyr out of the party in December 1846 and denounced National Reform. But party radicals—Smith A. Boughton, J. C. Allaben, and William S. Hawley—who supported Devyr's free soil and abolition ideas remained influential in the country. Their popularity with the rank and file prompted the *Freeholder* to warn tenants in October of "hangers-on" and "yes men" who had joined the Anti-Renters for political or pecuniary gain and called for their expulsion from the party.[83]

The destruction of the Anti-Rent coalition at the 1847 state convention resulted from political divisions that had been obscured by electoral victories in 1845 and 1846. Devyr, Evans, and Boughton made a final, desperate effort to unite the two movements. Thirty National Reformers asked to be seated with twenty-three Anti-Rent delegates. Stating that all National Reformers were Anti-Renters, but not vice versa, all but four Anti-Rent delegates voted against seating the National Reformers. The four dissenting voters represented the radical core of the movement: Boughton, Allaben, Hawley, and Elisha Hammond of Schoharie County.[84] The state party nominations widened the rift. Of the twelve offices under nomination, only four went to Democrats. Boughton, Hawley, and two other delegates protested. Boughton proclaimed that he would hold mass meetings in the manor districts to bolt the ticket. The breach became irreparable when *Young America,* the National Reform organ, endorsed Boughton and Allaben as "the real anti-renters." With Boughton—the symbolic and spiritual head of the movement—at odds with the political arm, the Equal Rights Party was doomed to defeat. The loss statewide was so bad that the *Freeholder* did not publish the returns.[85]

The electoral disaster had important ramifications. The legislature had worked for nearly two years on Anti-Rent legislation, though much of it had failed to become law. But two weeks after the 1847 election, the legislature retreated on Anti-Rent measures and voted down a bill that would have made it legal for a tenant to investigate his landlord's title. Divisions within the movement had borne fruit. Legislators could afford to ignore Anti-Rent demands.[86]

Tenants questioned the political choices that they had made in their attempt to restrict the landlords' power. In a revealing piece, Hawley, the editor of the Delhi *Voice of the People,* assessed the "gains" made by Delaware County tenants. The intention of the Anti-Rent Party had been to break the hold of the corrupt and monopolistic Delhi Regency, to curb economic and social injustice, and to spread office-holding among the towns. But, he admitted, Delhi residents occupied nine of the Equal Rights Party's fourteen elected or appointed county offices. He asked his readers searchingly:

> Is this the state of things which were to be brought about, to bring down the pride and lofty bearing of this place, where originated all the oppression and Posseism heaped upon the Tenantry of Delaware? Is this the reaction that was supposed to be brought about by an independent political organization? Is this the redress of long and sore grievances which was promised? Is this the end and ultimatum of the hopes and prayers of the oppressed? If so, then we are not surprised that the People have become lukewarm—that their zeal has subsided, and that we as a party are retrograding to the dark and blinded days of partyism! Shall we now lay by our sword and buckler, and calmly, quietly submit to the reign and dominion of a *New Regency,* more avaricious and gormandizing than that which now lies harmless and powerless at the feet of public indignation?

Hawley called upon tenants to "ORGANIZE and KEEP ORGANIZED" to revitalize the party. But his tone sounded more like a eulogy than a call to arms.[87]

For landlords, instability inside the Whig Party left them equally bereft of support. Jacob Livingston of Cherry Valley chafed at the willingness of Young, Seward, and Weed to embrace Anti-Renters. Livingston found himself unable to resist Anti-Rent attacks in his hometown and county and declared:

> As a Whig I do not intend to support any but a Whig but cannot & will not support Young he belongs to the Seward & Weed faction. In supporting them I am aiding and supporting armed insubordination to established law by voting for him who while a law maker supported those violators of established Law & order. That party was instrumental in calling the convention merely to Jeopardize every thing that they might have a new shuffle & if successful control the

> next Presidential Election & bring forward Seward as the candidate. . . . Young[,] Seward & Weed are determined to drive all those whom they cannot manage from the party & then to call them aristocrats & exclusioners & destroy their influence in plain language to set the poor against the rich & thus control the state by destroying the minority who ought to have as much protection as the majority.[88]

Having failed to prevent a takeover, Livingston confided in 1847, "I believe the Whigs have been instrumental in placing the State in its present unfortunate position. I am no longer a Whig and I am certainly not a democrat but would willingly come out of both parties & aid to form a third party to support Law & order & men from both parties could be found who would be able to keep the radicals of both parties in check. something must be done of that kind or we are gone."[89] His call for a third party demonstrated that the Whigs were too divided to provide Anti-Renters with a reliable power base to win elections. This, as much as any sellout of tenants by reform Whigs, ended any chance for continued agitation of the rent issue.[90]

Perhaps most important, Anti-Rent itself had become anachronistic in the process of forcing political change. Anti-Renters first intended to preserve a set of communal rights within a political economy that tolerated a semi-subsistence economy based on family labor, the sharing of labor with neighbors, and informal credit relationships. The calico Indians, Anti-Rent associations, and petition drives reflected the sensibilities of common New Yorkers who still believed in the notion that the community shared a common interest, that individual and community needs were one. But as Anti-Rent progressed, divisions within the community that had been evolving for years suddenly burst forth and rendered the notion of a common will meaningless. Tenants attacked landlords; merchants, farmers, and artisans chose sides; family members split; neighbors threatened each other with violence; and civil war reigned as the state militia and local lawmen combed Anti-Rent districts for Indians. Class, wealth, occupation, even gender now divided neighborhoods, towns, and counties and could no longer be ignored.[91]

The reemergence of the slavery issue in national politics in 1848 irritated these schisms and prevented a reconciliation between tenants and the Democratic Party, a development that would have lasting effects in central New York politics. Pennsylvania congressman David Wilmot had begun the furor in 1846 when he introduced a bill declaring that all territory taken in the Mexican War would be free territory. The Wilmot Proviso divided the major parties along sectional lines, causing the New York Democratic Party to split again along pro–Wilmot Barnburner and pro–southern Hunker lines. When the former joined the Free Soil Party in 1848, which nominated Martin Van Buren for president, local party alliances once again fractured. The agrarian platform of the Free Soil Party attracted Anti-Renters to its banner, despite the fact that Van Buren's Barnburner machine had violently put down the tenant insurgency. It was a natural choice for tenants to make, for the Whigs nominated southern slaveholder Zachary Taylor, while regular Democrats nominated Lewis Cass, whose doctrine of popular sovereignty would have allowed territories to choose whether to be slave or free. Comparing the gubernatorial returns in Delaware and Otsego Counties gives a good indication of the significance of the tenant vote in the 1848 election. In Delaware, Free Soiler John A. Dix received 64 percent of the total vote, while the regular Democratic candidate mustered a paltry 10 percent. But in Otsego, where Anti-Rent was weak, the Whig candidate carried the contest with 41 percent of the vote, with the Democrat polling 38 percent and Dix only 21 percent.[92] Though, as subsequent chapters will show, the Anti-Renters did not take a straight path toward the Republican Party between 1849 and 1856, their support for the Free Soil Party in 1848 confirmed their continued disillusionment with the Democrats. Other national developments escalated this partisan discord. When Irish famine refugees began flooding the countryside seeking work in the late 1840s, many Anti-Renters joined the nativist movement. By 1851, the last vestiges of the Anti-Rent vote in Delaware County, the fleeting Working-Men's Party, called for the exclusion of all immigrants, especially Catholics, and blamed them for the economic problems of the Republic. Forced to survive in the market economy, former Anti-Renters—once bound by a sense of mutual endeavor—now became defensive, lashing out at perceived competitors and even each other. This would be reflected in the political and economic choices they made in the 1850s.[93]

In the wake of the Anti-Rent movement, electoral politics in all three counties shifted perceptibly. The issues that animated politics in the 1850s—slavery, temperance, free soil, and nativism—primarily revolved around issues of class, ethnicity, or individual conscience, not localized democracy. All called for the government to intervene in the personal lives of citizens. Anti-Renters had entered the movement in 1839 largely loyal to the Democratic Party and its emphasis on personal liberty and small government. But in the partisan atmosphere of the 1840s, they had been forced to differentiate themselves from neighbors as political, social, and economic actors.

But for tenants, the most unsettling aspect of the demise of Anti-Rent as a political force was its failure to protect family-based semisubsistence agriculture. The state constitution of 1846 did not require land reform, leaving the issue to the legislature, the courts, and individual landlords and tenants.

A number of landlords began selling their property to tenants when it became clear that the state was willing to prosecute tenant claims against the gentry. Some, like John A. King, sold their land to tenants. The Van Rensselaers sold their interest in the land to speculator Walter Church, who after defeating the last Anti-Rent suits in 1859 continued to sue tenants for back rent. And George Clarke sold some land but primarily converted his estate to hops and dairy farming on shares. He would lose his estate only when he went bankrupt in 1885.[94]

Having fought desperately to gain title to their land, tenants now would have to compete in the open market place. Because they had not secured genuine economic power, which Devyr and others had warned them to pursue, they had little defense in the rapidly expanding economy of the mid-nineteenth century. Politically as well, there were few places to turn. In the 1850s, nativism, temperance, and slavery dominated political discourse, and—because the provisions of the 1846 constitution removed much of the power over the economy from the legislature and placed it in private hands or the courts—voter insurgencies like the Anti-Rent Equal Rights Party could not effectively alter economic policy. The Anti-Rent vision of a decentralized, democratically responsive government had been lost in the fires of partisan politics and economic change.

4. New Crops, New Challenges: Farm Life in the 1850s

> I hav not got enny monney nor I cannot git Enny at presant I have trid to hire it but I cannot git it monney appears to be ver cease at presant hear I sold my oxen and cart in may to pay a det of nintey nine Dollars I was Sude in April on anote of twenty two Dollars the jugmunt was given the 24 of April and remains unpaid yet I have Set out 1166 hils of hops this Spring I am in hopes if I can git along till they bar they will give me a start. . . .
>
> —Farmer Silas Wellman, 1852

At the end of the Anti-Rent crisis, central New York farmers—whether tenant or yeoman—had reason to believe that they had achieved the essential elements of the agrarian republic that their parents had set out to create. For one thing, the state constitution of 1846 had struck a substantial blow to the landed aristocracy of the state, preventing the further expansion of manor tenancy and enacting democratic political reforms. Second, several landlords had begun selling land to tenants, and the state attorney general initiated title suits against holdouts. The cornerstones of agrarian ideology—land ownership and political democracy—seemed about to be extended to virtually all farmers. But farmers did not foresee the resurgence of tenancy, albeit in a new guise; nor did they anticipate that their erstwhile allies, the middle class, would make common cause with the landed gentry to stimulate a rapid transformation of the agricultural sector in the 1850s. This effort centered on a new crop—hops—and, as the subsequent chapter will discuss, railroad development. Forced to enter cash crop production, many farmers would share the experience of Silas Wellman, who borrowed heavily from his brother to convert to hops farming. Their challenge was to do so without losing their land or the social world they valued so highly.[1]

Thanks to hops and increasing specialization in dairying, an integrated market economy developed in central New York in the 1850s, taking shape within a process of conflict and negotiation by farmers, landlords, merchants, and artisans. In particular, landlords, merchants, and wealthy farmers spearheaded efforts to increase production, specialization, and mechanization in agriculture, both by example and coercion. Yeoman and tenant farmers, for their part, attempted to avoid hops production, focusing instead on dairying, which fitted more closely with their ideal of the family farm and a historical experience that had demonstrated the benefits of limited commitment to cash crops. But this strategy proved untenable for most as both the national and local economies expanded. The family farm concurrently underwent a transformation of its own as new labor relations forced farm men and women to redefine their roles within it. When the Civil War broke out in 1861, the farm family was still in flux. Yet there was hope, for many farm men and women succeeded through hard work, determination, and cooperation. Whether agrarians could defend this emerging order as local and national issues such as the Albany & Susquehanna Railroad and slavery fractured political life remained to be seen.[2]

Farmers spent the years immediately following the Anti-Rent crisis preparing for life without the gentry. On the one hand, this meant that many more would own their land in fee. On the other, the manor system had provided farmers with cheap ground rents, easy credit, rights of occupancy, and access to the commons for timber and pasturage, all of which now had to be secured in other ways. They concentrated on land ownership between 1847 and 1850. In Schoharie County, for example, John A. King offered his land for sale to tenants on their terms in 1847, and virtually all purchased. For some, like James Burnet of Blenheim, this meant forestalling payments on other debts, as he explained to one creditor, Judge Lyman Sanford: "I have had the money once But paid it for the soil of my land to John A. king when he was out I expect to have some more next weak and I will Call and see you and pay it if I Dont get Dis apointed sad and am able to get about. . . . Dont make me cost if you can help it till I call."[3] By this process, tenancy rates for farmers dropped in the three counties from an estimated 30 to 40 percent in the 1830s. As the previous chapter indicated, landlessness dropped by 1850. Rates among farmers were 28 percent in Otsego, 14 percent in Schoharie, and 15 percent in Delaware. Even given the fact that many farmers owned only a portion of the land they tilled, these numbers represented an improvement over the past. And by 1855, landlessness among the entire population fell to 27 percent in Schoharie, 25 percent in Otsego, and 22 percent in Delaware. Yet tenancy persisted, and townships with some of the most valuable farmland had the lowest freehold rates per family. For example, 85 percent of George Clarke's tenants in Cherry Valley, Middlefield, and Milford both owned and rented land in 1850. It was perhaps this hidden tenancy that led newly elected attorney general Levi S. Chatfield to estimate in 1850 that nearly half of the land in the region remained under lease.[4]

Not all landlords proved as cooperative as King, however. The agent for Goldsbrow Banyar's estate, for example, traveled to Delaware, Schoharie, and Otsego in 1850 to assess the lots and fix a fair market price based on the value of the soil, timber, water, improvements, and production of each farm. The agent reported that it was difficult to crack community solidarity, as most residents were "disinclined to affix a price to our lands" other than the ten dollars per acre set by local Anti-Rent associations. With determination, however, he managed to persuade several to do so. The agent then offered to sell the land to the tenant or, failing that, to an outside buyer. If he could not sell the land, he offered a new lease. In a typical instance, the agent offered two tenants who lived in Otego, Otsego County, an opportunity to purchase their lot at fourteen dollars per acre. One of the two refused to pay above ten dollars. His partner, however, privately told the agent that he would rather purchase than have a stranger take the land. The agent decided "to serve notice to quit on Burges—& to then offer the lot to Jerome B. Y."[5] The new constitution had done nothing to force landlords to the bargaining table, and men like Burges and Jerome B. Y. thus faced a difficult choice: maintain a united front and possibly lose one's farm, or break ranks to save their interest in it.

Tenants who held life leases on George Clarke's lands confronted an even more unpleasant situation. The young landlord had taken control of his father's estate in 1843 determined to make it a going enterprise. His first order of business had been to collect back rents, which had accumulated after the death of his father in 1835 because of lax management by the estate's trustee, attorney Richard Cooper. Second, he instructed Cooper to investigate whether the individuals named in life leases—some of which had been in force for forty years—remained alive. If a tenant refused to pay back rents or if lives had expired, Cooper was to offer a new annual lease at higher rent or evict. Since individuals named in the leases often had died, many tenants had to vacate the land without compensation for improvements or else accept restructured contracts. At the urging of the Otsego and Montgomery County Anti-Rent associations, therefore, the state attorney general initiated a title suit against Clarke in 1849. Tenants refused to leave their homes while the suit was pending. After they won a favorable decision in a lower court, Clarke filed an appeal. Tenants feared the court would reverse the decision but dug in their heels, hoping to force a compromise by wearing Clarke down with lengthy, costly ejectment suits.[6] As agent Thomas Machin wrote in 1851, "Most of them are of that principal that they now have a perfict Right to Your Lands allthough the Lease have expired but so long as they can hold the Lands at one Shiling per anum they will all do so and as it falls on you to prove that all of the lives are dead after a notice has been served."[7]

While Anti-Rent associations pressured Clarke to sell out, individual tenants had to decide on the fly how to handle evictions. Improvements to farms represented at least two generations of sweat that had created wealth—in tenant minds the equivalent of savings that ought to be theirs. A number hoped to convince Clarke to sell farms by assuming a cooperative posture. George C. Clyde of Cherry Valley, Otsego County, asked Cooper to forgo a lawsuit against his brother's estate so that he could purchase the farm leases at auction. He would pay the arrears but wanted to purchase shortly afterward. Others did not have that option and tried to cut their losses. W. H. Coon of Milford could not pay Clarke's price for the family farm and implored Clarke to buy him out before the two surviving individuals named in the lease, both elderly, died and left him propertyless and shorn of his savings.[8] Finally, some tenants who failed to prevent eviction destroyed the improvements they had built with their own hands, poisoning livestock and tearing down or burning fences on vacant farms to prevent Clarke from enjoying the fruits of their labor.[9] These actions bespoke the desperation of men faced with poverty and dependence if the title suit failed.

While tenants fought to gain title to their farms, inheritance and family support remained the surest way for others to get and keep a farm. Farmers who inherited land, purchased it from their parents, or took a mortgage from a relative usually did not have to seek credit elsewhere, did not fall into dependence, and often prospered. Charles Harley of Roxbury inherited his father's 300-acre farm in 1843. Harley's father owned six rental properties, three of which he occupied between 1805 and 1835 as he worked himself up the economic ladder. The estate was worth $460 after debts. Using the income from the farms, Harley established a holding worth $1,555 above debts in 1849 and eventually became a successful merchant.[10] Moses F. Lyon's grandfather purchased a 150-acre farm in Bloomville, Delaware County, in 1813 for $1,000. Two years later, the grandfather deeded the land to Moses' father, Henry, for $800. Between 1838 and 1848, Moses taught school in Kortright and, while his father traveled, managed the farm from 1845 to 1850. Then Henry sold Moses 125 acres for $800, provided Moses would pay the balance of a mortgage held by Henry's cousin Seth. Henry gave Moses a mortgage for the price of the farm plus $600 to cover starting costs. Henry asked only $50 interest.[11] Freeholders like Harley and Lyon therefore entered the 1850s in an advantaged position compared to other farmers.

But what kind of world did central New York farmers hope to build in the wake of the Anti-Rent crisis? Census returns for 1850 suggested that both tenants and freeholders had made a decisive shift away from cash crops in Schoharie, Delaware, and Otsego Counties between 1840 and 1850. Wool production plummeted by approximately 30 percent, while wheat output decreased by 13 percent in Schoharie, 48 percent in Otsego, and 78 percent in Delaware. Dairy products declined slightly in Delaware and Otsego Counties and by a third in Schoharie, and the number of swine dropped by 40 percent in the three counties.[12]

Two explanations for this precipitous drop seem reasonable. First, low prices in the 1840s may have convinced farmers to stop raising sheep or planting crops that required high inputs without guaranteed returns. This explanation has limitations, for economic recovery began in 1847, yet cash crop production did not rebound. Given the long-standing connection farmers made between cash crops and dependence, a second inference has considerable merit. That is, farmers believed that they were in the process of extracting themselves from dependent relations on the landed elite and were attempting to put in place the more parsimonious agrarian republic they had advocated for decades. Increased production of food and fodder crops confirmed this: buckwheat output more than doubled; corn increased 72 percent in Schoharie, 137 percent in Otsego, and 157 percent in Delaware; and oat and rye harvests increased as much as one-third. And crops like buckwheat provided a cheap way to restore soil fertility. These changes occurred while improved acreage increased by 10 percent in Otsego, 12 percent in

Table 3. Cash Crop Output in Delaware, Otsego, and Schoharie Counties, 1840–60

County	Dairy Products (lb)	Hops (lb)	Wool (lb)	Wheat (bhl)
Delaware				
1840	3,931,758	0	235,032	94,120
1850	3,869,623	5,538	165,221	20,295
1860	5,010,895	116,568	127,128	57,619
Otsego				
1840	4,556,415	168,605	451,064	148,880
1850	4,547,544	1,132,052	325,598	76,652
1860	5,448,546	3,507,069	244,118	106,522
Schoharie				
1840	2,596,677	260	134,257	72,871
1850	1,702,438	10,587	95,185	63,241
1860	2,316,338	1,441,648	114,991	93,272

Sources: Sixth Census of the United States, 1840 (Washington, D.C.: Blair and Rives, 1840); *Seventh Census of the United States, 1850* (Washington, D.C.: Robert Armstrong, 1853); *Eighth Census of the United States, 1860* (Washington, D.C.: Government Printing Office, 1864).

Schoharie, and 57 percent in Delaware. Even as farmers cleared new fields, therefore, they shunned old cash crops.[13]

The steadfast refusal by central New York farmers to embrace notions of progress that had taken root in other sections of the North astounded outsiders. Melvin W. Hill, of Williamsburg, Massachusetts, found Charlottesville, Schoharie County, where he attended Methodist seminary, "an awful place" in 1853. Roads were primitive and the people sleepy and uninterested in making money. He was astonished when "a very large Comet" sent the village's "superstitious Inhabitants . . . running to and fro through the streets of their great and but a few hours previous, happy city, Seeking for a place of safety from the terrible fate which they imagined awaited them."[14] John W. Champlin encountered deplorable conditions in the Delaware County backcountry while surveying in 1853. Working near Middletown he noted, "The county here looks poor, as well as the inhabitants: farms poorly fenced; houses small and going to decay, people rough and ignorant." The southern section of the county was a tangle of primordial woods with few passable roads. Only a few villages—Delhi, Walton, Andes, Colcester—broke this pattern. Evidently new to the county, Champlin only chuckled when several menacing young men attempted to make him and his assistant shout "Down with the Rent."[15]

Landlords and merchants shared Hill's and Champlin's impressions and set out to modernize the backcountry by expanding cash crop production. The resurgence of the national economy after the discovery of California gold and the repeal of England's Corn Laws—which had limited that country's agricultural imports—at the end of the decade helped stimulate agricultural investment and innovation. And, beginning in 1853, the state court of appeals—starting with Clarke's case—reversed lower court rulings that would have vacated landlord titles. The state legislature meanwhile passed general laws allowing the free chartering of banks, railroads, and other corporations under the auspices of the 1846 constitution. These enabled individuals to form corporations merely by filing bylaws, a stockholder list, and a statement of the capitalization of the company with the secretary of state. These were very generous; under the General Railroad Act of 1850, only 10 percent of the capital stock had to be subscribed for the corporation to form. With these laissez-faire laws in place, local entrepreneurs eyed the countryside as a paradise ripe for development. Local farmers immediately felt pressure from both the gentry and merchants to reorient agricultural production toward cash crops through new contracts and credit relationships that curbed their autonomy.[16]

Above all, the agricultural transformation of the region took place thanks to the hop plant, and none promoted its growth more than George Clarke. Local and state agricultural societies had identified hops as one of several crops suited to the soil, climate, and topography of the region decades before, but hops had a small market in the United States before 1847. For one thing, Britain, a major consumer of beer and ale, had restricted the importation of hops under the Corn Laws. And, before German immigrants began to pour into the country in 1848, Americans preferred whiskey and rum to beer. Afterward, both the foreign and domestic markets expanded—so quickly, in fact, that demand for hops seemed limitless. Clarke thus renewed his father's effort to replace life with share leases that would require a mixture of hops and dairy production. He went so far as to sell off unwanted parcels to fund the purchase of new farms better suited to hops.[17]

Because hops production was highly specialized and costly, wealthy men like Clarke played a critical role in creating the boom economy of the 1850s. To begin with, the hop plant needed very particular care throughout the growing season, and haphazard methods spelled disaster for farmers. When a farmer established a hops yard, he had to set roots several feet apart in hills to allow for proper drainage, light, and cultivation. Then the farmer had to wait, for hops yielded in the second year, and only after three years would there be a full crop. Usually farmers planted corn or potatoes between the hills in the first year to keep down weeds and get some return on their yards. From first planting, hops yards had to be plowed, grubbed, and fertilized in the spring to ensure proper growth. In the second year, the process was repeated, then the shoots were trained around twenty-foot poles and then around rope strung between the poles to provide adequate light to the blossoms. At the end of that summer, picking commenced the moment the hops ripened to ensure that they had the proper oil content and flavor. They could not be harvested wet, either, for they would develop rust and spoil. And hops had to be carefully processed for market. Hops were laid out on canvas cloth that was placed across slats above stoves in hop kilns, dried slowly using sulfur to cure them, then pressed into large bales. Each step required specialized knowledge, and badly prepared hops had little value. A well-tended hops yard could produce for twenty years, but neglect could ruin years of hard work. Since few farmers knew how to grow hops in the early years, men like Clarke spent a great deal of time teaching them proper methods. Even then, Clarke did not trust tenants to dry the crop—the most important and difficult aspect of producing high-grade hops—requiring them instead to deliver his half of the crop to his own kilns.[18]

Starting costs were also high and, added to the three-year period before a full harvest, meant few could enter production without backing by entrepreneurs like Clarke. Springfield carpenter Brewster Conkling, for example, estimated in 1853 that a 140-foot barn would cost Clarke $2,600, a hop kiln and stove $780, and a forty-by-twenty-foot cheese house $350 on one share farm. The $3,730 merely outfitted the farm. Clarke still had to purchase poles, livestock, tools, hop roots, presses, and labor. A small producer like Silas Wellman had to make do with much less lavish expenditures. When he borrowed money from his brother William, it was only to purchase hop roots. Whether William could afford to wait three years for repayment is unknown, but in a countryside strapped for cash and credit, such dealings placed extended families at risk of collective failure. As a result, many would-be hops farmers, lured by inflated estimated returns of $250 per acre, took mortgages from local merchants to cover starting costs or took up share farms from men like Clarke. The latter was the most frequent gateway to hops farming but came at a cost. Clarke and other landlords retained all rights to crops in the fields until a final division was made and intervened directly in the management of the farm. Tenancy thus took on a new face in the 1850s.[19]

Clarke's achievement was more noteworthy given the fact that he staged a "cashless" agricultural revolution on his estate that then spread throughout the area. Much like southern planters during Reconstruction, Clarke recognized that his primary asset was land and that, in absence of cash, share farming would provide the labor needed to cultivate cash crops. But he needed credit, which was in short supply before the gold rush. As credit eased, Clarke mortgaged his lease properties to generate working capital. The fact that he held so much land enabled him to get credit from banks and insurance companies that did not deal with small holders. Hence Clarke acted as a credit conduit for the region. As he put it in 1864, "I am the big water wheel of the whole mill up here and If I . . . stop everything else stops."[20] He spent grandly on his share farms, providing a dwelling, a garden, timber for fuel and fences, hop roots, poles, and land, while the tenant supplied labor to grow, pick, and, in some cases, dry the crop "according to custom in Springfield."[21] Clarke usually stipulated that tenants keep a dairy stocked with his cows to produce butter or cheese on shares, which had the benefit of providing manure for needy hop plants. All other crops—apples, hay, oats—were divided equally; he retained all other timber, water, and mineral rights; and tenants made specific improvements to the holding. Clarke purchased supplies in volume, credited the goods to his tenants' accounts, and placed liens on crops to ensure that he received his half of the produce of the farm. Since tenant purchases ate up returns, he rarely paid out much cash at the end of the season, and many fell into debt to Clarke. Though his estate records do not list production on all of his farms in Otsego County, a sense of the scale of operations he instituted can be gleaned from the accounts he kept for his "home" farms near his mansion, Hyde Hall, in Springfield. Hops production there rose from 7,566 pounds on his two main farms in 1853, the first year he kept records, to 122,425 in 1858, when fifteen farms comprised his "home" operations.[22]

Clarke's lands surrounded Otsego Lake—with holdings in Springfield, Cherry Valley, Middlefield, Otsego, Milford, and Westford—and he consolidated his half of each harvest in great barns at Hyde Hall and in Milford. He sold his butter, cheese, and hops through merchant houses in New York and London. He encouraged tenants to market their portion of the crop through him, for a percentage, which helped him monopolize the trade. He became known by the end of the 1850s as the nation's preeminent hops grower.[23] In the span of fifteen years, 1845 to 1860, Clarke developed an integrated system for growing and marketing hops and dairy products. He bridged the gap between gentleman landlord and agricultural entrepreneur by personally collecting rents, negotiating new leases, and commencing suits against debtors. Along with other wealthy landowners like Cooperstown's William H. Averell, Clarke's drive to reap windfall profits led to the rapid development of sophisticated agricultural and marketing structures unknown in central New York before 1850.[24]

Hops production rose to dizzying heights in the 1850s. In 1850, Otsego County already produced nearly half of the total state crop, selling 1,132,052 of the 2,536,299 pounds of hops produced in New York. By 1855, the Cooperstown *Freeman's Journal* reported that Otsego County hops farmers expected to yield more than 800 pounds per acre and sell them at $.30 per pound, netting $240 per acre. Growers sold 2,000,000 pounds of hops for $600,000—$400,000 in profits—in 1854. The county produced 3,507,069 pounds by 1860.[25] The editor did not exaggerate when he said, "The *Hop* up in Otsego is a great institution."[26] Oneonta merchant Eliakim R. Ford revealed the deepening dependence of the local economy on hops when he asked fellow banker Lyman Sanford to take a six-month note for a debt because "living in a Hop district renders it inconvenient to raise funds outside of common business just at this time."[27]

The downside of hops, however, was price volatility, which encouraged speculation by producers, brokers, and consumers. For Clarke, the thrill was intoxicating, and he spent much of his time trying to outwit competitors by improvising new marketing methods or cornering the market. For example, Clarke attempted to circumvent American brokerage houses by selling direct to London brewers, using family and friends in England as sales agents. He also asked them to provide inside information on the state of the English and German crops so that he could predict price fluctuations at home. His speculative dreams included raising levels of production high enough that he could dictate hops prices, which prompted him to purchase more and more farms. William Averell's daughter, no stranger to wealth and luxury, exclaimed in 1856, "Mr. Clarke has a perfect mania for buying land and keeping fine horses—He has ten horses and more land then I can tell."[28] Clarke's cousin and overseer warned him of the pitfalls of his course: "I will simply say don't speculate again be content as a Planter make your produce and sell it. [D]on't hold particularly such an unsteady commodity as Hops. You have a *devilish* fine Property keep it intact and you will have no mean inheritance to leave your dear little children."[29] Clarke should have heeded the advice, for brewers held a distinct advantage. When hops supplies fell, they bought the lowest grade hops to save money, yet during gluts they played growers against each other and purchased high quality hops for little. Like many producers, Clarke often held his lots too long and either took a loss or had to sell his crop in competition with the next year's hops. He rarely reaped high enough profits to pull himself out of debt and continued to operate on credit.[30]

But others did make it big thanks to hops, especially merchants and brokers who carried none of the liabilities that came with growing the crop. Hops merchants appeared in the 1850s and, much to the dismay of growers, soon had a stranglehold on marketing and credit. David Wilber began his career as a speculator and financier by purchasing hops from his fellow small growers in Milford, Otsego County, in the late 1840s. He and partner John Eddy moved their hops, wool, leather, and butter brokerage to Oneonta in 1855 to connect with the projected A&S Railroad. By 1859, their business had assets of $81,934. Wilber boasted a personal estate worth $25,000 and Eddy one worth $50,000. Wilber became the premier hops dealer in the area by 1870, opened a bank in Oneonta in 1872, and was involved in railroad finance and the Republican Party. Eddy enjoyed a similar role in the community.[31]

Hops also drew outside entrepreneurs to the region in pursuit of big profits. Silas Dutcher left his job with a western New York railroad in 1854 to make his fortune as an agricultural speculator. He noted in 1856 that he had "thought much about the farming business and made up my mind I did not wish to farm unless I am able to do so without being obliged to labor myself." Instead, Dutcher spent several months learning the trade in Trenton, New Jersey, in early 1855 and then embarked across central New York, introducing himself to merchants and hops farmers. He became an agent for a New York brokerage in August and began purchasing hops in Otsego County with the help of an uncle who introduced him to local farmers. He then traveled to Philadelphia, visited local brewers, and left in December, "pretty well posted on the hop trade." Dutcher got his first break when he sold a consignment of hops at a tidy profit in February 1856. Inspired, he devised a plan for "managing the Hop business" to favor brewers and, having secured backers, reported in July that he sold $2,000 worth of hops in a single day. He made so much money that he retired a decade later a wealthy man. In the cold calculus of the market economy, the wealth generated by hops went to nearly everyone except the farmers who grew them.[32]

* * *

While hops occupied center stage, dairying expanded substantially during the 1850s, again thanks to the efforts of wealthy farmers. Some impetus came from landlords like Clarke, whose share leases included provisions that tenants care for a specific number of his cattle, properly breed his stock (usually with his bulls), and produce quality butter or cheese.[33] Many landlords eschewed hops and instead let dairy farms on shares. These demanded a premium on the market, for many farmers preferred dairying to hops farming. Landlords could thus require references that indicated a farmer's reliability and the profitability of his previous farms. Jedediah Miller of Schoharie County made such an inquiry of Lyman Sanford in 1858: "An application has been made to me for a farm by Mr. Fox who says he has carried on your farm for several years—Mine is a hill farm adapted to pasturage, there are about 300 acres, and to carry it on advantageously some force is to be made—and at some considerable expense is Mr. Fox the man and his family to take charge of 20 or more cows, and would the butter they shall make be fit for any table?—If he and his family should fail to do well, it would injure both—Frankly tell me, and all you shall communicate shall never be known."[34]

As with hops, innovation in dairying came from above. George D. Wheeler of Deposit was a member of one of Delaware County's leading families. Born in 1818, he began farming in 1855 after working as a lumberman, shopkeeper, forwarding agent, and mine manager. He leased Laurel Bank Farm in Deposit from his father for $350 in 1858 and applied business practices to agriculture, carefully investing in bred stock, crops, and equipment to improve rates of return. By 1863, he and partner Charles Daniel ran a large operation for that time, with 66 head of cattle, 103 sheep, 20 swine, 4 turkeys, and 2 horses. Laurel Bank Farm boasted a mower and reaper, two level and one sidehill plows, two lumber wagons, two harrows, one corn cultivator, a threshing machine, a fanning mill, sap pans, a road scraper, a wood machine, and hand tools. Two years later, Wheeler and Daniel hired two men to live on the farm and provided houses for both, in addition to taking on seasonal labor. Wheeler and Daniel produced for the market on as large a scale as they could afford and expanded whenever possible.[35]

Progressive farmers like Wheeler joined with local merchants to put in motion a reordering of farm labor that centered on mechanization and wage labor. They believed that the spirit of enterprise would more than offset displacement of poorer farmers by generating moral virtue in an otherwise backward countryside. For example, the Republican Schoharie *Patriot* wrote in 1857 that machines exerted a positive moral influence on farmers by freeing time for intellectual and social pursuits. Farm machinery companies, often run by local manufacturers, flooded stores with pamphlets extolling the virtues of hop presses, mowers, plows, and threshers. Testimonials by successful local farmers improved the prestige of the machines and implied that the machines embodied the same excellent qualities as their owners. Pamphlets contrasted well-tended, machine-rich farms with the overgrown, dilapidated farms of those who still practiced labor-intensive methods. One advertiser attempted to establish further a "mechanized" social order by naming his superior machine "Farmer" while calling a simpler, cheaper version "Yeoman."[36]

With the farm economy expanding, entrepreneurs began manufacturing farm implements and thus stimulated industrialization in rural townships. George Westinghouse of Central Bridge, Schoharie County, the father of the noted industrialist, began his career making farm implements before moving his operation to Schenectady. Cobleskill merchant, banker, and A&S promoter Minard Harder began manufacturing the Empire threshing machine in 1859 to take advantage of the large oat harvests produced following the demise of wool production in the area. Veteran Otsego County politician Levi Beardsley manufactured haying machines in South Edmeston. Oneonta merchant, banker, and railroad promoter Eliakim R. Ford developed a hops cultivator that earned a gold medal from the New York State Agricultural Society in 1867. These men also speculated in farm commodities, shipped crops, invested in railroads, ran general stores, or extended credit to farmers, allowing them to accumulate further capital for investment.[37]

In all of these ways, then, the expansion of hops and dairying put in motion a series of changes that made it difficult for farmers to remain outside of cash crop production. With large operators buying up property, land values rose. This, in turn, exerted pressure on tenants to cut deals with landlords who otherwise might sell or lease to outsiders. New farmers also had to pay more for land, which meant having to invest in cash crops to pay mortgages. And existing farmers put in an acre or two of hops to offset higher tax assessments that came with the boom and, after 1858, from township funding of the A&S. Meanwhile, mechanization undercut smaller producers who used traditional methods. The end result for ordinary New Yorkers was a world quite different from the one they had anticipated would emerge after 1846; rather, they found their families, livelihoods, and strategies for the future very much in flux.

* * *

Still, small farmers continued to praise a limited, mixed-crop farm strategy, believing that it would ensure prosperity. The Democratic Cooperstown *Freeman's Journal* enumerated the county's agricultural production in 1855 yet credited conservative strategies for its success: "The inhabitants of this inland, rural district, are comparatively independent. They have comfortable farm houses, well fenced farms, good roads, good schools, wagons and carriages to ride to church, and plenty to eat and drink and wherewithal clothe themselves; and are, happily free from pecuniary embarrassments and pinching want which afflict so many communities elsewhere." Avoiding debt and wild speculation would bring farmers independence. Thus the *Freeman's Journal* celebrated the social benefits of limited commercial production.[38]

The central underpinning of this system was the independent family farm, an ideal that had endured changes during the Anti-Rent years but remained recognizable in the 1850s. The primary development was the solidification of dairying as the main source of income for farm families. This partly reflected the decline of other economic pursuits. With declines in wool, wheat, and lumbering, men became involved in dairying, which had once been female labor. Women, too, focused more on the dairy, as home manufacture of cloth, woolens, and knitted goods declined sharply in Otsego, Delaware, and Schoharie Counties after 1835, falling from a total of 8.7 million yards to less than 1 million by 1855. Farm couples worked together much more closely than before, adjusting fairly well to what might have been a divisive process. Evidence from across the mid-Atlantic region indicated that men and women modified the traditional patriarchal household, forging a new set of gender relations labeled "mutuality" by historians. Men remained primarily responsible for field crops, the care and feeding of cattle, and so on, while women continued the skilled preparation of butter and cheese. Yet, since their labor now represented the main source of income, women asserted greater authority over the economic and social decisions of the household, with men going along in recognition of their wives' equal commitment to the maintenance of the family farm as the cornerstone of agrarian life.[39]

But farm families also chose dairying when possible because it offered a stable livelihood compared to hops culture. Though it brought modest annual returns, a dairy farm could be operated with family labor and without high expenses. Farm families traded most of their dairy products to local merchants, and since butter was used by all households, a ready market always existed for it, especially as cities like New York, Albany, Utica, and Syracuse grew in population. As well, standards for butter quality remained primitive before the Civil War, which allowed farmers to spurn purebred cows and to continue to allow animals to graze on unimproved meadows or in the woods. And since silos were not yet in use, most animals could not be fed a proper diet in winter, meaning that families made butter only in the summer. The rest of the year could be used for other pursuits. Farmers resisted improvements in breeding, partly because of the expense but also because the more specialized the breed, the more care it required—fencing, sanitation, feeding, and medicine. For farm families that did not want to become enslaved to the market, therefore, dairying seemed to offer a measure of independence.[40]

But such independence could be achieved only if farm families could weather pressure to expand market involvement. The Industrial Revolution set in motion forces in central New York that made a more modest, localized economy difficult to sustain. This came in several forms, but as early as 1851 farmers complained that their sons could not find good jobs outside of the family farm and had to leave the area.[41] The market also began to divide farmers and artisans as each struggled to survive. In 1853, Delaware County blacksmiths, forced to charge low prices because of competition from "convict or factory labor," staged a strike against farmers. They charged that farmers had abandoned them, seduced by cheap manufactured goods. Farmers countered by opening a cooperative blacksmith shop. Looking back at services he no longer performed, one blacksmith angrily wrote, "I ask them where they got their wagons, plows, ox chains, pitchforks, hoes, hinges, and in fact almost every article wanted by the farmer." He noted that farmers and blacksmiths once traded goods and pasturage to help each other out, but now farmers thought only of price. Pulled by external forces, farmers and mechanics thus lashed out at each other, damaging the unified front to which Anti-Rent had aspired in the 1840s.[42]

The gradual shift toward integration with the national economy made it difficult for marginal agricultural producers and laborers to make a living. The recession of 1854–55, for example, winnowed out many dirt farmers who were unwilling or unable to take up dairying or hops raising. Whereas the manor system had offered them a level of security—and some interest in protecting the value of their farms—poor tenants now abandoned their holdings. Delaware County landlord Ann E. Gould complained in 1855 that "those who have Contracts are leaving, after being on long enough to pul the [hemlock] bark and sell it—and are such miserable things that it would be useless to do any thing but let them go—as it is we are constantly involved in [legal] expenses without attaining anything."[43] Middletown attorney Matthew Griffin remarked that year that Delaware County laborers "are realy in a Suffering Condition as there is no work to be had and Provisions cannot be bot Short of Cash."[44] In the 1850s, therefore, old strategies of combining farming and outside work proved increasingly untenable. Rather, the economy required specialization, which in turn demanded financial resources beyond the means of people on the lower end of the economic ladder. One Schoharie County miscreant articulated the frustration many undoubtedly felt in 1853. He told his creditor, Lyman Sanford, that he was leaving for Canada and would not repay him; rather, "I think more of my bacon than I do of your money so keep it and fuck all you Can."[45]

The more fortunate likewise struggled to adjust to the new economy. In particular, the lack of circulating cash in the region—which reflected the relative underdevelopment of the economy—handicapped business transactions. Locals who produced for the market had to devise hybrid barter-cash transactions to counterbalance this dearth. Merchant-farmer John D. Shaul of Springfield, Otsego County, for example, agreed to sell his cheese to merchant R. Bomber for $7 per "hundred" in 1853. However, Shaul retained an option to buy out the contract for $.50 per hundred. The deal guaranteed Shaul a minimum return on his cheese, but if he found a better price, he could get out of the contract for a small fee. For Bomber, risk was minimal. He either received the cheese or netted a tidy profit from the option that would add to, rather than deplete, his cash supply. Ingenious as this arrangement was, it reflected the weakness of the local economy—even the most entrepreneurial citizens struggled to catch up commercially with the rest of the nation.[46]

For other farmers, barter and labor exchanges continued to be an alternative to cash outlays for goods and labor. Farmers in Hamden, Delaware County, for example, established a market day in 1859 to trade or sell goods and livestock—not with outsiders but with each other.[47] At the same time, farmers used machinery—often seen as a hallmark of capitalist agriculture—in ways other than manufacturers intended or hoped, often sharing mowing, reaping, haying, and harvesting equipment. If a man owned a threshing machine, he traveled to neighboring farms and threshed with several men. He received a portion of the crop and secured their help for his harvest. Working together, they could produce several hundred bushels per day and finish threshing in the neighborhood in two or three weeks. Farmers thus adapted new machinery to their social system in a way that illustrated the unwillingness of central New Yorkers to abandon cooperative, noncash labor exchanges that had worked well in the past.[48]

In the end, however, this did not prevent unemployment among both farmers and laborers, which made them vulnerable to new, exploitative labor arrangements. Thus, tenants who accepted George Clarke's share leases in the 1850s labored under terms that few would have accepted a generation before. John Chamberlain, who leased a Cherry Valley farm, signed a chattel mortgage for his rent "and also the value of seed grain furnished to me by said Clarke." Chamberlain assigned Clarke "thirty acres Oats on hill fifteen acres Oats on side hill back of house—and 2 acres of Oats opposite house—Eight acres Barley all of which said grain is now growing on the farm now in my occupation[.] Also all the hops growing on said premises." The mortgage would be satisfied if Chamberlain delivered 208 bushels of oats, 20 bushels of barley, and all of the hops "well dried cleaned and in good order" to Hyde Hall by November 1. Unlike life leases, which required minimal payments in cash or kind, the new leases demanded a scale of production that made growing food crops difficult and forced tenants into the cash crop economy.[49] Clarke asked for and received outrageous concessions from incoming farmers. In some cases, he required tenants to surrender any claims against him before he gave them new contracts; in others, he demanded that they pay the back rents owed by former tenants.[50]

Men like Chamberlain were caught in an all-too-common bind by the mid-1850s. Both tenants and yeomen increasingly had to borrow to maintain their farms, pay taxes, and invest in new crops. Escaping debt, whether to a landlord or merchant, proved difficult. One farmer explained to a creditor that "a farmers means are slow even if he is doing well." He would repay his fifty-dollar debt if his crops returned a profit that season: "I shall milke ten cows this season I shall begin to make butter that will be sailable to put in firkins about the first of may butter made in April will not be sailabal to send off but will sell to merchants about home for such things as we need in the family."[51] The more specialized the operation, the more catastrophic normal agricultural risks became, as one tenant of Cherry Valley merchant Jesse Sutliff discovered during a drought in 1848: "[H]e says the dairy farm whiche he hires and 45 cows which has usuly neeted him 150 dollars profit," Sutliff's agent reported, "has done nothing more than pay the rent."[52] Twenty years after purchasing his land from John A. King, James Burnet of Blenheim, Schoharie County, had accrued substantial debts to Lyman Sanford as he continued to try to expand hops production to make up annual shortfalls.

He asked Sanford to delay filing suit for a bad note in 1867: "I have sold a yoke of oxen to be taken away next weak for $220 am to have $120 on delivery and to wait till hoppicken for the balance if we must pay it I will haf to pay it out of that but wanted to by some cows with the money have simpathy judge for us poor fellers on the hill or as much as you can."[53] Years later, locals recalled that hops-growing townships were the poorest in the region, testimony to the fact that many had gambled on the crop and lost.[54]

Furthermore, farmers learned quickly that merchants and banks did not hesitate to foreclose on delinquents, which further undermined their independence. Lyman Sanford and his partners, for example, called in farmers' notes to capitalize a bank in Middleburgh in 1857. In addition to draining the countryside of credit, this action left those in arrears facing immediate foreclosure.[55] Some farmers found themselves in the precarious position of borrowing from one wealthy man to pay another. Eliakim R. Ford refinanced John A. Boyd's debt to Sanford in 1858, which gave Boyd several more months to pay but at higher interest. More important, he became dependent on Ford's goodwill if he could not repay the loan.[56] And men who had borrowed from relatives might find themselves indebted to financiers when kin were forced to consolidate resources. John Pindar of Gardnersville, Schoharie County, asked Lyman Sanford to purchase a $6,000 mortgage he had given to his son-in-law so that he might finance his son's move west. Pindar hoped Sanford would be lenient toward his son-in-law, who wanted to expand his hops farm to include a dairy. Forced to favor one child, Pindar chose his son and left his daughter and son-in-law's fate in Sanford's hands.[57]

In the face of these circumstances, farm families took special pains to avoid exposure to debt, which meant sticking together. George D. Taylor, of Jefferson, Schoharie County, wrote that his family "deliberately arranged in advance" the passing of the farm from father to son so that there would be no chance of losing the property in probate. Furthermore, he proudly noted, the family succeeded because they rigidly used only the farm's earnings, not "outside money," to finance improvements.[58] Blenheim Anti-Rent leader Benjamin Curtis's family exchanged labor and thus trimmed the costs of farming. Family members traveled great distances to help each other during peak seasons. This ethos rippled through the stern warning Taylor's father gave him: "Remember it's the Taylors against the world and the world against the Taylors."[59] When Henry Lyon of Bloomville, Delaware County, had a falling out with his son Moses in 1850, Henry's cousin and creditor Seth Lyon counseled him to repair the breach. "You have duties towards each other very binding," Seth warned, suggesting that since "Moses I suppose was home five years during the seasons you was away and was prospered in managing in your room," Henry now owed it to his son to "put an addition to the house and let him have a part of the Farm . . . try to lay aside self all you ought, and be yeilding to each other, and be sure not to fall out by the way."[60] Seth later granted Moses a mortgage at low interest and forestalled collections during hard times.[61]

But however much families attempted to reproduce agricultural stability, many young men and women discovered that the new economy prevented them from achieving the agrarian ideal of a freehold farm on which to raise their families. The expense necessary to do so was high, which made many young men have to work a number of jobs—school teaching, farm labor, share farming, mill work, teamstering, logging—in order to establish themselves financially before marrying and settling on a farm. Men who reached majority after 1850 increasingly failed to do either.[62]

The careers of two Otsego County men highlighted the changes the farm community underwent in the 1850s and 1860s. John N. Colburn and Lucius Bushnell each failed to become a successful farmer and had to take jobs that each felt were unequal to that of husbandman. Colburn grew up in Burlington and left the family farm to work at a paper mill in Franklin, Delaware County, in 1852. He returned in May to help his uncle farm. He listed Burlington Flats as his home in his next diary, written in 1856, where he worked for a second uncle while also teaching school. He kept up his spirits by attending Baptist meetings and assured himself that "a lazy man an idler is one, not designed by God to dwell on the Earth, he has placed him here to be active not active in vain and trifling amusements, but on the other hand, it is designed that his activity should be exercised, in doing good to his *fellow man.*" Colburn continued helping both uncles on their farms and taught school in Burlington Flats, Toddsville, and South New Berlin through 1858. Soon after, he bought a paper mill in Toddsville with his brother Leonard. They barely kept the mill running between shortages of rags and straw, the inability to maintain a solid force of rag sorters from among local farm women, and the vicissitudes of the market for rag paper. The brothers therefore spent much of their time either idling or working Leonard's farm, and Leonard cobbled shoes in the neighborhood. They periodically returned to Burlington to help an uncle to hay or harvest crops.[63] Colburn deplored manufacturing and intimated several times between 1861 and 1868 that he might leave the business. He quit in 1868, noting, "The abilities for financial management, with me, are very small. I deplore it very much, and my lank purse shows the effect." He moved to New Berlin to help his father farm full time. Nearly forty years old and still single, he felt unfulfilled because he did not own his own farm.[64]

Lucius Bushnell's experience resembled Colburn's. His father, Horace, worked as a carpenter and farmer in Gilbertsville. In 1857, Lucius turned eighteen and, like Moses Lyon, fought with his father over his place in the family. Horace refused to let Lucius leave home and establish himself independently. Angry, Lucius moved to Elmira, New York, then Stamford, Connecticut, and finally to Lansing, Iowa, working a number of agricultural and mill jobs, mostly for family members, before returning home in 1859 to work at the family sawmill. But Lucius remained discontented with carpentry and hoped to settle down on a farm with his fiancée, Rosetta Hammond. He worked constantly, mowing hay, setting hop roots, raising barns, harvesting corn, shingling houses, and speculating in sheep. But Lucius could not get ahead. He remarked sadly on November 15, 1859, "worked in shop. Guess I will always work in shop. Don't see anything else." Like Colburn, Bushnell did not succeed as a farmer. In his next diary entries in 1877, he worked chiefly as a carpenter, leaving Rosetta in charge of their general farm.[65]

Women endured similar frustrations, but these were compounded by their secondary status in rural society. Those who reached majority and could no longer live at home or who failed to marry suffered distinct disadvantages. Unmarried women living outside of the family often had to move between households seeking employment and endured perpetual dependence on others, since wages were low. Catherine Bartoe faced such a situation when she wrote to innkeeper G. H. Edgerton of Kortright, Delaware County, seeking work in 1854: "As I am out of bisiness I thought I would wright to you so if you Want any help I will come and work for you As I am left Alone I wish to get A place to work if you want help pleas wright me Direct Conesville Schoharie Co if you Should not want help if any of your neighbors wants help pleas Wright."[66] Catherine Wood of Butternuts, Otsego County, moved from job to job from 1857 to 1884 but never settled. Once in the cycle of dependency, she could not extricate herself, and camp meetings, church, and visiting could not relieve the terrible loneliness she experienced. Unable to fully enter local social networks that revolved around families, she felt deeply troubled and complained of headaches, sickness, and melancholy.[67]

The social impact of the evolving agricultural economy also altered intergenerational and gender relationships. On the one hand, the close interaction between family members when making butter was reminiscent of the unified approach to labor that had informed family economic strategies before the Anti-Rent years but entailed greater gender and generational cooperation. Rosetta Hammond, for example, proudly recorded that her father taught her how to milk cows on her fifteenth birthday in 1857. She milked three cows the first day and ten by the end of the week.[68] Farmers tried to mold hops to the same model. With hops, the main need for labor came in the fall, when hops matured and needed to be picked quickly to ensure quality. Farmers first tried to maintain the tradition of using only family or neighbors at harvest. Because of this, they generally restricted acreage to a maximum of five acres. But in the middle of the harvest season, labor was difficult to procure; hence, hops growers employed women and children to pick hops. Wage labor quickly came to dominate the harvest because exchanges between households could not meet demand. The scale of operations strained household resources. Growers bedded pickers in the farmhouse, barn, or outbuildings, and the farm wife had to work throughout the year to lay in enough foodstuffs to feed the pickers, baking bread and pies before the harvest, and then had to cook hot meals for the workers. She also continued her traditional chores in the house, garden, and dairy, a staggering load. Nonetheless, in the 1850s the harvest retained many rural traditions, including labor exchanges and "vistin'," and brought several weeks of busy activity and pleasant social life despite the hard work.[69]

But in other ways, hops culture fractured traditional relations, having a profound influence on young men and women and generating a new and at times exhilarating peer culture. Young women formed the bulk of the harvesting corps that entered the fields each year to strip the vines. The labor was divided so that a young man acted as box tender for four pickers, usually young women aged thirteen to twenty-five. The box tender pulled the poles from the ground laden with hops and placed them across a bar on the hops boxes. After the pickers filled the boxes, he took the contents to the kiln for drying. The harvesters received cash, even when families exchanged labor, and generally were allowed to pocket the money. Not surprisingly, perhaps, spinning girls from local mills formed the first groups of pickers. They had already broken free, to a degree, from the household—though fathers still controlled their work routines and remuneration.[70]

Young women welcomed hops-picking season because it offered relief from daily household routines. On September 6, 1857, Rosetta Hammond went to the farm of William H. Young to pick with a host of neighbors. The next day she noted, "Picked hops all day very tired. Picked two boxes in the eve Hannah & I went to village bought me a collar to embroider." She netted $4.09 "or 7 shillings"—about 12.5 cents per box—at the Youngs', averaging nearly two boxes per day. Her family exchanged harvest work with two other families. She bought treats in town, "a dress and a few notions," with her earnings.[71]

The harvest also gave Rosetta an opportunity to socialize. Friends visited her in the fields, and she attended a camp meeting and "had a good time saw lots of folks." Rosetta also "went a visiting to the Traceys had Pickle & Whiskey" and reported on September 12, "Picked hops all day had lots of company, in eve . . . went to hear brass band play." The few difficulties she encountered did not dampen her spirits. On September 9, "Had a real hurly burly about picking hops they wanted us to pick better but picked two boxes." Though Rosetta was not particularly concerned about the row, it provided evidence that hops growers were keenly aware that they needed to set firm production standards if they hoped to make a profit and that local pickers did not particularly wish to accommodate them. And, on September 15, she noted, "In the eve went to the hop house saw someone that I did not care about." The incident pointed to one of the greatest difficulties farmers had during harvest: keeping men (who generally slept in the hop house or barn) from making unwanted advances toward young women. Nonetheless, Rosetta had a good time and felt sad when she returned home. Hers was a common experience, and by the 1860s nearly all young women and many children in the countryside picked hops.[72]

Hops picking allowed young men and women to interact in the fields with far less adult supervision than before, created avenues for them to establish a degree of independence, and offered them community prestige not accorded them by other types of farming. The Bloomville *Mirror* proudly announced that Ray Kaple and Eunice Thompson of Decatur, Otsego County, each picked nearly eight boxes of hops in a day at a Schoharie County farm, declaring, "Beat this, and we will try again!"[73] Young men aspired to become box tenders as the job brought them into close contact with female pickers whom they could impress with the sheer physical strength required for the task, giving box tenders unmatched recognition among their peers. Folk customs such as the "kissing loop" developed from the casual contact between the sexes. A young man kissed the first girl he saw after finding that peculiar growth of the hop plant and would, tradition had it, marry her. Combined with hops dances and camp meetings, the harvest offered new ways for the young to meet, court, and marry.[74]

Generational changes could also be seen in courtship and marriage. When Lucius Bushnell and Rosetta Hammond decided to get engaged, his parents strongly opposed the union, but the two stubbornly overcame their resistance. And though Rosetta engaged in traditional pursuits leading up to her wedding day—quilting bees, sewing, dressmaking—she and Lucius read avidly from the *Marriage Guide* rather than rely on the counsel of friends and family. Rosetta remarked that she hoped that the guide would help her "to inform myself for the plaice I am to fulfil."[75] Traditions such as horning—the folk practice of serenading newlyweds with rough music on their wedding night—

also fell out of vogue. Horning remained common throughout the 1850s and 1860s as Lucius Bushnell, Rosetta Hammond, and Louise Nethaway recorded in their diaries, though many "respectable" folk found it vulgar. Pressure mounted from middle-class reformers to end the practice. The pro-temperance Bloomville *Mirror,* for example, expressed distaste for revelers in Long Eddy, Delaware County, charging them with drunkenness and incivility for joining in a horning.[76] Henry Conklin, who grew up in Blenheim, Schoharie County, and his bride ruined their horning in 1856 by taking part in the merrymaking: "This shamed them out and they soon went off home declaring they would never go to another horning."[77] Henry and his bride misunderstood the meaning of what had once been a vital community ritual intended to interrupt the couple's first night of conjugal bliss and, in the process, demonstrate that individual desires or needs were very much subordinate to the rest of the neighborhood.[78]

The potential for an almost complete breakdown of the family under the strain of the market economy, hops production, wage labor, and new gender and intergenerational relations could be seen most profoundly in a lawsuit brought by laborer Edwin H. Cass against his employer, Otsego County hops farmer Joseph Blanchard, for back pay in 1865. The proximate cause of the controversy came in the middle of the hops harvest, when Cass angrily stalked off the Blanchard farm following a confrontation with Blanchard's son George concerning where Cass sat at the dinner table. But preceding events indicated that the hops economy had shattered traditional patterns of authority within central New York households, the very foundation of the agrarian order.[79]

From the outside, the Blanchard family appeared to be a traditional patriarchal household. Joseph Blanchard, a machinist by trade, was seventy years old in 1864 and owned a 200-acre farm in the town of Maryland. Authority over agricultural and domestic labor was divided by gender. He oversaw farm operations, keeping two hops yards, with eight acres in production. His son Morey, forty-four years old and single, operated the farm and hired and managed the laborers, including his brother George, a Civil War veteran. Joseph's wife presided over household chores, but her eldest daughter, Olivia, managed daily affairs for her sickly mother and kept the family accounts. Olivia explained the arrangement to the court in the following way: "Pa has the farm—and the personal property—Morey works them under the same arrangement we all *do the work* & pay for *the place* & *to have* a place to live upon."

Though the Blanchards attempted to maintain family solidarity, stresses caused by commercial farming led them to act as individuals first, family members later. For one thing, Joseph and Morey did not coordinate farm strategies effectively, including the hiring of laborers. Misunderstandings abounded. Joseph claimed his son had hired Cass in 1862 without his knowledge, yet Cass reported that Morey had stated that he could not be hired until Joseph agreed. This led Cass to believe that the elder Blanchard had full charge of the farm, yet the father made no such claim. Cass, perhaps expecting a more traditional paternalist household, struggled to figure out who exercised ultimate authority throughout the period of his employment.

Confronting a vacuum, Cass tried to take advantage of the situation to improve his own position and soon ran afoul of the men in the family. Cass first signed on as a common hand during the 1862 hops harvest, then made a formal agreement in 1863 to work six to eight months at $15.50 per month plus room and board. Among his duties would be serving as box tender during the hops harvest, a desired job because he would have four girls working directly under him. Such a job carried prestige, but farmers also expected such men to behave honorably in order to prevent untoward advances on young women or incidents of favoritism that might demoralize others. The Blanchards, like other farm families, therefore attempted to regulate contact between men and women during the harvest. As with other families, males, including those of the host family, slept in the hop house or barn during harvest, while women shared the bedrooms of the farmhouse. The sexes also sat at separate ends of the table at supper.

Events turned on this issue. During the 1863 harvest, two of the young women working under Cass, the Smith sisters, struck his fancy, and he decided to sit at the head of the supper table between them. Knowing this defied custom, Cass asked for Olivia's permission. Though one suspects her mother might have acted differently, Olivia assented. At this point, George interfered, demanding that Cass return to the men's end of the table, and as the exchange escalated, Morey arrived and also instructed Cass to sit with the men. Cass countered that both Olivia and the Smith sisters had agreed to the arrangement and, further, that since Morey had no authority within the household, he would quit rather than be treated in such an arbitrary manner.

No one seemed able to resolve the mess. Rather, Morey prudently asked Cass to discuss the manner outside, inside the hop kiln. There he pleaded with Cass to stay on and asked him to be tolerant toward George, who had difficulty adjusting to civilian life after his service in the Union army. Joseph arrived and told Cass to ignore George and return to the table. Cass went inside and heard Joseph blaming Olivia for the "d——d fus." Cass interceded on her behalf. Joseph, conceding that he had little authority within the household, agreed to let the women decide where Cass should sit. They wanted him at the head of the table. The next day, Cass came to breakfast after milking and sat down where the women directed him. George, outraged, advanced and knocked him over, letting go of him only when others interfered in the scuffle. Joseph and George argued in the master bedroom, precipitating a second fistfight. The women asked Cass to intervene, but Joseph angrily told him to leave the bedroom, despite George's stranglehold on his neck. Cass left the farm and did not return. The Blanchards refused to pay Cass and charged that he enticed the Smith girls to leave to pick hops at another farm.

Several themes emerge from the events at the Blanchard farm. First, the family as an economic unit faced deep strains because of the demands of hops production. As titular head of the family, Joseph showed no ability to control it. Designed to keep kith and kin together, the arrangement had the opposite effect. Morey and Olivia acted confused over their authority within their spheres, which was complicated by Joseph's absolute but ineffective authority over them and their mother's illness. Second, the events showed how changing economics were challenging traditional gender relationships. The women forcefully asserted their independence. Olivia disregarded family ties and supported Cass's testimony against her father and brothers. In so doing, she challenged male power and authority within the family and stridently defended her control of the household. The Smith sisters, by leaving the farm, clearly acted contrary to the wishes of their own family and those entrusted with their care, the Blanchards. Their behavior suggested that they regarded themselves as free agents, especially rejecting male control over their labor and social lives.

The Cass case may have been one incident, but it outlined the disruptive changes in rural society brought on by the hops economy. By the time of the Civil War, the hops harvest involved so many people that few could not be touched by its influence. After the war, farmers would have to ship laborers into the countryside to handle the huge harvests. Many looked to dairy farming as the best alternative to the displacement that the Cass case highlighted. Conforming more closely to the yeoman ideal of the family farm, dairying came to embody the most effective vehicle for advancing republican sensibilities in the new capitalist economy.[80]

The years 1850 to 1865 therefore left a mixed legacy for central New York farmers. At the beginning of the period, they believed that they could achieve agricultural prosperity based on land ownership, limited market production, and the traditional patriarchal household. But the hops economy, and to a lesser extent dairying, commercialized the region with lightning speed. Remnants of the gentry and the merchant community used the failure of land reform and new economic powers granted after 1846 to leverage farmers into greater market investment, first through tenancy and later credit relations. Old social networks fractured under this assault, and as the Cass case demonstrated, instability extended into farm households as well. In this atmosphere, the debate over the A&S Railroad came to symbolize a broader conflict in the region over the proper political economy as farmers increasingly identified the project with a growing disordering of rural life. They would take their fight into the political arena in the 1850s yet, as the next chapter will show, found it difficult to be heard in a din of national issues that were absorbing the attention of the political parties.

5. Agrarianism Outflanked: Farmers, Railroads, and Politics, 1848–65

> The people will resist the tax, and if all the Courts in the country shall sanction it, the property holders will not submit to be thus despoiled of the hard earnings of their industry, to fill the pockets of a few leading men who are engaged in the project. A more unjust and oppressive power could not be assumed under the worst despotism on earth, than this Company are endeavoring to exercise—that is to lay the people under contribution for an object to which they are opposed, and must prove ruinous to them in the end.
>
> —"A Tax Payer," 1858

The decade of the 1850s proved as challenging politically as it was socially for central New York farmers. While the rapid transformation of the agricultural economy unsettled community and family structures—privileging the most market-oriented producers—those who embraced the Jacksonian notion of limited government and political democracy found themselves on the defensive during the decade. No issue better illustrated the defeat of the agrarian ideal in politics than the debate over the construction of the Albany & Susquehanna Railroad. Chartered in 1851, the company began as a private enterprise but soon sought municipal funding. Opposition erupted along the proposed route between Albany and Binghamton, with citizens organizing to block town bonding acts passed by the state legislature in 1856. Voicing a similar critique of corporations as Anti-Renters did a decade before, anti-railroad spokesmen like "A Tax Payer" placed faith in democratic institutions—the legislature, the courts, and political parties—to defeat the company and restore public accountability to government.[1] Yet opponents of the railroad were defeated—in part because the national crisis between the North and the South over slavery shattered the Democratic Party, their most natural allies, and denied them critical institutional backing. The A&S, however, had the resources to buy political support and did so liberally at the state and local level. The success of the A&S at controlling the political process illustrated the institutionalization of an entrepreneurial ethos of progress in government by the end of the antebellum period. In this new political economy, grassroots mobilization could not overcome the increasing dominance of big business in politics.

As this chapter argues, three main factors contributed to this outcome: the new constitutional framework adopted in 1846 granted wide powers to corporations to operate without public oversight, including by the legislature; political parties were increasingly absorbed with national issues at the expense of local issues; and factionalism undermined local party cohesion on issues like the A&S—a situation irritated by widespread corruption. Still, the most interesting aspect of the transformation was that farmers did not abandon the agrarian ideal despite being shouldered out of party politics. Rather, they would attempt to configure a virtuous republican rural order after the Civil War through nonpartisan activism, education, and moral reform. With the Republican Party increasingly committed to business and federalism during the Civil War era and Democrats hamstrung by charges of disloyalty, farmers had little choice but to pursue such a course.

* * *

The constitution of 1846 and a set of general incorporation laws passed soon after—including the General Railroad Act of 1850—initiated a fundamental reordering of the political economy of New York State that would have significant consequences for agrarians. First, as previously noted, the constitution expanded democracy by making virtually all offices elective, a reform welcomed by all but the most rock-ribbed conservatives in the state. Equally important for understanding the A&S debate, however, was that the constitution weakened the legislature by limiting its power to grant corporate charters and established the principle that general laws, not special legislation, provided the only democratic path toward economic development. A spate of general incorporation laws, including the General Railroad Act, were passed in the next several years. This law enabled a railway company to form simply by filing articles of association with the secretary of state and providing evidence that 10 percent of its capital stock had been subscribed. It also created a set of basic regulations over charges, capitalization, and dividends but left oversight largely to the courts. Additionally, the law provided limited liability for stockholders for all but labor costs. Last, the legislature granted the right of eminent domain to railway companies, placing the burden of proof on individual land owners to prevent seizure. Combined with other general laws in banking and manufacturing, New York by 1850 boasted the most liberal constitutional regime in the country.[2]

A product of Barnburner, Whig, and Anti-Rent criticism of distributive politics, these reforms enjoyed wide support, though Barnburners' anxiety that corporations would abuse their nearly unchecked powers produced the few regulations contained in the General Railroad Act. But overall, each faction was pleased. Whigs rejoiced that they had removed the government as a drag on entrepreneurship; Barnburners considered the reforms a victory for small government; and Anti-Renters celebrated the vanquishing of the old aristocracy and its influence in Albany. Even the biggest losers in the process, Hunkers, found a silver lining; with the legislature stripped of power in both the economic and political spheres, the more-numerous Barnburners would gain little patronage to keep themselves in power. The honeymoon lasted but a few years, however, when it became clear that unrestrained enterprise corrupted politics.[3]

In the meantime, the political parties turned their attention toward reunifying coalitions splintered by a decade of crisis. This was a formidable undertaking, given the sharp splits in both Democratic and Whig ranks. Democrats had the unenviable task of reunifying Barnburners, Hunkers, Free Soilers, and Anti-Renters. The new state constitution and general laws helped here, for limited democratic government seemed to have triumphed—something each group supported. And many who had jumped to the Anti-Rent or Free Soil cause did not wish to permanently strengthen the Whigs and were willing to negotiate a settlement. But once the crisis over the admission of California as a free state began brewing in Congress, New York Democrats discovered that the issue of slavery continued to divide the party. In 1849, the majority of Democrats—either moderate Hunkers or Barnburners who had not bolted to the Free Soil party in 1848—agreed to allow Free Soilers back into the party without demanding that they accept the national party's policy of noninterference in slavery. Conservative Hunkers such as William C. Bouck resisted amalgamation, arguing that the party could not win without southern support and that true Democrats would not trifle with property relations of any kind—slave or otherwise. Thus, an updated set of Democratic factions were born: "Soft Shells" tolerated Free Soilers, "Hard Shells" resisted reconciliation. Popular support for the Compromise of 1850 helped Softs—who did not want disunion but also had little regard for the slave system—gain control of the state party machinery despite Hard truculence.[4]

The Whig Party, too, suffered divisions between reformers and conservatives over Anti-Rent and slavery. Whigs held together their unstable coalition by championing positive government. But this agenda was losing force. Reformers, led at the state level by William H. Seward, had agitated on behalf of tenants and embraced temperance, antislavery, and other middle-class reforms. Landlord Jacob Livingston of Cherry Valley, Otsego County, summarized conservative frustrations in 1850, noting that radicals supported "free schools . . . they also court the Anti rent men . . . they are abolitionist of the worst kind that question is settled but they still want to agitate it merely for their political advancement."[5] Seward opposed the Compromise of 1850 and survived a conservative attempt to seize control of the party that year. Pledged to noninterference in slavery, conservatives created a splinter organization, the "Silver Greys," to rally behind President Millard Fillmore and the compromise. Often Silver Greys shared interests with Hard Democrats—including support for internal improvements and nativism—yet their forthright advocacy of a strong central government and the prerogatives of the wealthy prevented a union. These schisms would prove fatal to the Whig Party.[6]

At the local level, the Democratic Party set out in 1849 to rekindle the Jacksonian flame in Delaware, Otsego, and Schoharie Counties, despite Anti-Rent's continued influence over the electorate and the Democracy's poor record on behalf of tenants. To overcome this, the party melded traditional Jacksonian values with support for the new constitution and the compromise. They preached a message that had played well in the backcountry, calling for limited, decentralized government, local autonomy, and each individual's right to a level playing field. They stressed that the Whigs intended to use the powers of government to dictate economic and social policy, with reform "isms" leading the way. In Otsego County, this strategy reconciled the competing Democratic factions in 1849. There, the dominant Softs ceded half of the nominations to their rivals and proclaimed that the party had achieved "a union, honorable because it does not exact humiliating concessions from either side."[7] Democrats swept the fall election. In 1850—amidst the turmoil of the sectional crisis—the pro-compromise Soft gubernatorial candidate Horatio Seymour carried Otsego, which now seemed securely back in Democratic hands.[8]

In Schoharie and Delaware, however, Democrats confronted deeper fissures and achieved mixed results. In Schoharie, Hard kingpin Lyman Sanford and his semiretired father-in-law, William C. Bouck, had regained control of the county after 1847, headed off Free Soil in 1848, and kept the party unified enough to carry most offices for Democrats in 1849 and 1850. Nonetheless, the situation in Schoharie revealed the depth of disagreement between Softs and Hards on key issues. Sanford found heretical the Softs' rigid stand on reduced government expenditures on internal improvements and their noncommittal position on slavery. Inclined to see politics in pragmatic terms—as a means of accumulating power, patronage, and pork—Sanford considered Soft adherence to principle naive. To best the Whigs, in his mind, Democrats had to keep control of the White House, Congress, and state government, thereby preserving the flow of perks down to the local level. Government funding of canals and railroads and cooperation with the southern wing of the party offered the only means to achieve this end. Because Sanford and Bouck had a record of helping Whigs win if it meant keeping control of the county Democratic machinery, Democrats were fortunate that Schoharie Whigs were badly divided themselves.[9] In 1848, Anti-Rent Whig leader Thomas Smith of Cobleskill warned that conservatives "have been very much troubled about matters in our county" and blocked a conservative convention designed to purge radicals. But his party, never strong in the county to begin with, could not replicate its successes of the 1840s once Sanford brought Anti-Rent Democrats back into the party.[10]

In 1849, Delaware presented Democrats the greatest obstacle to mending the party. Soft leaders charged that successive Whig governors had expanded the public debt in violation of the spirit and letter of the 1846 constitution and demanded a rapid contraction of government:

> We again repeat our unshaken belief in the necessity of untiring vigilance against the encroachment of delegated power upon the rights of the masses and of individuals; the strict accountability of public agents; and the strict construction of the State and National Constitutions; that we demand equal taxation and cheap postage for the people; a retrenchment of the expenses and patronage of the National Government; the abolition of all unnecessary offices and salaries; the election by the People of all civil officers in the services of General Government . . . the individual liability of stockholders in all monied, trading and manufacturing corporations; and that we are opposed to public debts, funding schemes, high taxes, standing armies, profuse expenditures of the public treasury. . . .

This language echoed the state party's agenda and laid the ideological foundation for later Soft opposition to the A&S but was not sufficient to rally wayward Free Soilers. Instead, Delaware Softs had to accept the Free Soilers' proclamation "that as Democrats and disciples of Jefferson, we deplore the existence of slavery in our country, and are wholly and totally opposed to its extension over territory now free" and that western lands should be reserved for actual settlers at minimal cost. Even among Softs, such sentiments bordered on heresy—and were likely to cause Hards to shun the party that November—but they recognized that victory required Free Soil support.[11]

Democratic efforts paid few dividends at the polls, however. In 1849, the only Democrat to outpoll a Whig in the state races in Delaware was Hard attorney general candidate Levi S. Chatfield of Otsego County, who promised to initiate title suits against landlords. Tenants simply did not trust the old Barnburner Delhi Regency—many of whom had been the judges, prosecutors, and lawmen who had put down the Anti-Rent rebellion—that directed the reunified party. The following year, despite Democratic efforts to shift focus to retrenchment, support for the compromise, and the commencement of suits by Chatfield, Anti-Renters again voted Whig, with reform candidate Washington Hunt beating Horatio Seymour by a 600-vote margin in the county. Though far more sympathetic to Anti-Rent calls for localism, democracy, and limited government than Whigs, old grievances continued to plague Democrats.[12]

In this context, with Democrats trying to right their fortunes by attacking Whig notions of political economy and with the local agricultural economy beginning a rapid transition, a group of entrepreneurs chartered the A&S Railroad Company. The conflict surrounding this enterprise would test just how well Democrats could address the needs of the farmers they purported to represent in the changing economic climate of the 1850s. In the process, the fight over the railroad would establish the limits of agrarian politics on the eve of the Civil War.

* * *

No project better illustrated the coming of age of central New York entrepreneurs as a distinctive group than the A&S. This process had begun as early as the panic of 1837, when many had concluded that the Jacksonians no longer represented their interests, that farmers would consign them to bankruptcy, and that righting their fortunes meant overhauling and rationalizing the credit system, commercial transactions, and government. Anti-Rent denunciations of all debts, not just rents, alienated many more. The merchants of the region had resolutely backed the new constitution, believing that easing government restrictions on incorporation, banking, and other aspects of commerce would enable businessmen to develop existing and new markets freely. Some abandoned the region, notably Oneonta merchant Collis P. Huntington, who departed Otsego County for San Francisco in 1849 with $20,000 worth of goods to make his fortune outfitting gold miners and later earn fame as a railroad builder. Others financed expeditions by local men, hoping to generate a tidy profit to invest locally. Those who stayed at home contemplated ways to modernize their communities.[13]

Among these, many were smitten by the buoyant ethos of progress articulated by the Seward wing of the Whig Party. Undoubtedly some had heard Governor Seward speak at Cherry Valley a decade before, when he had cautioned the audience not to exalt the past, since history had been fraught with ignorance, tyranny, and human suffering. "Ours is a country in which all that is old is yet new," he declared, and as a nation without a past, America could remake human society. Christianity, science, industry, and free labor would help citizens abolish the tyranny of hierarchy that afflicted the monarchies of Europe—and, vestigially, the United States. He offered a set of principles to guide this experiment: "That peace is indispensable to the improvement and happiness of man, that improvement is his highest duty, and arts, not arms, his right occupation. That Republican Government resting upon equal and universal suffrage can only secure an exemption from the ambition of conquest and the popular discontents which involve nations in foreign wars and civil commotions [and] can only be maintained in a community where education is universally enjoyed, and where internal improvements bind together the various portions of the country in a community of interest and affection." Should Americans take up his call, they could create a universal nation, freed of the constraining influences of locality, class, ethnicity, and faction.[14]

Far fewer central New Yorkers responded to the governor's call than he would have wished in 1840, but a decade later, advocates of a new order had gained considerable ground. Local merchants and professionals lauded the urban middle class and its ideals—temperance, abolition, and evangelical religion—as they sought solutions to the problems that they believed plagued the countryside. Erastus Crafts, a Whig merchant of Laurens, Otsego County, whose enterprises included sheepherding, a tannery, a distillery, and grist and woolen mills, bubbled with enthusiasm when he compared urban with rural life in a letter to New York City attorney and politician William W. Campbell—who had been raised in Cherry Valley—in 1851: "There exists among your '*Merchant Princes*' so much enterprize industry & public virtue, so much intelligence & perserverance, so high a sense of integrity & honor in their dealings that I think they will understand their mission in the city which is now the emporium of Capital of the 'new world' is destined to be carried forward *under providence* by them & their successors untill it shall become the emporium of Capital & Commerce of the Whole World."[15] This spirit of entrepreneurship, so different from the skepticism of radical Anti-Renters and Soft Shell Democrats toward unrestrained capital, had come to define and motivate the emerging middle class in Otsego, Delaware, and Schoharie Counties. Furthermore, merchants had concluded that for the good of the community, it was their duty to bring this world to the countryside.[16]

For many merchants, the first step would be to build a railroad to link the region with the outside world and rationalize agricultural trade by lowering their cost of funneling local produce to Albany or New York markets. Additionally, finished goods brought into the region could be sold at lower prices and in greater quantities than before. With credit eased by the gold rush and the hops economy showing promise, promoters had reason to believe a railroad might succeed. And, if all went well, the road might be profitable enough to net tidy dividends for stockholders—a tantalizing prospect for would-be entrepreneurs in a cash-starved region.[17] Sure that their ideals and the needs of the community were one and that society would suffer without such improvements, they dismissed opposition arguments as "old fogyism."[18]

Local entrepreneurs had yearned for a railroad for two decades, but they had never managed to get previously chartered projects beyond the planning stages. The latest attempt had come in 1849, when Oneonta merchants had secured a charter for the Schenectady & Susquehanna from the legislature, which would have run along the approximate route of the A&S. But without the power of eminent domain or limited liability for shareholders, investors declined to risk capital in an enterprise that might not generate immediate returns. When the General Railroad Act eliminated these impediments, the Oneontans once again tried their hand at railroad building. Led by merchant Eliakim R. Ford, a Free Soil Democrat, railroad boosters sent Hard Democrat William W. Snow of Oneonta, a career politician and lobbyist, to meet with prominent Albany businessmen to discuss funding the A&S. Several, including Erastus Corning, Edward C. Delavan, Robert H. Pruyn, and Ezra P. Prentice, the president of the New York State Agricultural Society, agreed to support the undertaking. They called a formal meeting to organize the corporation on April 2, 1851, at Oneonta, the midpoint on the proposed line.[19]

The Albany agreement was a coup for Snow and the country promoters but came at a cost. Albany capitalists had a different set of motives than the Oneontans. Ford had a decidedly local purpose: he hoped to boost his village and arrest his county's economic and political slide. However, the Albany men were playing a higher-stakes game. The city had grown into one of the nation's leading industrial and commercial centers as the eastern terminus of both the Erie Canal and the New York Central Railroad—run by Corning, who was also involved in iron, banking, and other enterprises—and served as a hub connecting the west to New York City, New England, and Canada. Yet in 1851, the New York & Erie Railroad was about to open and would bypass Albany by connecting the Great Lakes to New York City directly across the southern tier of the state. Furthermore, the Delaware & Hudson Canal Company had a virtual monopoly on the coal and lumber trade out of northeastern Pennsylvania. Industrialization and urban growth made both commodities indispensable in the Northeast, and Albany would be well served to wrest that trade from Kingston, the terminus of the D&H Canal. For the Albany men, the local needs of central New York were useful rhetorically to demonstrate the public utility of the railroad but were peripheral to the enterprise. Albany capital would come to the countryside on its own terms.[20]

Nevertheless, the promoters presented the road to the people of central New York as a civic enterprise that resulted from popular demand for improved transportation. The Oneonta meeting was organized to give the impression that sentiment in favor of the A&S was spontaneous and the company thoroughly republican in its character. Boosters canvassed the townships along the right of way seeking investors and secured pledges for 1,402 shares from prominent citizens. On April 2, "delegates" from towns along the proposed route descended on Oneonta to place the A&S before the people. They elected officers to preside over the meeting (giving it the democratic air of a political convention) and then "drew up" articles of association before a throng estimated at 1,500 people. Ford and Delavan ceremoniously subscribed to 100 shares each, and 321 preselected "volunteers" stepped forward to subscribe to 1,211 more shares. The meeting approved the structure of the company and authorized a capital stock of $1,400,000 at $100 per share (under the provisions of the General Railroad Act, a company had to capitalize at $10,000 per mile to receive a charter, and the A&S would be 140 miles long). Subscribers would pay for their shares in $10 installments on demand of the board of directors. The subscribers then "elected" the board of directors. Delavan served as president, William V. Many of Albany, vice president; Samuel B. Beach of Oneonta, secretary; and Robert Pruyn of Albany, treasurer. The company established its headquarters at Albany.[21]

The men who filed to the speaker's podium to subscribe to shares were wealthier, more commercially oriented citizens. Farmers made up 65 percent and merchants and tradesmen 31 percent of the 213 rural subscribers who can be traced in the 1850 census, the latter figure double that of the whole population. These men were substantially better off than their neighbors, owning on average $3,400 of real estate. The farmers in the group averaged $3,780, well above the $2,500 average value of farms in the three counties. With sixty-five Albany businessmen buying 41 percent of the shares subscribed at the meeting, though they constituted only 21 percent of the men

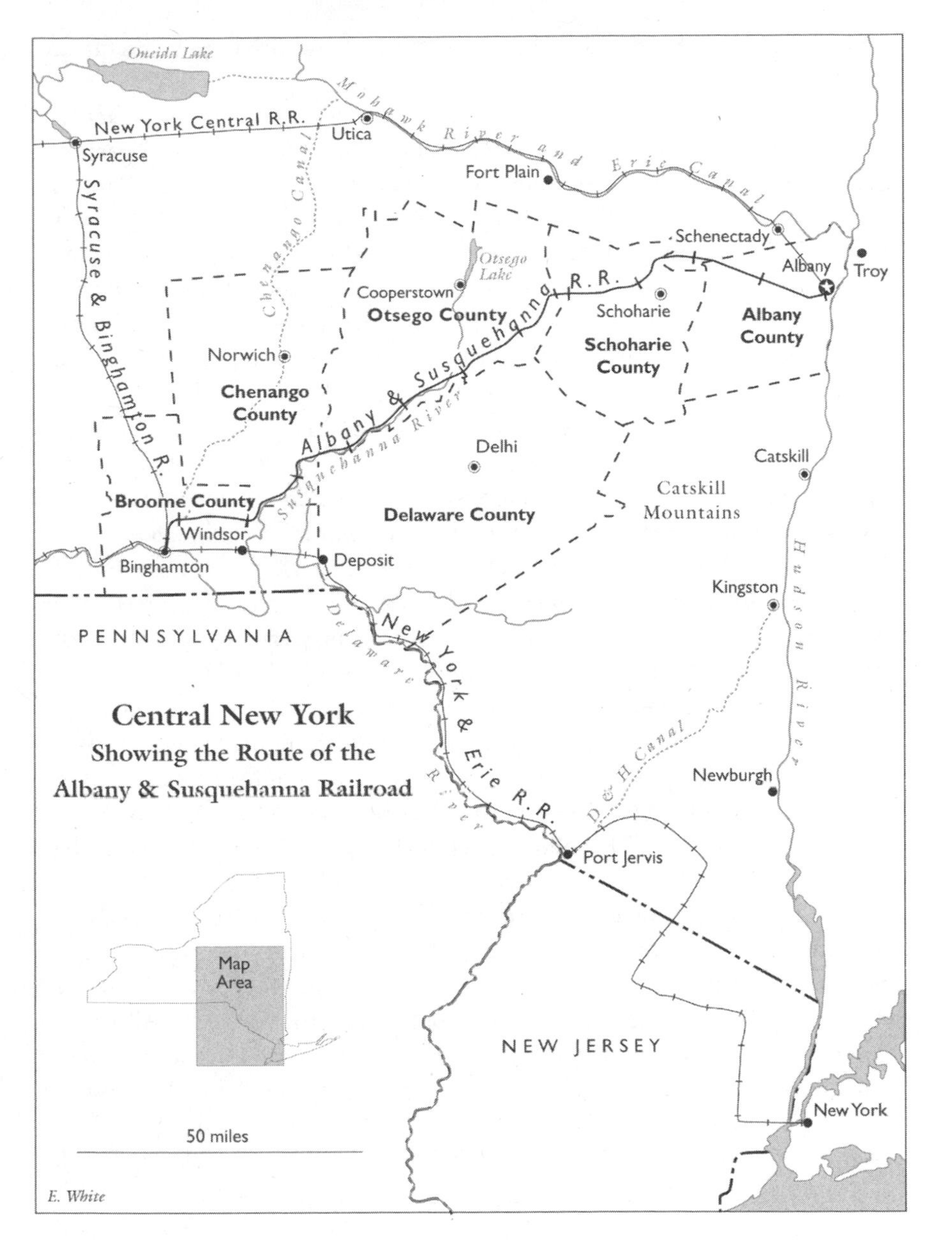

Map of the Route of the Albany & Susquehanna Railroad.

who signed up, the A&S forged a union of rural and urban proponents of progress.[22]

The most eager country advocates of the railway engaged in manufacturing, finance, and speculation and were drawn from both political parties. Agricultural implement manufacturer George Westinghouse of Central Bridge, Schoharie County, was an original subscriber. Soon after the company was formed, he volunteered to meet the additional cost of building the line to his village in order to ship his machines.[23] Merchant Charles Courter of neighboring Cobleskill, a Hard Shell Democrat, chartered several local banks, including one in Cobleskill with Whig attorney and politician Joseph H. Ramsey. The Mercantile Agency reported that Courter was "interested in everything"—local and western real estate and railroads, insurance, manufacturing, and selling Singer sewing machines with partner and A&S backer Minard Harder.[24] Oneonta merchants were so unified in their promotion of the A&S that the Cooperstown *Freeman's Journal* marveled, "Oneonta is growing rapidly in population and business, and bids fair to become one of the largest—if not *the* largest village—in the county. It is already the residence of several very enterprising businessmen, who are prepared to make the most of the Susquehanna Railroad for the benefit of their village."[25]

Having successfully mobilized a corps of entrepreneurs behind the A&S, the directors worked to convince farmers living along the right of way to purchase shares to make up the remaining 90 percent of the capital stock of the road. Chief engineer Gideon Hotchkiss optimistically wrote company president Delavan that "with Som exception, Stock will be taken liberally by the inhabitants after the preliminary Survey & all are Solicitous to know when the party will be on."[26] The directors visited or corresponded with leading citizens in each town along the line to discover the primary crops and manufactured goods produced there, as well as what new businesses might be expected to develop as a result of the construction of the A&S. The directors also circulated rumors that the right of way would pass through townships that most strongly supported the road, an outright falsehood.[27]

This pattern of misrepresentation would generate increasing opposition to the A&S, especially among farmers, whose political loyalties shifted back toward the Democrats and their anticorporate message in 1851. In late October, the Soft Delaware *Gazette* warned, "Should the whig party succeed in maintaining itself in power . . . extravagance, taxation and the federal system of class legislation will continue. . . . It is the party that believes and has always believed that the powers of government are . . . 'to take care of the rich, that the rich might take care of the poor.'" The paper further cautioned Anti-Renters and Free Soilers that the Whigs were "in close embrace with the cotton aristocracy of New York City" and would abandon land reform and free soil. The paper asked whether farmers would sustain Whig class warfare or "have an assurance that economy will again pervade the management of public affairs, and that laws may be passed equal in their character and adapted to the spirit of the age." The message was straightforward: a virtuous, modern agrarian republic was only attainable through the Democratic Party.[28] "Democracy," the paper concluded, "delights in simple, frugal government without complicity of laws by which the rich are made richer and the poor poorer. . . . It is a primary principle of democracy to guard with ceaseless vigilance the rights of all and to banish from its ranks every one who trafficks in the spoils of office, or makes merchandize of the people's money." The editor excoriated Whigs for granting dozens of special charters to corporations, a practice supposedly put to rest under the new constitution, and charged that they intended to continue to pass "unconstitutional laws to aid unnecessary works" and would use charters to purchase the support of competing party factions.[29]

The 1851 election returns demonstrated that the Democrats had gauged correctly the mood of the electorate. Democrats won nearly all state offices—including seats on the state canal board that offered lucrative patronage—and regained control of the senate. In Delaware, Democrats won state contests by healthy margins, losing only a few county races to Whig/Anti-Rent candidates. The Anti-Renters, when confronted with candidates chosen by the state party organizations, clearly trusted Democrats and their message far more than the Whigs. Taken as a whole, the 1851 election suggested that the Whig ethos of progress had run aground on fears that it masked the selfish, federalist motives of monied men.[30]

With this Democratic resurgence, the A&S promoters feared that the citizens of Delaware, Otsego, and Schoharie Counties might be no more interested in personally investing in the railroad than having the government do so. In April 1851, President Delavan urged the board to commence the subscription drive no later than the fall because, when it came to convincing farmers to invest, "Slow will be the tug."[31] But not until February 1852 did the company hold mass meetings "in churches, hotels, and town halls . . . a number of the directors traveled the length of the line, sometimes, in certain parts of the country, on foot and from house to house, 'tagging' everybody possible," where directors distributed literature and urged prominent citizens to subscribe to induce others to follow suit.[32] Promoters played upon local pride, arguing that the quality of the soil and character of the people of the three counties had enabled them to maintain population despite better transportation facilities available in other parts of the state. The railroad, they declared, would once again place the counties at the forefront of the state.[33]

The A&S promoters articulated a message of civic enterprise, but their actions suggested less lofty motives. At many meetings, for example, the company arranged to have local elites take substantial stock subscriptions to entice others to do the same; however, many of these were reduced or nullified afterward.[34] Despite such attempts to manipulate audiences, however, Delavan's prediction proved all too accurate. Whatever initial curiosity farmers had about the railroad, they had no intention of ponying up even $100 to see it built. Director Edward Tompkins, traveling through Chenango, Otsego, and Delaware Counties in February 1852, lamented, "We drag along day & night through the mud to fulfil our appointments, without expecting to find any body to meet us—In the last particular we are not disappointed."[35] Two years later, the Whig Oneonta *Herald* concluded in frustration, "The Rail Road which is to be built thro' this valley, seems rather to scare some of them; and we are of the opinion that they will be found standing '*still,*' unless they do different than they have been in the habit of doing, in years past, and act more united on matters of public improvement."[36]

The directors, however, barely hesitated at the poor showing and instead pursued municipal aid in Binghamton and Albany. Promoters lobbied the state legislature to obtain special legislation in March 1852 to allow the village of Binghamton to subscribe to the capital stock of the road by plebiscite. Director Tompkins managed the voting with Machiavellian precision, as he recounted later:

> The opponents were quite as busy & if there had been the least chance left for them when the polls were opened, the vote against us would have been large. But the battle had then *been fought,* & all that was necessary was to follow up the victory. It cost some money for teams &c to get the voters out, as the law might be so construed as to require two thirds *of all the tax payers* in its favor, whether any voted against it or not—I insisted on that construction, as it stimulated our friends, and kept our opponents away from the polls. If we were compelled to get 2/3, their names counts as much as their vote; Hence *all but two stayed at home;* and the vote goes forth as unanimous: No *effect* is better than if it had only been a majority.[37]

The village took $50,000 in stock. Then Delavan launched a two-phase initiative to get Albany to fund the road. First, the company held a subscription meeting on the steps of the capitol in June 1852 and sold 1,846 shares, with seven leading city capitalists involved in the A&S taking a third of the total. The company then convinced the legislature to grant the city authority to extend a $1 million loan to the company by claiming that the June meeting had demonstrated public support for the enterprise. A number of prominent Albany citizens petitioned against the measure because no work had been done on the road, no construction contract was in the offing, and the construction estimates had climbed from $4 million to $7 million. Yet the company carried the referendum, managing to poll 2,591 favorable votes in the city's first and second wards, where the total number of registered voters approximated 1,400.[38]

As Softs had warned, the bad old days of distributive politics seemed to have returned, but now in more troubling guise. Though it limited the legislature's right to grant charters, the state constitution had not explicitly taken away its ability to promulgate laws favorable to particular corporations, counties, or municipalities. Combined with the General Railroad Act's failure to provide adequate regulatory enforcement of corporations, New Yorkers had unleashed a mad scramble in the capital for government dollars. The A&S directors would have to take full advantage of this situation because the Albany loan was structured to protect the city in case the railway went bankrupt. The A&S would receive city bonds in $250,000 increments and would only get successive issues once it had generated an equal amount from other sources (stock sales or revenues) and could show that it had been spent on construction. With less than $700,000 subscribed and only 10 percent paid in by stockholders, the first bond issue represented the company's primary source of working capital—and the bonds still had to be sold on the market. Without further sales of stock, which seemed unlikely, the A&S would not even be able to collect the full amount of the loan.[39] Director George W. Chase of Maryland, Otsego County, observed in July 1852, "The RR feeling has considerably died away since our meetings—our vicinity has heretofore been pushed about as far as could be for stocks."[40]

Success, then, depended on further government aid, which in turn depended on electing men favorable to the road. The A&S dropped any pretense to nonpartisanship and turned to Whigs and Hard Shell Democrats to elect pro-railroad candidates. In the fall election, despite Whig failures across the state, Chase, a Seward Whig, defeated Soft Samuel Gordon of Delaware County for Congress from Delaware and Otsego Counties.[41] Thus, while the Softs railed against a "foul morass of aristocratic benefits" that included "protective tariffs, sumtuary laws, limitations upon suffrage, property qualifications for office, alien laws, test acts, monopolies, special charters, and every other species of class legislation,"[42] conservative Democrats accepted Chase's overtures. After all, they had been artful practitioners of distributive politics and had supported internal improvements since Jacksonian times when these sprang from local initiatives; they now had few moral qualms about the logrolling, horse-trading, and corruption that defined Albany politics in the railroad age. The 1852 election thus left Democrats in a quandary. On one hand, they evinced an ideology that central New Yorkers generally embraced, especially farmers, yet on the other, continuing factional disputes within their own party prevented them from winning critical races that might help them put their program into law. And in many ways, 1852 would be their last opportunity, for the Hards and Softs who joined to support Franklin Pierce for president would split again within a year.[43]

While the A&S was beginning to put in place a network of sympathetic politicians, rumblings among stockholders began in early 1853 because the company had not let a contract yet was making new calls on subscriptions.[44] Under the gun, and needing to collect the rest of the Albany loan, the company concluded an agreement with a New York firm—Morris, Miller, Baker & Co.—in May to construct the road. But once again, the road's actions were tainted with scandal. As the Senate Committee on Railroads would later report, "At the time of the making of the contract, only eight hundred thousand dollars had been subscribed to the capital stock of the company, and in order to fill up the capital, as specified in the articles of association, they required Morris, Miller, Baker & Co. to subscribe for seven hundred thousand of the stock, and, as the contractors allege under oath, loaned them sixty thousand dollars out of the funds of the company to pay in the ten per cent on six hundred thousand dollars." The fraudulent subscriptions did not represent actual capital invested in the company and therefore violated the General Railroad Act's requirement that a company's full capital stock be subscribed before a contract could be let.[45]

The directors then overplayed a bad hand by planting newspaper stories praising the contract and claiming that construction would be under way in the summer. The pro-railroad Oneonta *Herald* opined that "the contract recently made for the completion of your Road with the strongest Company that has ever undertaken a similar work in America, has removed it entirely from among the doubtful projects of the day, and rendered its construction a fixed fact."[46] The paper carried on a relentless campaign to convince farmers to subscribe, while one piece of A&S literature even advised farmers to borrow money to subscribe to stock, claiming it was a sure investment.[47] But new subscribers did not materialize, and when the questionable dealings of the directors began to surface, many angry subscribers refused to pay their premiums amid charges of fraud, stock watering, and lies.[48]

National political events prevented an effective Democratic attack on the A&S, despite this widespread outrage. President Pierce reopened factional splits in the New York Democratic Party in 1853 by awarding the choicest patronage to Softs, despite the fact that Hards were more sympathetic to his policy of conciliation with the party's southern wing. In Schoharie, Softs seized the opportunity to attack the pro-railroad Hard machine of Lyman Sanford. Director Charles Courter, Sanford's confederate in Cobleskill, feared for the road's future because "disaffected Stockholders, and the opponents of our project are Industriously Reporting, that they do not believe that the road will be built, they admit that one & a half miles are under contract, but say it is only to gull the Stockholders, & get the Second 10 per Ct., They represent it to be one of the Cheapest Sections on the Road, and if the contractors were in Earnest, they would be driving the work more vigorously, now unless this feeling Can be allayed and Confidence restored it will be very tedious work to collect here."[49] When Softs won most of state and local races in the three counties in the fall 1853 elections, they believed that their message of limited government was the reason. Yet, New York State voted Whig that year, which meant that they would have plenty of patronage to dispense, and Whigs by and large supported the A&S.[50]

What could have been seen as a political victory for the A&S was immediately threatened in early 1854. First, stockholders discovered that very little work had been completed on the line.[51] Claiming they had purchased stock on the premise that "the road could be built for about three millions of dollars, and that [the directors] 'believed it to be the very best and most profitable investment we could make of our money,'" sixty Albany stockholders who owned 661 shares of stock petitioned the board of directors in April to dissolve the company to avoid further losses. They urged "the company to avoid 'an enterprise which, in our honest judgement, must result in a loss in our stock, increase of city taxes, and may involve us in heavy liabilities beyond our stock.'"[52]

Then MMB & Co., which had barely commenced grading, announced to the board in May that it wished to terminate its contract because men "whose opinions as to the value of railroad securities of the State of New York is almost conclusive here and in Europe" had no faith in the A&S.[53] As the two sides negotiated throughout the summer, work stalled in August, and the parties took the disagreement to court. Compounding matters, in June 1854 Wall Street learned that Robert Schuyler, who had close ties to MMB & Co., had defrauded the New York & New Haven and the New York & Harlem Railroads. Investors stopped buying railroad securities, especially of companies connected to Schuyler. Meanwhile, the complex, illegal relationship between the A&S and its contractor finally became public.[54]

Goaded by Soft newspapers like the *Freeman's Journal*, which called for the termination of the company, bitter subscribers refused to pay their second installment on their stock—a tactic that had helped kill many turnpikes under the old constitutional system. Promoter Harvey Baker of Oneonta recalled, "The failure of the contractors had a demoralizing effect upon all the hasty element among the stockholders of the road and gave its opponents, who were no insignificant number, an opportunity to do the best for its defeat."[55] The company decried this "fault-finding spirit" and attempted to restore faith in the directors with the now-familiar falsehoods: "The more thoroughly the action of the board is known, the more satisfied, we believe, will the public be, that both caution and discretion has guided it in putting the work under contract, and the more willing will the stockholders be to sustain the board and the president cheerfully, and meet the calls made on them with as little delay and grudging as may be possible."[56] The directors took aggressive action, drawing up threatening form letters dated simply "185_." Almost absurdly, considering that the directors intended to send the letters out over a period of several years, the letters stated that "$12,000 is being spent each month," that "rails were to be laid in the Spring to Schoharie," and that the directors "would deprecate the necessity of resorting to compulsory measures, yet having no alternative, they beg leave most respectively to say, that unless the ______ call on your stock is paid on or before the ______ it must of necessity be put to suit."[57] One can only imagine how recipients of these letters reacted to the obvious deception.

And so the company began a series of suits against delinquent subscribers in 1854 that would reveal to central New Yorkers that the political economy of their state had changed in ways they could not have expected in 1846. Three cases demonstrated that judicial oversight of corporations did not protect the public from corrupt or unethical business practices. The first, heard before Justice Andrew G. Shaw in Franklin, Delaware County, revealed that electing judges did not eradicate elite influence in the court system, as the framers of the 1846 constitution had intended. An original stockholder, Shaw did not recuse himself. On the contrary, he dismissed defense evidence that indicated that company agents had misrepresented the financial health of the company and had employed prominent citizens to falsely subscribe to stock to induce subscriptions. Shaw accepted at face value the denials of the company agent who sold the delinquents their stock and ruled in favor of the A&S. The Oneonta *Herald* crowed, "Gentlemen, you had better pay up your ten per cent now, if you do not want to pay it and also the cost of a suit. The directors are determined to prosecute all who do not pay."[58]

Moreover, since the General Railroad Act made an a priori assumption that railways *always* provided an essential public service, the courts could rule only on whether corporations acted according to the letter of the law, not a road's public utility. The A&S suit against Abraham M. Stanton of Laurens in 1854 laid this out clearly. The defendant refused to comply with the company's demand for his second payment on his stock in 1853 because, he claimed, company sales agent Eliakim R. Ford knowingly had made two fraudulent statements: first, that the road could be built for $4 million, and second, that several specific Albany financiers had taken stock in the company. Later, Stanton discovered that the projected cost of the road was $7 million and that the Albany men had not purchased shares. Stanton charged that Ford's prevarications indicated that the A&S was nothing more than a company put together by a small group of local entrepreneurs who intended to bilk rural investors of their hard-earned cash. The judge in this case stated that caveat emptor applied and accepted Ford's testimony that he had only expressed "opinions" about the company's financial prospects, its stockholder list, and its capitalization.[59] The judge further ruled that the defendant's allegations of fraud, even if true, did not outweigh the public utility of the project. Unlike eighteenth-century British bubble schemes, he wrote, the A&S was "not a visionary project of a few rash adventurers, seeking to strip honest men of their earnings, but is that of an incorporated association of gentlemen of wealth, talents, and public spirit, formed under the laws of the State, for the commendable object of constructing a railroad through an interesting portion of the country."[60] In effect, this ruling meant that *any* railroad, no matter how unpopular, corrupt, or poorly managed, was de facto in the public interest and therefore free to conduct itself as it saw fit, except in instances where it specifically violated provisions of the General Railroad Act. The notions of personal integrity, honesty, honor, and accountability held by Stanton were irrelevant.[61]

The ruling handed down in a third A&S suit, against Edward E. Kendrick of Albany, extended further judicial protection to the venture. Kendrick also refused to pay his second installment, but he based his defense on evidence that company directors had used the names of third parties without their approval or payment to subscribe to stock at the Oneonta meeting in 1851. With the subscriber list thus padded, they then filed articles of incorporation with the secretary of state. Kendrick contended—quite reasonably—that this constituted fraud and, more important, negated the corporation's original charter. He therefore could not be held liable for a debt to a company that did not legally exist. The judge ruled that by purchasing stock in the road, Kendrick de facto approved the charter and bylaws of the company; he could not later claim that the corporation did not exist. How one might be protected from stock fraud under this ruling is not clear. Kendrick's second argument—that the subsequent misconduct of the directors in letting the contract fraudulently to MMB & Co. nullified any claims against him—also did not impress the judge, who wrote simply, "As to the transactions of the company with third parties, he and the other members, are equally bound by the acts of their common agents [the directors]." Stockholders thus had no remedy at the bar for the directors' misdeeds, for stockholders and the company were one.[62]

Stockholders refused to submit to the rulings and continued to ignore calls for payment, steadfastly maintaining that they, not the company, had been injured. Even with the favorable rulings, therefore, the directors considered shutting down the enterprise. However, several proposed that the A&S seek municipal funding and hired Cobleskill Whig attorney Joseph H. Ramsey to study the question. After conversations with a number of politicians and financiers, he reported to the board that such a move would be possible with careful legislative management. Ramsey then secured the Whig nomination for assembly from his district so that he could usher the legislation through personally. Whether this bid would succeed depended on the ability of A&S supporters to weather the political storms bearing down on the region from across the state and nation.[63]

The proximate cause was the passage of the Kansas-Nebraska Act in 1854, which reopened the debate over slavery in the territories. Free Soilers feared that the doctrine of "popular sovereignty" enshrined by the act would extend slavery northward, thus limiting the expansion of free territory. Softs were in a bind, for they depended on Free Soil Democrats to maintain their coalition yet were also beholden to President Pierce for patronage, and he supported the law. Hards, for their part, came out squarely in favor of the Nebraska bill in the hope that they could discredit their adversaries among southern Democrats. They nominated a separate slate in 1854. Thus the party's energies were almost completely absorbed with divisive issues that pushed its more effective message, limited government and individual liberties, to the background.[64]

In central New York, Democrats kept the Kansas-Nebraska Act from atomizing the party by the slimmest of margins. Hards attacked Softs for trying to finesse the Nebraska issue, claiming that "small, sincere and fanatical band of abolitionists" had split the party when there should only "be but two organizations in the State; one would be the national democratic organization, and the other an anti-slavery organization, occupying the place and taking the platform this year adopted by the whigs. There was not, and would not be, any halfway house."[65] But Softs deflected the slavery issue locally, arguing that the Democrats must unite against Whig temperance legislation, which would hurt the local economy: "Orchards must be cut down, hop yards ploughed up, and barley fields laid waste, to save the dear people from self-destruction."[66] In Otsego, where Soft gubernatorial candidate Horatio Seymour's pro-liquor stance played well among hops growers, Democrats won every state and local race.[67]

But the election returns in Delaware and Schoharie revealed Democratic instability. In Delaware County, Seymour just edged Whig Myron Clark while the Whig congressional candidate decisively defeated his Soft opponent. Still, Soft tenant leader John McDonald's victory in the race for county supervisor of the poor suggested that Softs still had a chance to win the agrarian vote.[68] In Schoharie, however, boss Lyman Sanford's championing of the Nebraska act caused rank-and-file Democrats to rebel. Even though Sanford had his newspaper, the Schoharie *Republican,* replace the Hard state ticket with the Softs' just before election day, Clark carried the county in a tight contest. In the race for the assembly in the first district, respected Anti-Rent leader and Hard John Mayham lost to his Whig opponent. The election was important for the fortunes of the A&S because the chaotic political situation allowed Whig Joseph Ramsey to win Schoharie County's second assembly district, while Whig victories statewide ensured that in Albany, at least, the road would get a favorable hearing. Even though Softs had won the most races in the three counties, their defeat across the state kept from office men who might fight against the A&S.[69]

The maneuvering to obtain municipal funding for the A&S began as soon as Ramsey entered the New York Assembly in 1855. Whereas municipal support for railways was not new, arranging for towns to purchase stock via bond issues was unusual. Since company stock sold at 10 to 15 percent of par and its bonds sold at a discount, any town bonding bill would be sure to elicit stiff resistance, especially from anti-tax Softs. Ramsey therefore crafted the legislation much like the Binghamton law. The bill stated that approval of the measure by two-thirds of the taxpayers owning two-thirds of the taxable property in each township would enable two town commissioners, appointed by the county judge, to sell bonds to purchase A&S stock. The proposed law had several advantages for the A&S. First, only taxpayers could vote, which minimized the possibility that Softs could mobilize tenants, tradesmen, laborers, and others who did not own real estate against the bill. Second, the A&S had cultivated relations with local justices and believed they could count on favorable commissioners to persuade wavering voters during the drive. Third, the two-thirds clause was ambiguous: it could be read to mean that two-thirds of *all* taxpayers had to approve the road *and* that they also had to represent two-thirds of the taxable property. This would be difficult to achieve. On the other hand, the law could be read quite differently, calling for a two-thirds majority only among those whose property added up to two-thirds of the taxable property in each township. The latter would be much more attainable. For the moment, none of this mattered as outrage throughout the state against railroad corruption had provoked the legislature to form a Board of Railroad Commissioners to enforce the regulations contained in the General Railroad Act. When 156 prominent Albany businessmen published a petition decrying the town bonding act as a ploy to salvage the investment of the promoters without actually constructing the road, the legislature backed away from Ramsey's bill.[70]

Within a short time, however, the state Whig Party fused with Free Soil Democrats to form the Republican Party. Whigs dominated the new party in New York, and Whig economic policies in particular survived the merger. Fusion condemned to failure the Democratic Party's bid to reestablish itself as the voice of the agrarian community. The party's devotion to the principle of small government and its refusal to put forward "class legislation" blinded it to the need farmers felt for positive legislation to protect their interests, whether in the form of federal guarantees that western lands would be left open to free settlement, or stricter regulation of corporations and other legislation to ensure that capital did not grow too powerful within the polity. Softs continued to decry "spend-thrift legislation" and "the harpies that infest the Capitol" but failed to offer a specific reform agenda to voters.[71] In New York's realignment, Republicans thus wrested a substantial portion of the agrarian vote from Democrats on the slavery issue, even though Free Soilers, by and large, disliked the holdover Whig economic agenda. Absorption of agrarians into the Republican camp would help blunt partisan attacks on the A&S in the next five years.

In the meantime, factional upheaval gave A&S supporters time to line up votes for their candidates. In Schoharie County, Sanford and the A&S faction worked to ensure favorable results for both the Hards and the railroad. A&S backers engineered the key nominations of Ramsey for state senator on the Republican ticket and Sanford for county judge on the Democratic slate. If both were elected, Ramsey could guide the town bonding bill through the senate, and Sanford would be in position to appoint sympathetic town railroad commissioners to manage the approval process. Sanford and director Charles Courter struck a secret deal with Ramsey in which Courter would publicly endorse the Republican candidate for county judge, Henry Smith, to trick Softs into believing that Smith was the A&S's candidate and thus vote for Sanford. Sanford would then deliver key Hard votes to Ramsey. The bargain worked handsomely. Sanford won easily, while Ramsey, whose district included Delaware and Schoharie, polled enough votes in each to win the election without carrying either in a three-way race. Meantime, Schoharie delivered its usual Democratic vote in most other races, enabling Sanford to claim that his management had kept the county in party hands. It was a model political intrigue.[72]

Though the Otsego County election did not turn on the A&S issue in 1855, it illustrated well the price Softs paid for refusing to promulgate policies that would benefit farmer constituents. By 1855, hops production had soared in the county, yet when pro-liquor Democrats attempted to create a fusion party that would bring Softs, Hards, and Know-Nothings into an alliance that would promote the interests of hops growers, the Soft leadership balked. Softs could not bring themselves to back what they termed special legislation.[73] The entire Soft state ticket lost. Republicans won all seven state races and three local ones, while the pro-liquor Know-Nothings scored victories in three county contests. Once again, the Democratic Party had offered farmers little evidence that it would actively protect agrarian interests.[74]

With political winds blowing in the railroad's favor, Ramsey reintroduced the town bonding legislation, this time from the powerful position as chairman of the Senate Committee on Railroads. A&S supporters made a second push for municipal funding in 1856. Reaction from Soft opponents was swift. Over 250 Albany residents published a petition in the Soft Albany *Argus* renewing charges that the directors intended to use the money to salvage their investment without building the road. A group of prominent Albany capitalists led by Soft politician and New York Central Railroad president Erastus Corning also attacked the municipal bonding law because it allowed the company several more years to comply with the terms of the Albany loan. As stockholders of the A&S, they requested the directors to cease operations to avoid further losses.[75]

The directors fought back viciously in the legislature. On the floor of the senate, Edward Tompkins denounced the opponents as "Shylocks, Jews, old fogies—frightened timid Dutchmen, selfish mercenary chaps." Tompkins charged that Corning was angry because the directors refused to buy iron from him. Thanks to Ramsey's careful lobbying, the bill passed despite the opposition, and the company ordered construction to resume in December 1857.[76] Charles Courter jubilantly exclaimed, "And now the question is Shall we build our *Road; or* Shall we always, Stay here in the *Woods* with out any thorough fare."[77]

The A&S again staged a massive misinformation campaign to hoodwink townspeople considering the measure. Tompkins believed that farmers could be gulled into approving the bonding acts if the company appealed to their foolish greed:

> It is quite important that the explanation should be very *minute* and plain. In the country such documents are read more carefully than in the city, and there is not always the same basis of general information that would render the demonstration of axioms unnecessary—The farmer family lives in the kitchen, and at the evening one of the boys who has been to school more than his parents had the opportunity to do, will read it aloud and the kitchen cabinet will discuss it, as it appears to them in all its bearings. I am as times go, a very good sort of Democrat with a full party share of confidence in the masses, yet I would rather trust a genuine, well turned sophism, stated with the utmost confidence and frankness, and made very *plain,* to influence the *crowd,* than all the half-told truths that have been enunciated since tongues and pens came into general use.[78]

Contemptuous of the rural folk they targeted, the directors held meetings in every township, where they emphasized civic duty and progress, asserted that the railroad would earn a profit, and downplayed the fact that if the company failed, or if the towns could not meet the interest payments on the bonds, taxpayers would be forced to make good the loss.[79]

In an atmosphere of increasing criticism of railroad corruption in Albany, the directors' legislative victory did not check resistance to the A&S. Indeed, reformers from both Democratic and Republican ranks attacked the town bonding law as yet another example of money overriding public interest in politics. Softs focused on the issue of cleaning up government by quashing the influence of special interests in Albany. Local newspapers printed blistering editorials. The *Freeman's Journal* declared that the A&S was yet another undertaking that corrupted the legislative process and cloaked the fleecing of the people in civic garb. Connecting members of the A&S board with a series of corporate scandals, the editor remarked that "the public have had enough of *their* service."[80] While Softs traditionally had opposed the road, the A&S also faced an assault by reformist Republicans. One contributor to the Republican Schoharie *Patriot,* for example, observed that no one had objected to "Capitalists" investing in and building the A&S. However, he noted, mechanics and farmers, who had small means, could not be expected—nor should they try—to invest in the railroad, even collectively. The writer lambasted the directors for engaging in petty corruption and selfishly foisting the cost of the project on the people and for promoting a railroad that "would, like all similar projects, benefit some, while it ruined thousands of others, or depreciated their property in value."[81] Ramsey labeled opponents "old fogeys," but anti-railroad forces gathered strength nonetheless.[82]

The parties did not take a formal stand on the railroad at this point, but opponents organized in each township and refused to sign petitions authorizing town subscriptions to stock. In all but a few towns, anti-railroad forces triumphed over the company.[83] They received aid from outside the region, much of it from the Soft-led New York Central Railroad, which sent agents to wavering towns to encourage dissenters and to build Democratic support around the railroad issue. In Otsego County, Know-Nothing assemblyman T. D. Bailey complained that the Central, disgruntled shareholders, and local opponents were winning over taxpayers by claiming that the railroad would benefit only a chosen few citizens at the expense of others. Having helped the A&S secure the bonding law in the assembly, he now requested a substantial bankroll to avert disaster for the road.[84]

Where the company did manage to get a favorable vote, opponents turned to the courts. In Cobleskill, when newly elected county judge Lyman Sanford appointed Director Charles Courter town railroad commissioner and Courter promptly secured approval and purchased stock in the company, opposition leader Richard J. Grant sued Courter to prevent the sale. Grant's interpretation of the state constitution echoed the Softs' strict constructionism and their privileging of individual over corporate rights. He contended that the 1846 constitution had limited the power of towns to extend their credit to private corporations in order to stop abuses practiced under the old government. He contended that the legislature must "*restrict* cities and incorporated villages in their *power of taxation; in contracting debts and in loaning their credit*" in order to ensure that government did not allow elites to impose their will on the majority or to take public funds for private use. He was especially perturbed at the assumption inherent in the law that the road would serve the public interest. The bonding acts would effectively levy a $1 million tax, not for "any *public* necessity; but merely to furnish the necessary capital to enable a private company, for their own private emolument, to engage in business as *common carriers.*" The acts represented "a fearful and dangerous enlargement" of the right of the municipalities to tax citizens yet contained no democratic checks on "disastrous speculations" that would "involve whole communities in poverty and bankruptcy." In the end, he argued, "Whatever advantages may be anticipated from the construction of the road there is not any such *public necessity* for its construction, as should justify a resort to *taxation* for that purpose." Worse, he hinted that approval in Cobleskill had been improperly obtained, further evidence that the bonding law could only hurt democracy.[85]

Grant perceived an imminent threat to political democracy and representative government in the law, first because the town bonding act undermined the authority of the legislature as the only body that had the power to tax. To delegate this power to any other public or private entity would violate the due process clause by removing the power of taxation from the legislature and thus from the people as a whole. Since railroads—like banks, turnpikes, and insurance companies—were private corporations, government could not convey its powers to them. Second, Grant took no comfort in the approval process outlined in the law because, by definition, townships could not judge the common interest of the people of the region or state, and the wealthy might easily manipulate the vote at the local level. Finally, democracy was betrayed because the people could not prevent friendly county judges from appointing commissioners favorable to the A&S. The town bonding act, according to Grant, created an unfair "lien" on his property by allowing select individuals to seize his property to fund their own enterprise while making property rights contingent on the whim of a dubious majority: "The more property he has, the more he is required to pay, and the greater the lien; if he does not pay that lien, *his property* shall be sold . . . to pay a debt contracted for the benefit of a *private corporation*."[86] Grant therefore hoped to reverse the law by appealing to the Softs' interpretation of the constitution and General Railroad Act, each of which, they believed, uncoupled private enterprise and government.

Republican judge William B. Wright rejected Grant's contention that the road was merely a private concern. Though he questioned "the wisdom of a town in a largely agricultural county, where there is no unusual concentration of wealth burdening itself even temporarily with a debt, and becoming a stockholder in a Rail Road Corporation," Wright ruled that because railroad commissioners acted in the public interest, no constitutional restraints prevented town subscriptions.[87] He dismissed Grant's argument that the legislature could not make such a law, concluding that the 1856 town bonding act was "the fruit of the legitimate exercise of legislative power" because the railroad was "a work of conceded necessity to its inhabitants, and calculated to augment the value of its taxable property." He also rejected Grant's contention that the 1856 law violated "the *spirit*" of the constitution because "an otherwise valid exercise of the law-making power, cannot be unconstitutional, for the reason that the law is antagonistic, in spirit, to certain provisions of the constitution, provided it not be in direct or necessary conflict with them."[88] Wright's decision gave the company judicial protection from the general public unless the company violated the letter of the law.

While A&S opponents awaited Wright's decision, national events again intervened in local affairs. The Kansas controversy spread fears across New York of southern radicalism that would fuel a Republican triumph in the 1856 presidential election in the state. When South Carolina's Preston Brooks caned Charles Sumner on the floor of the U.S. Senate in May, Cooperstown protesters mobbed the home of U.S. Supreme Court justice Samuel Nelson, a Hard Shell who supported the southern wing of the Democratic Party.[89] Republicans launched a campaign to attract disaffected Democrats and Know-Nothings.[90] "The question at issue," according to the Republican Oneonta *Herald*, was not "whether the black race shall be enslaved in Kansas, but whether the white man shall also be enslaved in all our governmental territory." The Democrats, it warned, were "doing all in their power to subdue the white man to that same base servitude."[91] A rally at Cooperstown drew a crowd estimated at five thousand to twenty-five thousand,[92] and Jane Russell Averell, daughter of a Republican boss, concluded that "there is not the slightest doubt as to which party 'Otsego' belongs."[93] When meetings, speeches, and invective failed, Republicans simply spent money to influence voters. The Democratic campaign had nothing new to offer voters except a refusion of the Softs and Hards, which elicited little enthusiasm.[94]

The results were, first, that Otsego and Delaware Counties rallied to the Republican Party. Free soil was popular in both counties, and the Republican gubernatorial candidate, John A. King, a former landlord, had embraced free soil and probably was less offensive to voters than the Soft candidate, Amasa J. Parker, who was unrepentant about the harsh sentences he had handed down to Anti-Rent insurgents in the trials that had followed the Steele murder in 1845.[95] But in Schoharie County, Hard Democrats kept the county out of Republican hands by once again catering to pro-railroad Republicans. Because Schoharie County previously had resisted reformist appeals, Republicans muted their antislavery message, while Hards used the A&S to defeat their Soft rivals. State Soft leaders sent emissaries to Cobleskill laden with cash to buy votes for their candidates, but Sanford, Courter, and Ramsey swung pro-railroad Republicans behind Hard candidates. Local offices by and large went to Hards or Republican friends of the A&S while the county delivered a Democratic plurality for president and governor.[96]

The local context thus was so volatile that it precluded any successful effort by Softs to translate statewide calls for tax reduction into a local campaign against municipal spending on internal improvements. Hards controlled Schoharie and supported the railroad; in Otsego, agitation against internal improvements would undermine the fragile entente between Softs and Hards; and Delaware's political system was still too much in flux to risk such a plan backfiring. Faced with well-funded, organized, and politically connected promoters, opponents suffered repeated setbacks. First, the railroad directors secured an amendment to the town bonding acts in 1857 requiring approval by only a simple majority of taxpayers holding a majority of the property in each town. Then political leaders in anti-railroad townships unwittingly placed trust in colleagues already in the railroad's pocket. Soft Carlisle town clerk John H. Angle, for example, confided to secret A&S supporter and county judge Lyman Sanford that "there is getting to be considerable Strife in this Town about the Loan to the Albany & Susquehanna Railroad" and boasted that his town would not be browbeaten into supporting the railroad. Since Sanford would appoint the town commissioners, Angle requested that Sanford notify him as to whom the A&S wished to appoint so that he could counter the move. No reply came, and Sanford appointed pro-railroad men to the post.[97]

The 1857 elections illustrated that Softs had too little strength to battle the entrenched A&S political coalition. In Schoharie County, Hards again solicited Republican help. When a coalition of "Anti-railroad, high pressure, Know Nothing Democrats" nominated Softs for town supervisor, justice, and clerk, Charles Courter reported, "It is said they had a terrible time at the Polls, Rum flowed in torrents," but with his help the Republicans won all three offices.[98] Hards kept control of the party and easily managed the outcome of the fall election. Democrats won in state and local races as Republicans conceded Schoharie to Hards in exchange for the railroad.[99] Delaware County Republicans won all seven state offices, but their vote total fell by nearly 40 percent, while Democrats retained enough voters to elect their anti-railroad candidate for state senate, Edward I. Burhans, in a close local contest.[100] Glimmers of a Democratic resurgence appeared in Otsego, where Hards and Softs united to battle Republicans on the temperance issue—a formidable alliance in a hops-growing county. Democrats won six close races in local contests primarily by arguing against prohibition and noting that Republican zeal for free soil and antislavery had dampened after the party had carried the state in 1856. But, while Burhans would prove a formidable anti-railroad spokesman in the legislature and Democratic success in local elections was hopeful, statewide Republican victories prevented any broad reamalgamation of the agrarian vote.[101]

Thus, by 1857, A&S opponents had been defeated in the courts and could not command enough votes to mount an electoral assault on the road, despite success mobilizing at the township level to block town bonding. Their recourse was to petition the legislature to repeal the town bonding acts, sending seven petitions to the senate and more to the assembly by March 1858.[102] Several key themes emerged in this campaign that emphasized Soft skepticism toward entrusting corporations with public powers. "H," a Laurens Democrat, admitted, "I should be benefitted by its construction—but I have seen too much mismanagement in regard to it, to put too much power in the hands of the Directors. I don't want my town saddled with debt for *nothing*—and before I vote to issue them town bonds, I want to know *who* will take them at par, and to be assured that, with their aid, the road *certainly will be built.*"[103] "O" told the Republican Schoharie *Patriot* that he did not oppose the construction of the railroad but strongly objected to public financing. Unlike roads and turnpikes, he wrote, railroads "are a species of 'close corporations,' where great facility for speculation and fraud is given, and where, as we have too often seen, stockholders have been cheated and swindled." He called municipal funding an "arbitrary" tax on those least likely to benefit from the road. The merchant elite, he wrote, had instituted "unequal" and "oppressive" laws that flouted the democratic spirit of the constitution.[104] The opposition was strong enough to lead one supporter of the A&S to conclude that the road "is one of those things only talked of, but not to be. Petitions in great number, and petitioners almost innumerable, ask the repeal of the Act of 1856 & the amendment thereto in 1857, Authorizing towns to subscribe for stock, and in each case representing a great majority of the assessed valuation of the towns. The prospect is extremely dark."[105]

Soft state senator Edward I. Burhans introduced legislation to repeal the acts in February 1858, arguing that they represented the worst feature of a misguided, profligate state policy on railroads: "We know how much it costs to make Rail Roads, and what they are worth when constructed. We have learned, at severe cost, how utterly unreliable are the *estimates* of Engineers."[106] Even the Republican Albany *Evening Journal* observed that "the worthless stock" of numerous railroads statewide "shows how even roads which are completed are prone to failure." The editor wrote, "While the delusion lasted, they were regarded with such favor, that thousands of Farmers along their Lines paid their hard earned dollars for stocks which are now not worth the Paper and Ink they cost." With the railroad still having only half of the funds needed to complete the line, any policy other than repeal was "inexcusable."[107]

Opponents of the A&S played upon statewide fears of the growing influence of the railroad lobby in Albany to help convince reform-minded Republicans to take a stand against corporate corruption of politics. Otsego's Republican state senator, A. H. Laflin, declared that the railroad power in the state "has become a power hardly second to that of the State itself. . . . When we consider how mighty are the interest of capital and labor under railroad control," he continued, it was not surprising that "many regarded [it] as a most dangerous institution, as an '*imperium in imperio.*'" Though not an endorsement of repeal of the A&S town bonding acts, Laflin's speech accepted the legitimacy of "mutterings of discontent . . . heard among the people" about collusion between lawmakers and railroad managers, especially when farmers saw legislators receiving free railroad passes. He predicted that the people would rise "independent of all political leaders" and break "the fetters . . . imposed by a Railroad company, by a subsidized press, or a corrupt and bribed Legislature."[108]

With increasing optimism, opponents flocked to Albany on February 24, 1858, to attend senate hearings on repeal of the bonding acts. Railroad promoter Harvey Baker, who rode the stage to Albany, recalled that "from Schoharie a majority of the passengers were men from that county who were bent on the repeal of the law, and were on the way to Albany for that purpose. I rode all the way in silence, listening to the talk of these anti-railroad cranks. It required quite an exercise of will power to say nothing in defense of our road, but I said nothing." Opponents who were led by National Bank of Albany cashier R. C. Martin—a former Schoharie County resident with interest in the Schoharie Plank Road Company—and Cobleskill Republican Herman Becker argued that company directors consistently acted against community interests by misrepresenting its financial affairs, encouraging local elites to falsely subscribe to stock to induce others to purchase shares, and illegally letting a contract in 1853. The directors, they charged, had operated the company for their own emolument and intended to injure, not aid, the public. As for claims that the railroad would benefit farmers, Martin tartly said, "The counties of Delaware, Otsego, and Schoharie did not make enough butter to grease the wheels."[109]

But the opponents were outgunned. Railroad supporters Baker, Ramsey, Jedediah Miller, Thomas Lawyer, and General S. S. Burnside presented the railroad's case. Baker recalled that he played the decisive role in the hearings by posing, according to plan, as a farmer who supported the road, despite the fact that he was a millwright and land speculator in partnership with A&S directors Eliakim R. Ford and Jared Goodyear. The railroad, he pleaded, was necessary to lift the darkness that hung over the three counties because of inadequate transportation. The head of the committee congratulated Baker on his speech and wrote the majority opinion in favor of the town bonding acts. The minority report agreed with opponents of the road on nearly every count. Senator A. Hubbell criticized the directors of the A&S for subverting the governmental process that safeguarded the welfare of its citizens. The report concluded that while the government should provide aid to internal improvements that benefited the entire population, the A&S

served a narrow interest and should not receive public funds. Nonetheless, the full senate, strongly influenced by Ramsey's lobbying, supported the majority opinion.[110]

Opponents made a final effort to defeat the measure. One Milford man followed Baker "from place to place to try to induce his neighbors not to sign the consent to take stock." He threatened that his firm, Steere & Windsor, would leave town if forced to pay taxes to meet interest on the bonds. Democrat John Edgerton of Franklin, Delaware County, sent men to parts of town that Baker had not visited to firm up opposition. Franklin rejected the bond issue.[111] In Oneonta, company president Richard Franchot held a meeting to allay fears among supporters that the project would fail. Despite such maneuvering, Decatur residents obtained an injunction to prevent its commissioners from selling town bonds in August 1858. Maryland, Otego, Unadilla, Davenport, Sidney, Bainbridge, and Colesville subscribed to stock only on the condition that the company expend the funds in the township. The company defeated both initiatives in court.[112]

Yet the bitterness many community members continued to feel toward the A&S reinforced a popular distrust of corporations, especially railroads, in the countryside. "A Tax Payer" from Schoharie County called for public meetings and memorials to the legislature: "It is time they should be stopped, and the people have it in their power to break up this system of corruption, which has been going on for years, if they will take the matter in hand, and resolve no longer to submit to it."[113] His faith in democratic mobilization seemed stunningly naive in the wake of the road's success at defeating such efforts.

During the 1860 election, when New York Democrats unified against the Republican Party's sectional agenda, Democrats finally took a stand against the railroad. And Otsego and Delaware Democrats continued to battle public financing of the A&S and railroads in general. They attacked the Republican candidate for Congress, A&S president Richard Franchot, for his part in its "mismanagement." But it was too little too late, since the looming crisis with the South absorbed the attention of the electorate. Moreover, Democrats had failed to recognize that the A&S directors now operated within an integrated governmental and judicial system that encouraged the growth of capitalist enterprise in the name of the public interest. Thus, when they attacked Franchot for transgressing the old republican order, like "A Tax Payer" two years before, they did not question the constitutional structures that now allowed or encouraged the company to operate without restraint. They did not fully understand railroad politics or the special protection accorded corporations by the courts. As the A&S debate demonstrated, the decentralization of authority within the state government diminished the public's ability to affect change through electoral means.[114]

Democrats appealed to voters to reject as unrepublican the emerging political economy that Republicans advocated. When the editor of the *Freeman's Journal* charged that the Republican convention exuded "an atmosphere of bribery and corruption," he was holding to the Jacksonian idea of a sharp dividing line between government and business. Factions, he charged, offered "sums of from one to two hundred dollars" to sway single delegates and "much larger amounts for a single town." Thus, "THE MONEY POWER," not the people, ran the party.[115] If Republicans nominated Franchot to convince pro-railroad Democrats to bolt, "Why should they do so? Did Mr. F. personally subscribe a dollar toward that road? Has he not made money out of his connexion with it—receiving $2,500 per annum for his services?"[116] The editor wondered whether the A&S had spent as much money on constructing the railroad as it had on salaries, lobbying, and electioneering and asked Franchot to disclose the number of suits initiated against delinquent subscribers and explain why those cases had recently been discontinued. Lastly, "A Stockholder" asked Franchot to account for the litany of lies about the letting of contracts, the amount of construction actually commenced, and the expected date of the line's completion.[117]

But secession and the simple fact that the political economy of central New York had shifted decisively in favor of the new middle-class, industrial ethos embodied by the supporters of the railroad determined politics. Edgar Ryder of Barnerville, Schoharie County, recalled that in 1860, the Republican Wide-Awakes kept up a steady campaign for Lincoln: "Every little village had a campaign marching club with uniforms and torches and almost any evening all during the Fall from my father's home on the hill, we could see torchlight processions marching along the highways, preceded by fife and drum corps." The Democrats remained too disorganized to counter the Republicans. In both Delaware and Otsego Counties, Republicans emerged victorious; Democrats drew fewer than 150 Republican votes from Franchot. Even in Schoharie, where Democrats had held off the Republican challenge, Ryder wrote, "A thrill went over the land as the announcement was made" that Lincoln was elected even though "many of the best people shook their heads and expressed the fear that the country had made a mistake." National issues had risen to primacy in people's minds.[118]

The Civil War continued the Democratic Party's tendency to factionalize. In what surely stands out as one of the great ironies of a realignment already marked by odd alliances, former Hard Democrats, who had defended the South's right to hold slaves and deplored Republican radicalism, now supported Lincoln's war effort, while former Softs, who had tolerated free soil, opposed the Lincoln administration's rapid expansion of the powers of the federal state. Recognizing the continued importance of the railroad issue in local politics helps explain this switch. Republicans and Hards—now styling themselves "Union" Democrats—spent the years 1860 to 1865 fighting in Albany for state grants to complete the road. Softs, meanwhile, continued to adhere to a recognizably Jacksonian message of limited, decentralized government, individual liberties, and local democracy. This final Democratic split wrecked the party in all but Schoharie. Tarred with charges of treason, Democrats were effectively neutralized as a vehicle for promoting agrarian politics after the war, despite voicing an ideology sympathetic to farmers' plight.[119]

* * *

But the political saga of the road was not yet complete. The A&S needed additional governmental assistance to complete the railway and went to the legislature for grants several times during the Civil War years. There was little opposition, for after being saddled with the worthless stocks of the road, residents could avoid taxes only if the A&S was completed and turned a profit. The company had secured $700,700 from individual subscribers, $950,000 from municipal subscriptions, and a $1,000,000 loan from the city of Albany, a total of $2,650,700. Thanks to Ramsey's relentless lobbying—and not a little corruption—lawmakers passed eight bills granting aid to the Albany & Susquehanna between 1859 and 1867. Two bills, in 1863 and 1867, brought the railroad an additional $750,000. The total capital accumulated by the company thus reached $3,400,700—more than double its original capitalization.[120] Government aid, carefully sought after and lobbied for by the promoters of the railroad, allowed the railroad to be built without wide support among the population.[121]

The fate of the A&S symbolized the shifting roles of socioeconomic groups in the three counties. The middle class emerged as the defining social group in rural New York. The economic power they garnered from the commercial expansion of agriculture in the 1850s placed them on a par with the gentry in wealth and political power. For their part, the worlds of petty proprietors, artisans, and tenants were being dramatically altered by this new economic power. These latter groups had reacted to changing circumstances by breaking old party loyalties, seeking a party that would help reassert older republican concepts of limited government interference in economic affairs and a social order based on reciprocal bonds. But by making the needs of farmers a Republican issue, the middle class had successfully converted their agenda of temperance, antislavery, and capitalist expansion into the domi-

Construction on a trolley line in Index, south of Cooperstown. Such railways were built to connect outlying towns with the Albany & Susquehanna after the Civil War, creating a web of freight and passenger lines to link the region with the outside world. Photo #5-1056, Smith-Telfer Photographic Collection. Courtesy of the Fenimore Art Museum, Cooperstown, New York.

nant political force in central New York in the 1850s. In the process, they had reshaped local party politics in a way that made democratic insurgency less viable while joining in a statewide reframing of government institutions that again dampened dissent. The challenge for the rural people for the rest of the century would be to fashion new institutions and ideas that could advance their interests in this changed political economy.

6. The Maturation of the Market Economy: Dairy and Hops Culture, 1870–1900

The Civil War left a mixed legacy in central New York. Many farmers took advantage of wartime inflation to pay off outstanding debts, especially those who produced butter. But the war's effect on the long-term economic health of Otsego, Delaware, and Schoharie Counties remained unclear in 1870, when the first effects of economic depression appeared in the countryside. Hops farmers endured disastrous aphid attacks in 1863 and 1864 that reduced the crop to one-quarter of its former output. The depression of the 1870s would create an economic uncertainty that persisted throughout the century. Though butter prices fluctuated little during that time, hops prices underwent severe swings. Farmers caught in the uncertainties of an expanding market-based economy did not just roll with the punches, however. They fought back, and two interesting questions have yet to be asked for central New York. The first is rooted in an understanding of the past. That is, how did the region's history influence the responses of late-nineteenth-century farmers? The second is, how does their translation of their past provide us with an understanding of why they chose to act in certain ways in the decades following the Civil War?[1]

In his 1933 study of the Granger movement, Solon Buck argued that the Grange developed not out of economic hardship but from a sense of frustration among farmers that other sectors were growing more rapidly. Lee Benson, on the other hand, later contended that the increased capital investment that the war had encouraged left farmers badly extended when prices did not rise high enough to meet mortgage and loan demands. He believed that farm organizations acted primarily as political pressure groups for farmers who sought greater profits to meet the high fixed capital costs. Writing two decades after Benson, Jeffrey Williamson found that prices for agricultural products stabilized in the late nineteenth century, but mortgage rates and rents increased beyond the ability of farmers to meet their obligations. Without closely examining the particular economic situation of central New York farmers, though, we cannot assess the validity of these disparate ideas or understand what farmers hoped to accomplish in their organizations.[2]

* * *

Most farmers practiced mixed farming and produced either small amounts of hops or butter or both. The former formed a vital part of the economies of Otsego and Schoharie Counties from 1870 to 1900. As explained earlier, large producers and landlords had introduced the crop in the 1840s and 1850s, but by 1870, all levels of the agricultural sector engaged in its production. Gross output increased by 160 percent in Otsego County and 195 percent in Schoharie County between 1870 and 1890, despite dramatic price fluctuations and the appearance in 1870 of a worm that ate hop vines down to the root. Per acre production increased as well. On average, an Otsego County farmer expected to grow over 600 pounds per acre in 1890, roughly 120 pounds more than in 1880 (see table 4). The sheer intensity of production statewide, in which Otsego remained the primary producer-county throughout the century, kept New York the nation's leader. In 1860, New York had produced nearly 90 percent of the nation's hops; two decades later, it still dominated with a total output of 21,628,931 pounds that accounted for 81 percent of the nation's production (26,546,378 pounds).[3]

Table 4. Hops Production in Delaware, Otsego, and Schoharie Counties and in New York State, 1870–1900

	Delaware	Otsego	Schoharie	N.Y. State
Total output (lb)				
1870	307,431	2,919,629	1,610,457	17,558,457
1880	190,793	4,441,029	2,982,873	21,628,931
1890	29,047	4,698,687	3,148,855	20,063,029
1900	n/a	4,115,300	3,752,700	17,332,340
Total acres				
1880	569	9,118	5,871	39,072
1890	71	7,749	5,563	36,670
1900	n/a	7,035	5,966	27,512
Yield per acre (lb)				
1880	335	487	508	554
1890	409	606	566	547
1900	n/a	585	629	630

Sources: Ninth Census of the United States, 1870 (Washington, D.C.: Government Printing Office, 1873); *Tenth Census of the United States, 1880* (Washington, D.C.: Government Printing Office, 1883); *Eleventh Census of the United States, 1890* (Washington, D.C.: Government Printing Office, 1896); *Twelfth Census of the United States, 1900* (Washington, D.C.: Government Printing Office, 1901).

A slow decline of the hops sector, however, began in 1886, when prices remained low for the fourth consecutive year and one of the largest growers, George Clarke, went bankrupt. In Otsego County, acreage devoted to the crop fell during the last two decades of the nineteenth century, although production generally increased, indicating that small, inefficient farmers ceased producing. The 1880s and 1890s brought western competition as a variety of pests and diseases that afflicted the New York crop did not trouble growers in Washington, Wisconsin, and California. Pacific hops also yielded in the first year and could be grown on a scale unheard of in the hilly upcountry of New York. George Hyde Clarke, the son of George Sr., estimated in 1884 that one acre of hops cost $101.75 to grow, and at current prices, 1,000 pounds—a very good yield in Otsego County—grossed only $120, a small profit, especially if grown on shares. Otsego County growers started to replace hops with other crops in the 1890s.[4]

Dairy farming had expanded throughout the century, partly because the terrain of the mountainous sections of Schoharie and Delaware Counties offered poor land for raising grain. Part of the push came from merchant-landlords who bought up farms and required tenants to keep dairy cattle. Dairying also promised more stable prices than hops, and farmers could maintain other crops to help support their families. Butter production rose 156 percent in Delaware, 106 percent in Otsego, and 139 percent in Schoharie between 1870 and 1890 (see table 5).

As with hops, butter output decreased during the 1890s but for somewhat different reasons. Urban demand prompted central New Yorkers to begin shipping fresh milk to New York and Albany after the completion of the A&S Railroad in 1868. Numerous feeder lines built in the 1870s linked the backcountry with the main line, and the Ulster & Delaware and the Ontario & Western tapped Delaware and Schoharie farms by traversing the Catskill Mountains to connect the region with trunk lines in the Hudson Valley.

Table 5. Butter Production in Pounds in Delaware, Otsego, and Schoharie Counties and in New York State, 1870–1900

	Delaware	Otsego	Schoharie	N.Y. State
1870	6,135,715	3,566,283	2,190,668	107,147,526
1880	7,732,028	4,578,784	2,729,633	111,922,423
1890	9,590,349	3,774,292	3,042,583	98,241,813
1900	5,920,095	2,412,218	2,616,214	74,714,376

Sources: Ninth Census of the United States, 1870 (Washington, D.C.: Government Printing Office, 1873); *Tenth Census of the United States, 1880* (Washington, D.C.: Government Printing Office, 1883); *Eleventh Census of the United States, 1890* (Washington, D.C.: Government Printing Office, 1896); *Twelfth Census of the United States, 1900* (Washington, D.C.: Government Printing Office, 1901).

Railroads aggressively sought the milk trade by building creameries next to their lines. By 1900, each county sold large quantities of fresh milk to urban consumers: Delaware sent to market 46 percent of the total milk produced on its farms, Otsego 60 percent, and Schoharie 25 percent. Nearly all farmers in the three counties sent some dairy products to market. While 87 percent of farmers statewide sold dairy products, 92 percent did in Delaware and Schoharie Counties, and 90 percent did in Otsego. Between hops and dairy products, the economy of central New York depended by 1900 almost completely on national and international consumers, despite attempts by farmers to maintain at least a modicum of independence.[5]

The reliance on hops and dairying also altered landholding patterns in the late nineteenth century. In 1860, 16 percent of all northern farmers were tenants—ranging from 55 percent in Maryland to 8 percent in New York. But by 1900, New York tenancy levels rose to 24 percent, putting it on a par with other mid-Atlantic states and the Midwest. In Delaware, Otsego, and Schoharie Counties, tenancy was slightly above state averages, but freeholds rose substantially between 1855 and 1880, from 75 to 80 percent in Otsego County in 1855, 73 to 82 percent in Schoharie, and 78 to 87 percent in Delaware.[6]

Otsego and Schoharie Counties had a higher percentage of share tenants than Delaware, most likely because of the prominence of hops raising. But the economic stresses of the 1880s, when hops prices remained low and dairy farmers confronted competition from oleomargarine manufacturers, caused a decline in freeholds that slowly eroded the gains of the previous two decades (see table 6). Between 1880 and 1900, freeholding fell in all three counties, though more dramatically in Otsego and Schoharie, until by 1900, total tenancy levels had reached or exceeded post-Anti-Rent levels, with Otsego and Schoharie above the state average. The increase in tenancy accompanied a decline in the number of farms. Between 1880 and 1900, Otsego County farms dropped 8 percent, from 6,096 to 5,634, and Schoharie fell 12 percent, from 3,892 to 3,437. Even in Delaware County, with its expanding milk industry and greater unimproved acreage, the number of farms fell slightly, from 5,264 to 5,232.[7] Good economic times, the Civil War years, and the 1880s had provided enough stimulus to local agriculture for farmers to purchase and work their own land. But when prices declined, many hops farmers in particular were forced to cease production or even abandon their farms, leading contemporaries to observe ironically that "those towns were most prosperous that grew fewest hops."[8]

While the number of farms decreased between 1870 and 1900, the size of farms increased. Otsego farms on average increased from 101 to 109 acres, Schoharie farms from 100 to 107 acres, and Delaware farms from 146 to 152

Table 6. Land Tenure in Delaware, Otsego, and Schoharie Counties, and in New York State, 1880–1900

	Delaware	Otsego	Schoharie	N.Y. State
Total farms				
1880	5,264	6,096	3,892	241,058
1890	5,468	5,854	3,693	226,223
1900	5,232	5,634	3,437	226,720
Freehold rate				
1880	87%	80%	82%	83%
1890	86%	74%	80%	80%
1900	78%	64%	69%	67%
Cash tenancy rate				
1880	8%	8%	7%	8%
1890	9%	9%	8%	9%
1900	13%	10%	8%	11%
Share tenancy rate				
1880	5%	12%	10%	9%
1890	5%	18%	13%	12%
1900	4%	19%	17%	13%

Sources: Tenth Census of the United States, 1880 (Washington, D.C.: Government Printing Office, 1883); *Eleventh Census of the United States, 1890* (Washington, D.C.: Government Printing Office, 1896); *Twelfth Census of the United States, 1900* (Washington, D.C.: Government Printing Office, 1901).

Note: In 1900, the census listed a separate category, "Other," which included part-owners, owner-tenant farms, and farms with managers. Delaware listed 4% "Other," Otsego 7%, and Schoharie 6%. Figures may not add up exactly to 100%.

acres. In Otsego and Schoharie, 67 to 68 percent of farms in 1870 had fewer than 100 acres, while in Delaware 60 percent of the farms fell below that figure. By 1900, only 32 percent of farms in Delaware, 46 percent in Schoharie, and 49 percent in Otsego comprised fewer than 100 acres (see table 7). Statewide, 86 percent of all farms had fewer than 175 acres, with 86 percent in Schoharie, 84 percent in Otsego, and 67 percent in Delaware. During the years 1870 to 1900, population fell in both Otsego and Schoharie Counties while Delaware enjoyed a modest increase. In short, many farmers sold or abandoned their farms in response to the need for larger, more efficient operations as the hops and dairy economies were rationalized at the end of the century.[9]

Changes in central New York agriculture did not benefit all townships nor most small freeholders. Those towns, for example, that sat off the line of the A&S underwent economic decline in the years 1868–80, with shops closing and farmers struggling to make ends meet. Freeholds above 100 acres decreased by 8 percent in Delaware, 5 percent in Schoharie, and 4 percent in Otsego County between 1880 and 1890, while tenant farms grew in size. Moreover, the aggregate number of freeholds declined as the number of

Table 7. Distribution of Farms by Acreage in Delaware, Otsego, and Schoharie Counties and in New York State, 1870–1900

	Total	0–49 acres	50–99 acres	100–499 acres	500+ acres
Delaware					
1870	5,120	25%	35%	39%	1%
1900	5,232	13%	19%	66%	1%
Otsego					
1870	5,717	30%	37%	32%	1%
1900	5,634	23%	26%	51%	1%
Schoharie					
1870	3,618	29%	39%	31%	0%
1900	3,437	19%	27%	54%	0%
New York State					
1870	216,253	40%	34%	26%	0%
1900	226,720	30%	28%	41%	1%

Sources: Ninth Census of the United States, 1870 (Washington, D.C.: Government Printing Office, 1873); *Twelfth Census of the United States, 1900* (Washington, D.C.: Government Printing Office, 1901).

Note: Figures may not add up exactly to 100%.

tenant farms rose during the last two decades of the nineteenth century, partly due to merchants buying up struggling farms in outlying towns. By 1900, 17 percent of Delaware, 25 percent of Schoharie, and 29 percent of Otsego County farms operated under lease, with the number of tenant farms let on shares increasing markedly in the hops growing counties of Otsego (46 percent) and Schoharie (96 percent). Tenancy thus appears to have been one of the consequences of the devolution of the hops economy at the end of the century. Though overall tenancy remained low compared to other areas of the country, central New York farmers expressed a great deal of anxiety about the long-term implications of these trends.[10]

All these figures suggest that the agricultural sector of the three counties—as was the case across the nation—underwent important shifts in the terms and scale of farm operations that related directly to market forces. Yet the ways in which these changes altered individual lives and the fabric of rural life in the region have not been examined from the perspective suggested by Jackson Lears over fifteen years ago: shifting crops, production for consumers, and changing land patterns produced dislocation and dependence that challenged cherished notions of a good republican way of life and, in turn, caused central New York farmers to seek ways to protect important elements of that ideal rural order.[11]

* * *

Despite these being years of declining prices and increased economic dislocation, Schoharie County resident Jared Van Wagenen Jr., born in 1871, remembered the late 1800s as a rosier era: "Letting my memory run back across the years, it seems to me as if that era in actuality was not as hard as it appears in retrospect. Wonderful to tell, in the better farm regions rural society did not disintegrate. . . . Rural society might have been undergoing a season of testing, but it could not be said to be sick unto death. The established farmer proved himself as a hardy perennial and very difficult to kill. When the darkest clouds lifted, he as a rule was discovered yet on his farm and his culture still intact." Voicing the community's enduring notion of a farmers' republic, he wrote, "I hope for an agricultural civilization made up of free men on their own land, with crossroads hamlets not too different from the one where I have always lived, so that the sound of the church bell may float across the fields of a summer morning." But Van Wagenen came from an elite family and underestimated the adjustments in education and investment he and his peers made to perpetuate the family farm. He could not see that few central New York farmers made that transition as easily. Hops culture moved inexorably toward a time and labor system that more closely resembled that of industrial America than antebellum farming, and dairying became more competitive, in each case undercutting the ability of central New York farm families to maintain themselves.[12]

Large-scale hops growers constituted the elite class of local farmers who either employed share farmers or hired wage laborers to plant, cultivate, and harvest their crops. They had the largest capital investments and earned the largest profits of all farmers. For growers like George Clarke, share farming remained the labor form of choice in the 1870s and 1880s, though Clarke employed wage laborers on his own farm at Hyde Hall. In 1881, Clarke produced huge quantities of hops on the twenty-five "home" share farms that he personally oversaw and at Hyde Hall, his output that year reaching 76,794 pounds, over 12,250 pounds at Hyde Hall alone. His extensive share farming operation radiated outward from his home farms around Otsego Lake and down the Susquehanna to Milford, where his massive barn for collecting the annual produce of his tenant farms impressed locals and travelers alike.[13]

The share system on Clarke's estate evinced many similarities with sharecropping in the South, as his son explained: "The idea is this; we supply the land & they the labor & in the labor is included all farming utensils & work horses. All fixtures on the land are supplied by the landlord such as buildings fences & hop poles. Expenditure means such things as seed, hop roots & twine butter tubs, taxes. We also advance a certain percentage in cash for paying hop pickers. Cattle, sheep, swine or any thing in fact that can grow &

is wanted must be supplied jointly & so all the returns from thus are shared equally."[14] The Clarkes provided their tenants' marketing and credit needs by supplying goods needed to maintain the farm on credit and settling at the end of the season. Despite the Clarkes' insistence that they made contracts with "equals," share leases provided landlords with liens on the growing crops and considerable control over the productive process. Clarke especially tried to curb overproduction by controlling the output of his tenants: "Hops require so much attention that farmers very often let the rest of the farm 'go' & give all their attention to the hops, so when hops bring but little they have little of any thing else to fall back on. For that reason on our estate we limit the tenants to a very small percentage of land for hops." Like most northern landlords, George Clarke or his overseers visited tenant farms during critical periods to ensure proper planting, cultivation, and harvesting.[15]

The Clarke estate retained important links to its colonial past but had changed considerably to suit the new conditions of succeeding generations. George Hyde Clarke observed that "father's life work" had been "to release his Estate from the entanglements of a deceased policy, namely the old Colonial, mine will be to adapt it to modern ideas & uses." His father relied on reciprocal relationships between landlord and tenant to run the estate. As a result, many renters clung to old farming methods. The younger Clarke was determined to retain the habits of the gentry and infuse a scientific, rational approach on the farms, as he explained to his fiancée in 1885: "All the plans for the cultivation of the soil must proceed from headquarters whether it be in view of dairying, hop raising, grain growing or beef cattle & sheep or any other method. Besides all this a personal example has to be set. Vigorous action, being prompt, fairness & encouragement must be imported as well as guarding against those who have sinister motives. I leave it to you to see where a womans influence can be brought to bear. People in lower stations of life are more easily influenced & are brought more into contact with their superiors in the country than elsewhere & so I think you can see as well as I where our duty will lie."[16] Hyde, as he called himself, internalized his father's notions of noblesse oblige yet understood finance differently. Though George Sr. recognized that he had to reduce his holdings if he hoped to preserve his estate, he delayed selling parcels even as mortgage payments threatened to destroy the estate because he believed that the land, not the income derived from it, gave him social status. Hyde, on the other hand, wanted to sell outlying lands to pay off bad debts and place the estate on a sound financial footing.[17] Yet the elder Clarke clung to his estate even though a tenant uprising in the 1870s confirmed to him that the tenantry could not be relied upon to produce in the market economy: "The unsettled nature of our people is such that they will not follow the steady line incidental to the European peasants education and consequently become convulsive in their undertakings, expanding under the influences of hope into extravagence, and in such years as the past shrinking through fear into morbid inactivity, accompanied with the worst of practices."[18] His son often complained that tenants who could "afford to supply his share . . . are not easy to find."[19]

Other large growers turned to wage labor to streamline production on their farms. James F. Clark of Cooperstown began producing hops in 1876, establishing himself by 1881 as the "Hop King," while locals called his 150-acre farm "Hop City." Clark employed 1,000 people, 600 of whom were boarded during the harvest. The farm boasted a mayor, sheriff, jail, barbershop, blacksmith, cobbler, general store, and a hall for hops dances and religious services. His seven hop houses, each fed by two kilns, ran twenty-four hours a day during the harvest. He invested heavily in fertilizers and produced 210,000 to 250,000 pounds per year. Clark owned outlying lands but centralized the harvesting process much more systematically than other growers. Clark brought laborers—often unskilled immigrants—from Albany to Cooperstown on trains and sent them to an assigned farm. Clark's system closely

Otsego County hops pickers. This photograph captures the "ideal" harvest labor force—families—at work in the post–Civil War era. Note the women picking hops from the vine while male box tenders supply vine-laden poles to them. Negative ph-1127, Francis Ward Local History Collection. Courtesy of the New York State Historical Association, Cooperstown, New York.

approximated that of his Pacific competitors, and he survived longer than the less efficient Clarkes.[20]

Merchants like brothers Clifford and Zenas France of Cobleskill hired warehouse laborers for an hourly wage and created new arrangements that ensured delivery of hops without the personal obligations Clarke had with his tenants. The Frances' Fordton Hop Fields resembled Clark's "Hop City." The men employed permanent laborers to work either at the farm or at its warehouse next to the A&S. They established a network of suppliers in nearby townships to purchase commodities and ship them to Cobleskill on the railroad. They then sent the products to New York brokers or directly to the consumer. Like other hops dealers, the Frances jealously guarded information about local, national, and international crops and prices and devised codes to communicate the information by telegraph.[21]

The dictates of the international market for hops also altered traditional intercourse between local merchants and farmers. Merchants attempted to meet new demands by undercutting local producers, sometimes acting in concert with outsiders. Secret deals like one made between John D. Shaul, a substantial merchant, landlord, and politician from Springfield, Otsego County, and G. W. Wilkins of New York demonstrated the merchant community's desire to keep farmers uninformed and disunited:

> I have taken the pains to speak to my neighbors about consigning their hops to you & they seem a fraid to consign to any one as so many have been slaughtered or in other words they have not realized what they expected for their hops & have come to the conclusion to sell them at home or by sample Now if you have any moneyed men in your place that wish to handle hops I think they would do well to slip out to my house & I will qietly slip around with them and show them where they can find hops & plenty of them & they can sort them by lots & I have no doubt but it will pay well to come & I shall be glad to show them all about & by doing so it will have a tendency to have our people become acquainted with your dealers & it may prove a great benefit to you & us both.

Shaul told Wilkins that anyone who came to see him must ask only other merchants where to find him: "Do not let him say hops to any one for as soon as hops are mentioned the news will spread like wild fire that there is a start in hops & everyone will ask more than anyone would be willin to pay."[22] Merchants did not stop at undercutting local prices. At the height of the 1885 hops crisis, merchants and large growers speculated in Pacific Coast hops to drive down local prices and ensure their own financial welfare independent of the interests of local farmers. It may be understandable that a caterpillar that infested hops fields became derisively known in local parlance as "the hop merchant."[23]

Clifford and Zenas France also established "hop contracts" with local growers to prevent other dealers from invading their "territory" and to tie freeholders to hops production. In return for a prearranged price for their hops for several years, farmers agreed to cultivate a specified number of acres in hops for three years, "to keep said number of acres under good and proper cultivation," and to deliver the annual crop "cleanly picked, properly cured and put in good, new sacking, and in merchantable order and in bales of about two hundred pounds each" of a specific quality. The Frances paid farmers "less seven pounds per bale tare [a reduction of the sale price of each bale of hops to make up for the weight of packaging] for sacking . . . after customary inspection." If a grower produced poor hops, the Frances deducted the difference between the grade specified in the contract. The farmer delivered the hops "free of charge on ten days notice at the purchasers option" and insured the hops at full value, payable to the Frances, who reserved the right to supervise the harvest and curing of the crop. Finally, the brothers advanced a farmer cash to pay pickers and required him to purchase baling cloth at their store, then deducted these costs from the sale price. Hops contracts such as these tied freeholders to hops production in a way similar to share tenants. The system ensured reasonable stability in production and grade, a continuity of labor for hops merchants, and, at best, a moderate return for farmers.[24]

There were independent hops growers, of course, who took to the commercialization of farming. William H. Brooks of Cooperstown purchased a modest 100-acre farm for $3,000 in 1863 with five acres of hops and added two more in 1866. Between 1863 and 1884, a period of high prices, he grossed $38,180, an average profit of $150 per acre or $1,050 per year. He produced 1,300 pounds per acre, almost three times the Otsego County average, a testimony to his use of fertilizers and improved machinery. Brooks demonstrated that meticulous care of hops yards could yield a good return, but any decrease in productivity or price threatened even careful farmers with ruin. Careless or capital-poor husbandmen faced certain failure during lulls in the market. The *Delaware Dairyman* reported in March 1886 that farmers in Oneonta had been badly shaken by the recent low prices: "Some have plowed up their yards, others have suffered from the open winter, and a great many are prophesying that hops are never to pay well again." Most independent farmers who continued to grow hops after 1886 succeeded only by limiting cultivation to several acres and growing other crops to meet family needs. Hops grown on this scale would neither make them rich nor ruin them.[25]

Large tenant estates proved unable to weather economic downturns. Competition from the Pacific Coast, England, and Germany lowered prices steadily from 1882, when hops had reached $1.20 per pound, to 1886. George Clarke, whose estate had survived the Revolution and the Anti-Rent years intact, now went bankrupt. Clarke had accumulated much of his post-Anti-Rent property by mortgaging his Otsego County farms, his most valuable holdings. He stubbornly refused to sell land in other counties to pay his debts, instead servicing an extremely complex set of ad hoc financial arrangements that would allow him, he thought, to retain the entire estate. His creditors began to close in on him in 1885, though his son Hyde remarked, "The Govey arrived home, is looking well & doesn't seem a bit worried by being sued by nearly every body in Otsego County."[26] He failed in 1886, and his assignee, George Barnard, began selling his property to pay his creditors. Barnard terminated share leases and demanded that all new contracts be let on cash terms.[27]

Other hops growers survived the 1886 debacle, but Otsego, Schoharie, and Delaware Counties witnessed a steady decline in hops production. James F. Clark went bankrupt in 1897 when he tried to circumvent American brokers by shipping his hops directly to England, a tactic George Clarke had also found disastrous in the 1850s. Those who survived limited production to a few farms. DeWitt Mitchell of Schuyler Lake, Otsego County, averaged 1,000 pounds per acre on his forty-acre farm. George Hyde Clarke maintained nine farms at the turn of the century by combining dairy and hops production, though on a much smaller scale than his father. Still, blue mold and other pests repeatedly ruined crops. By 1920, few hops farms remained in the countryside; even fewer plebeians mourned their passing.[28]

Share farming as an institution thus evolved with changes in the local economy. The different forms of share leases in central New York reflected the wide range of individual needs of community members and the changing requirements of the marketplace. People worked plots as small as gardens or as large as Clarke's hops and dairy farms on shares. One contract required that the owner receive all of the produce of the farm, a highly unusual demand. Most approximated the experience of John Boocock, who leased a farm in 1876 from his employer of twenty-two years, John D. Shaul, for two-fifths of the dairy products and one-half of the hops and grain he raised. Boocock gave the crop to Shaul to sell. When Shaul completed the sale and subtracted Boocock's debts, Boocock received his share in cash. The Becker family leased their Roseboom, Otsego County, hops farm from 1874 to 1901 on shares and for cash. The lease first went to two men on shares in 1874. The following year, A. H. Becker leased the farm to Edith Becker for $300 for three years. In 1884, Edith purchased her own land and commenced hops production. When she gained title to the original farm near the turn of the century, she let it on shares as a dairy farm. By 1889, landlords and tenants commonly signed long-term contracts to increase stability, and landlords routinely placed liens on the crops to ensure payment. Most leases had been converted to cash by 1905, perhaps because of the unsuitability of milk production to share terms.[29]

Many yeomen still entered hops production to solve their financial woes. The process began in the 1850s, but the proliferation of farms that grew hops bespoke of the hold the crop had on the local economy. Even when hops prices crashed during the Civil War and natural forces ruined crops, farmers continued to grow the crop. Most farms advertised for sale boasted several acres of hops and outbuildings to entice buyers. Contemporaries remembered that most farmers who raised the crop cultivated five to ten acres. Communities with fairly low tenancy rates thus became more enmeshed in the cash crop. Finally, arrangements like the Frances' hops contracts tied freeholders into hops production for the first time in the 1880s. One farmer recalled the motivation behind this strategy: "A few farmers claimed to have made money if they got a good price; but by the time they got their bills paid, I don't think they were very wealthy. But there was always a chance that you would make good."[30] By 1897, however, better Otsego County farmers expected to gross only $5.40 per acre of hops, perhaps a 13 percent return on their investment. Scarce labor, poor weather, and price fluctuations could easily absorb that margin.[31]

Rising tenancy rates and hops contract arrangements disturbed the rural producer traditions of central New Yorkers, but the introduction of migrant labor and the scale of operations achieved by men like James Clark likewise disturbed the social order. New York State Agricultural Society president Marsena R. Patrick devoted a considerable portion of his 1867 address to the declining quality of labor and industry on farms in the state. He welcomed mechanization and the "factory system" of making butter and cheese but deplored the resulting proletarianization of farm labor. Too many young native American men and women, he said, left the farm for the glitter of the commercial world. The only workers available to farmers were immigrants, and "unmarried men of that class are unsettled in their habits, roving, and without local ties to bind them; ready to leave for trifling causes, or small inducements in the way of higher wages; the question becomes a very serious one, indeed, 'What are we to do for reliable Farm Laborers?'"[32]

For Patrick and many fellow New Yorkers, the labor issue involved more than whether children remained at home or immigrants formed the labor pool. It reflected their notions of republican society and all its nuances—family, gender, patriotism, and the overall social order. Capitalist expansion and immigration threatened social decay. Ontario & Western Railroad laborers marching menacingly from Hancock to Walton in Delaware County to demand back pay in 1873 and their sabotage of track, locomotives, and switches in 1875 brought labor unrest uncomfortably close to home.[33] The Democratic Cooperstown *Freeman's Journal* printed a scathing attack on industry and immigration in 1886: "The recent trials of anarchists in Milwaukee and in Chicago, and of the boycotters in New York, have revealed the existence in our midst of a large body of unnaturalized foreigners, who labor not to build up but to destroy the Republic." The Republican Party's tariff system, the paper opined, inflated wages and encouraged immigration. Once on American shores, immigrants took advantage of liberal American laws to instigate labor unrest. To prevent further moral decay, "the strong arm of the law must be put forth energetically to defend society against these pests." That such a sentiment could emanate from a Democratic paper—which generally railed against exclusion in American society—testified to the strength of a nativism tied to earlier notions of republican virtue and community to which predominantly white Protestant areas of central New York clung.[34]

These sentiments manifested themselves in different ways. Some simply looked upon immigrants, particularly the Irish, with little favor. Hyde Clarke reported proudly that he had secured native-born laborers to work his home farm at Hyde Hall in Springfield in 1885. Patrick offered hope for their reclamation and suggested that middling farmers adopt the "cottage system" of housing a hired man and his family rather than employing a number of single males. Such a system, he believed, would give immigrant laborers a view of the benefits of republican society and elevate them to respectability and family stability. Patrick's suggestions paralleled the gentry's understanding of their role in share system but called on all farmers to participate in the education process.[35]

The troubling concerns about rural society that Patrick voiced in 1867 foreshadowed developments in Otsego, Delaware, and Schoharie Counties that stimulated the formation of the Patrons of Husbandry. Prior to the Civil War, neighbors and young people drawn from nearby towns had picked hops in a festive neighborhood activity that reinforced ties of mutuality. Although hops culture over time set forces to work within the household that challenged traditional generational and gender roles, as described earlier, antebellum farm families had not readily perceived that family and community values were threatened. The first "foreign" pickers were urban working families, and many returned to the same farms each year. Local families felt comfortable with such laborers, and some struck up lasting friendships.[36]

Tenant farms on James F. Clark's "Hop City" estate south of Cooperstown in the late nineteenth century. This image shows dairy cattle grazing in the foreground with hops fields stretching over much of the landscape. Hop kilns are centrally located between farms in the center of the photo. Negative ph-1152, Francis Ward Local History Collection. Courtesy of the New York State Historical Association, Cooperstown, New York.

As late as 1870, the Morris *Chronicle* could report cheerily: "We are reminded that hop-picking is about to commence by seeing wagons loaded with girls en route for, to them, the delightful hop yards."[37]

After the Civil War, however, as the demand for pickers outstripped the countryside's ability to furnish them, unskilled laborers or "hoboes" arrived in the area in time for the harvest, lived in the woods, often raided local farms for chickens or grain, and engaged in drunken reveling. Townspeople and dairy farmers disliked this annual influx of the itinerants. The *Chronicle*, just one month after noting with joy the sight of young local women heading to the fields, portrayed the appearance of migrant harvesters quite differently: "A large number of seedy looking tramps of both sexes pass through the county daily. They are not afraid of padlocks and manage to get into cellars and out-houses with but little difficulty. Farmers had better rub the rust off their guns."[38] Other newspapers deplored the hiring of "city scum" and demanded that they be kept on farms and out of local taverns. The *Delaware Dairyman* of Franklin, Delaware County, remarked in 1886 that nothing better could have happened to central New York than the crash of hops prices. Noting that "the outlook for another year is not very favorable" and that many believed "hops are never to pay well again," the editor wrote, "In fact we agree with them, and add that we do not believe they ever did, counting the costs to the county people for alms house, jails and penitentiaries. Figuring the matter down fine as it will bear, they will find it was never a paying business, and Otsego Co., would have been richer today had there never been a hop yard within its limits."[39] Hops culture increased immorality, contributed to drunkenness, and tempted farmers to engage in production for a quick profit, thus inducing indigence. It brought landless laborers into the area who did not fit in well with republican society. For the editor of the *Dairyman*, only its eradication would cleanse central New York.[40]

Hops farmers thereby faced a dilemma over the social cost of commercial farming: either hire neighbors at high wages (if they could secure enough of them) or take on less expensive urban workers and migrants. A man's success in the straitened market of the late nineteenth century often depended on hiring the cheapest available labor. Farmers welcomed Iroquois from the Onondaga Reservation as pickers because they brought their own provisions, kept to themselves, and worked for low wages, but they made up a very limited portion of the available workforce. Beginning in the 1870s, male and female hoboes collected in large numbers during haying season and stayed through the hops harvest. One observer noted that "the woods was full of them." They performed informal jobs for farmers, usually earning their pay in vegetables and sundries, and were not allowed to sleep on the farm. Central New Yorkers farmers acted cautiously toward them. Women and children would not work in the same fields as "tramps" and avoided walking alone.[41]

The Cooperstown *Freeman's Journal* warned of deeper moral danger from such indigent outsiders in 1878: "This reckless vagabondage, commencing immediately after the Civil War, and increasing since the season of commercial depression, has grown to alarming proportions." This "dangerous class" threatened republican society: "The tendency of this evil is to spread and become deep-rooted, unless some legislation shall crush it before it shall reach a terrible extent. If now the tramp is a public nuisance, he will soon be an outlaw, defying society, if he does not get a severe check. The example of men, women and children, roaming over the country living off the gains of others, herding together in a promiscuous way cannot fail to exercise a most deleterious influence upon general society. It is of no use to attempt moral suasion with these persons. They do not place themselves under any such influence. They have no sensibilities, no religion, no nothing."[42] Methods of "recruiting" laborers exacerbated such fears. Large farmers hired urban labor brokers to recruit workers, but the agents many times seized unskilled laborers, who often spoke little English, from taverns, loaded them onto cattle cars, and placed a tag on them to ensure that they reached their destination. Such laborers proved particularly unruly and caused considerable tension between hops farmers and their neighbors. Cooperstown established a twenty-four-hour watch in the village and along Otsego Lake and enacted vagrancy laws to ward off transients. Other laws required farmers to police their own labor force strictly. Most paid such workers at the end of the harvest to avoid drinking problems.[43]

The new labor pool also altered the routines of host families and the pickers, together with the culture of the harvest. So long as urban pickers came from the trades, farmers continued to hold many of the traditional activities. But when hoboes harvested their crop, farmers separated local and "foreign" laborers and curbed drinking, dancing, and chicken roasts—where local boys stole chickens for the fire. Because food and lodging were essential, farm wives spent much of the year preparing meat, vegetables, breads, and cakes for the harvest. Some families, after laboring in the heat of the kitchen, fed workers in the cool of the basement. Men slept in the outbuildings while women occupied the indoor rooms. Sometimes farmers housed workers in dormitories away from the farmhouse if they distrusted the urban folk, and hoboes were left to fend for themselves. Farmers rarely served alcohol or beer to their workers, offering water or lemonade instead. Although few hops growers supported prohibition (for obvious reasons), they identified alcohol with the disorder and decay of the cities and wanted no part of either in their communities. In fact, many locals came to see previously welcome events like hops dances as sites at which the rabble congregated, defied the law, and threatened the social order.[44]

Overall, then, the presence of large numbers of migrant day laborers was one more element that symbolized the social changes that came with the commercialization of agriculture. In the 1890s, former Anti-Renters lamented the passing of the manor system because, they belatedly concluded, it had shielded them from the market economy. After they gained title to their land, they had to produce saleable commodities like hops that required high capital investment and cyclical debt. Western competition for those products soon wiped out the hill farmers. They had little choice but to emigrate. Many of those who remained in the area but could not compete because of high costs became permanent wage laborers. Many mill workers and farm laborers (laborers usually performed both tasks) lived in multiple-family houses and could provide little for their families. Elbridge Hunter of Downsville, Delaware County, typified this class of landless laborers. Forty-four years old in 1875, Hunter worked as a farm laborer and sawmill worker. He plied several other trades in the 1880s, including selling Bibles, and secured a mail route—in combination with general teamstering—in 1889, when he was fifty-eight years old. He rented a house from his brother and kept a garden, chickens, a cow, and, sometimes, horses. His daughters performed much of the household and farm labor. Freeholders remarked on the deplorable conditions within the households of local wage earners and yearned for a time when industry would cease the overproduction of finished goods and free workers to buy land to raise subsistence crops and sell small food surpluses.[45]

Farm families also attributed poverty and landlessness to commercial agriculture, especially hops, which unbalanced the gender and generational relationships that had formed the basis of their economic and social lives. To minimize these changes, some farm families practiced mixed farming with low capital investment in machinery. Dairy production grew in intensity because it least threatened the ideal of the traditional family farm. Cows required little care, certainly by the rudimentary standards of early dairy farmers, and permitted farmers to raise other crops. A farmer's family easily provided the labor to produce butter and cheese. Each could be consumed at home or sold for local use in small amounts and did not require delicate handling. The family thus functioned as a self-contained economic unit with the father in general control of the workings of the household. Tenant Walter E. Bard of New Berlin, Chenango County—just across the border of Morris, Otsego County—was a typical farmer in many ways. In 1876, he owned five cows and one heifer and produced two to four tubs of butter per month, which he either sold locally or shipped by rail when prices warranted. He made cheese less often, and also raised sheep, oats, corn, potatoes, and apples. He exchanged commodities for items he could not produce, regularly purchasing alcohol, tobacco, and cloth at local stores. Bard, his family, and their neighbors frequently exchanged work. But, like most other small producers, he was compelled to keep up with the needs of the marketplace and spent much of his time building an "arch" to bring cold water into his dairy to help prepare butter and fresh milk. The market increasingly influenced his farming strategies.[46]

The republican emphasis on protecting the lineal family, hard work, and community service remained strong despite the enticements of embracing liberal capitalist ideals. Evidence suggests that gender relations within farm households encouraged the adoption and maintenance of this strategy. Historian Nancy Grey Osterud has demonstrated that although men retained control of the productive process within dairy households, the relative flexibility of the gender division of labor fostered a sharing of decision-making between husbands and wives unseen in other agricultural pursuits. From fertility to economic choices, central New York farm couples relied heavily upon each other's counsel and acted, as nearly as possible within the patriarchal household, as equal partners. Likewise, Osterud argues, relationships within the household augmented tendencies toward mutuality in the neighborhood that contrasted with the male-oriented, individualistic ideology of middle-class America. It is not surprising that dairy farmers and their wives exhibited a strong tendency to unite in organizations like the Grange to safeguard their group interests.[47]

The complex set of community relationships that informed both economic and social life throughout the antebellum years thus persisted most clearly in dairy areas. Providing for one's children remained a primary concern. One Schoharie County farmer lamented in 1869, "Visit the school to day—long neglected duty—and upon going there find that it has been a loss to my family. Some things that should have been corrected before. Ella should have studied mental Arithmetic—hope it may no longer be neglected—Fannie & Ada troublesome. Fannie needs training in reading, & all need more Pains at home."[48] His sense of responsibility for his children found parallels elsewhere. In 1882, M. W. Frisbee of Bloomville, Delaware County, spent his days helping his sons work on the farms he had bought years earlier to give to them when they reached majority. Both owed him considerable amounts of money, but he remained philosophical about his role as father, confiding in his diary, "I have to pay for everything for family use so it goes."[49]

Central New York farmers participated in a wide range of social and economic exchanges that allowed them to maintain at least an air of community mutuality. Charles Broughton of Bloomville and his neighbors still bartered goods and labor, working the fields with those who could be considered social equals on an informal basis. In some years, Broughton aided neighbors more than they helped him; in other years, it was the reverse. At times, two or three men joined together to work gratis for elderly neighbors. At the end of the year the men settled with each other, taking into account the wide range of labor exchanges, goods, and credit extended to each other during the year. Broughton and his wife enjoyed a seemingly endless round of "visitin'" with such men and their wives.[50]

At the same time, however, Broughton hired men to work on his farm with whom he made more formal arrangements. Most of them worked for wages and all settled with him after their term of service, though Broughton gave them cash advances or credit for goods while they worked for him. An 1876 letter from Addison J. Blumberg, who always appeared as "Add" in Broughton's journal, illustrated the complexity of Broughton's relationship with his workers:

> Friend Charley
> Dear Sir
> As I have a chance to hire out by the month I want to know if you will give me off in haying perhaps you can do better as they talk wages down fearful How is your wife and Babies and Pets folks and Zeras folks and how does the fight go off under the hill since I come away has Lib got any hired girl yet be sure and answer this so I will get it before the first of June.[51]

Blumberg, whose formal arrangement with Broughton had ended in the spring, wanted Broughton not to be offended by his search for another job and to offer Broughton a chance to retain him for a fair price. This appeared to be a relatively simple labor arrangement. But Blumberg's reference to Pete Conroe and Zera Preston—a former Anti-Renter convicted of conspiracy in the murder of Osman Steele—and the ambiguous "Friend Charley / Dear Sir" revealed the complex relationship between the two men. Broughton treated the younger Blumberg as an apprentice. He took Add to church; visited Conroe, Preston, and others with Add along, sometimes with Lib Broughton and sometimes when just men were involved; and Broughton and his wife visited Blumberg's family. He also sent Blumberg to Pennsylvania for eleven days later in the year to conduct business for him. In later years, Broughton and Blumberg visited each other, cementing their bonds further, but not as equals.[52]

The two forms of labor exchange hinged on the position of individuals within the community wherein established men such as Broughton acted as patrons. The diary of Lucius Bushnell of Gilbertsville, Otsego County, reveals a similar pattern. In 1859, nineteen-year-old Bushnell, whose father ran a sawmill in that village, toiled at a wide array of tasks. He worked closely with his father but also cut wood, labored as a carpenter, took in hay for several individuals, and did a variety of other jobs. In 1878, when he had established himself within the community as a carpenter and farmer, Bushnell acted much like Broughton, extending credit to his laborers, ensuring that his hired girl received an education, and seeing that older members of the community had sufficient labor to run their farms. And, like Broughton, he participated in local government, serving as road commissioner.[53]

The long journals of Broughton and Bushnell, both stretching over twenty-five years, offer comprehensive accounts of a process hinted at in the scattered journals of other individuals. Each exhibited important characteristics of the republican ideal of the antebellum years. Both earned stature through hard work, virtue, and civic duty. Each gave back to the community in kind what had been offered him. They at once ensured the success and survival of their families while participating in a community-wide network of labor exchanges, mutual support activities, and the nurturing of younger men and women.

Yet their careers also highlighted changes in the republican ideal. Bushnell became increasingly involved in what could be termed small-scale capitalist activities after 1880. He speculated in local livestock and operated a local cheese factory while maintaining his carpentry business. Although it is hard to characterize him as a capitalist, his speculations bordered the fine line between mutuality of interest with and exploitation of local farmers.[54] Broughton's intensified production, careful breeding, and high capital investment at the end of the century symbolized the response of local dairy farmers to the milk trade with New York City. Shortly before he died in 1901, Broughton spent his time establishing the cooperative "Patrons' Creamery" in Delhi, a milk cooperative formed by local farmers as a result of "inshurance" and "milk" meetings. His actions—like Bushnell's—were not entirely altruistic, however, and he helped several local businessmen establish dairies.[55]

The level of business acumen and capital investment necessary to succeed increased rapidly in the years 1870–1900 as the market for dairy products became more competitive. State agricultural society president Marsena R. Patrick noted in 1867 that dairy farming had undergone considerable changes during the Civil War, the most important innovation being the "factory system" of butter and cheese production. Local entrepreneurs opened some of the factories, but farmers generally pooled resources in order to produce butter and cheese without consuming important work hours at home. Usually farmers took stock in a "factory," held meetings to discuss strategy, and selected one man to oversee operations. They carefully tabulated the amount of milk brought to the factory by each member of the corporation and paid dividends at the end of the year. Cheese and butter manufacture on tenant farms worked differently; generally each farm had its own facility.[56]

The factory system, too, represented a compromise between market demands and the desire to maintain important relationships of mutuality and labor exchange. Successful dairy farmers integrated limited market production with a fairly high degree of self-sufficiency early in the period. One Schoharie County man recalled that in the 1860s and 1870s, "there was little real concern for the effects of national trends upon the farmer's mode of life or upon his pocket book. His economic status was still pretty much under his own control, and the practice of thrift was an assurance of reasonable success." His father grossed $867.23 in 1871, with a net profit of $142.63, and lodged no complaints.[57] Other farmers eschewed making purely economic choices. Charles Broughton purchased new haying and dairy implements, but rather than increase production, he spent his newly found free time "loafing" or traveling to town. A. G. Beardslee of Pittsfield, Otsego County, maintained the same scale of dairy operations for over a decade, from 1881 to 1892. In July 1881, he produced 19,846 pounds of milk, which was made into 102 pounds of cheese at a local butter and cheese factory, netting him $202.43. Total production at the facility was 30,255 pounds of cheese and 4,214 pounds of butter worth $3,623.47. In July 1892, he took 18,299 pounds of milk to the factory, produced 85 pounds of cheese, and received $155.57. The total output of the factory was 29,463 pounds of cheese and 8,954 pounds of butter from 632,603 pounds of milk, or $5,586.00. The factory had expanded production, but Beardslee had not. If Beardslee provided a typical case, more farmers produced milk, but few expanded production beyond providing a competence for their families.[58]

The oleomargarine controversy of the 1880s, however, appears to have accelerated the demise of independent butter producers. By 1875, prices for butter and cheese peaked, leaving farmers with high fixed costs extremely sensitive to even minor price changes.[59] Farmers called for governmental intervention to stop the production of oleomargarine. New York merchants opposed the legislation, and the battle intensified in the 1880s when butter producers endured a series of price decreases. Delaware, Otsego, and Schoharie County dairymen faced losing their land if they did not increase production, but greater output would only lower prices. They failed to prohibit the manufacture of oleomargarine or to stop economic decline of butter and cheese production, but they did secure legislation that prevented oleomargarine from being sold as butter. Farmers continued to struggle, however, and found themselves with dairy cattle and no markets in 1895, when prices dropped and grasshoppers stripped local pastures bare.[60]

Farmers then compromised to meet the demands of the marketplace. The rail connections that had been built throughout the counties to connect isolated towns with the A&S, New York & Erie, New York Central, and West Shore Railroads provided rapid, efficient transportation to New York City.[61] Farmers in Delaware, Schoharie, and Otsego Counties thus switched from butter to milk production in the 1890s. The milk industry carried stronger imperatives for top breeding stock, clean scientific methods, machinery, and year-round production, something that had been absent on most dairy farms in the 1880s. George D. Taylor of Jefferson, born in 1888, recalled, "As I first remember the farm, the economy was still comparatively simple. During my boyhood there were introduced the grain mill, the grain binder, the corn binder, the threshing machine, the ensilage blower, the power saw, the sap evaporator, and the gas engine."[62] Many of these technologies had been available at midcentury and had been ignored by the yeomanry. Taylor's father had been one farmer who commercialized early, rapidly expanding his farm by purchasing adjoining lots and trying numerous innovations in an effort to get ahead. He accepted capital-intensive methods as the only way to succeed in a competitive market. In 1900, he doubled his herd to sixty milch cows, built a modern barn, and commenced year-round production of milk. He paid a heavy price, suffering a nervous breakdown at fifty when the finances of the farm finally fell apart during World War I, but the farm survived.[63]

Adapting to the demands of the milk industry created significant structural and economic changes on local farms. Each productive cow required forty pounds of quality feed per day to keep her working. Silos for storing fodder appeared in 1890 to enable farmers to produce milk throughout the year. And as dairymen turned their attention to improved stock and larger output, other machinery, such as grain binders, replaced more rudimentary farm machines by 1895.[64] The change in agricultural rhythms could be seen on Charles Broughton's farm. In the decade between 1886 and 1896, Broughton devoted increased attention to breeding stock and spent less time planting peripheral crops. He maintained his social circle but spent less time exchanging labor with fellow farmers. He did not hire much help and had become financially secure enough to lend money to his neighbors at interest. His dairy farm had become a business.[65]

The immersion of farmers in the commercial nexus had created notable changes in rural life by the end of the century. Taylor wrote that before the milk industry, "the farmer lived on what he raised. His cash income was an incidental but not vital consideration. To be sure, the smart farmers had managed to convert their surplus into some cash, but such cash was not regular enough or dependable enough to inspire any well-defined or concerted plan for it." He viewed cash as a "magic" influence on the farmer: "Almost immediately he began to change the order of his plans, aiming his whole production plan toward the mouth of the cow, with the assurance that the avails would come out in cash from her underside." Still, he recalled, national price fluctuations deeply affected the family's fortunes. The hard work required to realize a profit, the difficulty of acquiring adequate business knowledge, and the imperative to reinvest constantly in the capital resources of the farm made profits difficult to accumulate.[66]

The changing marketplace forced farmers to reassess their crop strategies, labor relations, family structure, and political loyalties. "Subsistence" still remained the byword of diversified farmers, but most found it untenable. Under a great deal of pressure to produce saleable goods, hops farmers questioned the social costs of production, and by the end of the century, smaller producers had begun to withdraw from market production rather than risk economic or social decline. Dairy farmers believed they had secured a fitting way of life for themselves and their families through mixed production, but economic difficulties stretched their resources to the limit. When butter production became unprofitable, they too faced stresses within the household as men took over the dairying process to supply fresh milk to urban markets.

By describing economic changes such as adaptation to the commercial economy, I am not suggesting, however, that farmers reacted passively to top-down forces. More than three decades ago in his seminal work, *The Search for Order,* Robert Wiebe hypothesized a crisis in rural communities occasioned by a "widespread loss of confidence in the powers of community." But his analysis that this crisis provoked a response that struck out against "whatever enemies their view of the world allowed them to see," for all its acute understanding that anxiety provokes human beings to act in seemingly "irrational" ways, failed to capture *how* rural producers responded, not just from fear, but in an attempt to preserve and refashion a way of life they valued and believed valuable.[67] The next chapters take a closer look at how the changing economic circumstances of the late nineteenth century caused farmers in central New York to form new organizations—the Grange, farmers' clubs, and other agricultural associations—through which they hoped to reinvigorate this way of life, even while they were becoming irrevocably connected to the market economy. Social stratification among them ultimately would stand in the way of those efforts but not before they made a final attempt to redirect their own lives to fit more closely with the Jeffersonian ideal of the independent farmer and to pressure the government into advancing agrarian reforms that they felt were necessary to save not only the farm sector but also the nation as a whole.

7. Tenant Unrest and Elite Cooperation: Responses to the New Economy

Historians generally agree that the late nineteenth century witnessed a fundamental shift in American political culture. They have stressed the decline of popular participation in the electoral process, the rise of reform and regulation, and the realignment that marked the birth of the fourth party system in the 1890s. Scholars have arrived at widely different conclusions about the role of farmers in this process. Most historians have come to view movements such as the Greenback Party, the Farmers' Alliance, and the Populists as liberal or even radical ones that sought social justice, a level playing field, or even quasi-socialistic reforms. But central New York followed a different path, despite facing an economic decline similar to other regions. In the 1870s, Otsego and Montgomery County tenant farmers turned to arson, anonymous threats, and the destruction of machinery in an effort to preserve paternalist relations as a bulwark against market participation. Meanwhile, large hops and dairy producers used agricultural societies and clubs to promote cooperative buying and selling in order to preserve their place within the market economy and eliminate smaller producers.[1]

Previous scholars have studied political participation by examining specific electoral campaigns, thereby obscuring unrest within farm communities such as the ones within these three counties. Richard L. McCormick, in his study of political reform in New York State in the 1890s, for example, argued that "rural citizens tended to be satisfied with party politics and to make relatively few demands on public authorities." Lee Benson went so far as to assert that farmers' calls for reform were "so opportunistic and narrowly viewed in terms of self-interest that it is impossible to discern any comprehensive, logically consistent ideological approaches" to the major political issues of the period. To both McCormick and Benson, northeastern farmers, in contrast to their peers in other regions, did not feel the need to organize collectively to promote their individual or group needs.[2]

Indeed, no single organization emerged that voiced the concerns of all classes of farmers. Yet, far from being complacent, central New York farmers exhibited a deep and abiding interest in controlling change to better suit the political, social, and economic demands of farm communities. Farmers' movements were multifaceted affairs that reflected struggles among different rural groups as well as with nonagricultural interests. Thus, while "Grangers" are best known for legislative assaults on railroads in the Midwest during the 1870s, the Grange membership and program differed regionally and locally from 1874 to 1900. Farmers also formed tenant groups, agricultural societies, and their own clubs. In central New York, tenants, middling farmers, and large producers confronted significantly different problems and found little common ground. So the failure of farmers in this region to unite in a broad consensus reflected a general differentiation of interests between hops and dairy farmers, tenants and commercial farmers, laborers and freeholders, not blanket support for the existing political system.[3]

* * *

As central New York farm families confronted the effects of commercial agriculture, national markets, and western competition described in the previous chapter, they formed several farm organizations through which they attempted to redefine the parameters of the market economy. In Otsego, Delaware, and Schoharie Counties, these were tenants' associations, wealthy farmers' clubs, and the Grange. Farmers' groups agreed that the family farm presented a social model for the rest of American society. But tenant, "middle-class," and wealthy farmers diverged over the methods and goals of farming and the social structure they hoped would emerge from their efforts to preserve the family farm.[4]

Tenants' first attempts to counter the effects of the postwar market on their lives were accompanied by violence. Central New York society in the 1870s and 1880s maintained a vestige of the aristocratic splendor of the manor system that had marked the Anti-Rent years, despite increasing freeholds. Understanding this local context is crucial for understanding farmer actions in these years. George Clarke's Hyde Hall on Otsego Lake commanded a vast estate that included much of the southern section of Montgomery County and the eastern half of Otsego County. Clarke's sister Anna Pell and her family occupied neighboring Swanswick. William H. Averell, the Cooperstown attorney, miller, and hops grower, owned Holt-Averell, which stood grandly at the southern end of the lake. Socially, economically, and politically, these families dominated their neighbors. Though large estates were increasingly rare, those that survived preserved many features of the traditional reciprocity and patterns of deference that marked antebellum relations between the gentry and the tenants.[5]

The gentry treated tenants essentially as peasants—indeed, they found American tenants wanting in industry and tractability compared to their European counterparts—and conducted themselves as masters of local society. When the wife of Hyde Clarke's coachman, for example, came to Hyde "for protection after [the coachman] had been horse whipping her for flirting with a hop picker," the young gentleman refused "because I thought she deserved it." A later row with a New York City hackney driver confirmed Hyde's confidence that he set the moral agenda for the lower orders: "I lost my temper with the cabman who of course tried to charge me more than his fare whereupon I called him & his class a set of thieves & he said he would have to be very smart to make anything out of me & I told him he could take anything he could get out of me & if he didn't shut up he would have to take a licking & that I had just 2 minutes to give it to him before the train left, which ended the altercation much to the amusement of the porters and baggage master."[6] The gentry regarded the discipline of their inferiors as a personal duty. When, while visiting Cooperstown, a "bagman" (salesman) wagered his tavern mates that he could walk beside a passing woman, then proceeded to do so, Hyde intercepted him and threatened all of the men with a horsewhipping if they attempted to dishonor a lady again.[7] Hyde summed up the gentry's sense of position when he spoke of his unmarried sisters' hopes for a good marriage: "The girls will have a much better chance there [Cooperstown] for here [England] although they have friends they have no real position & I think you will agree with me that they should have that which is naturally their right."[8]

The central New York landholding system encouraged the persistence of such paternalist relations. Clarke let land to tenants as a reward for service or loyalty. He enforced cooperation by evicting those who gave him trouble. And he preserved his power to act summarily by making personal or "customary" oral contracts with tenants. The informality of the arrangements encouraged negotiation and discouraged lawsuits that, in the absence of documentation, almost always favored men of higher station. But customary leases also tied Clarke and other members of the community together within a hierarchical system and, ideally, avoided confrontations. By employing local agents to collect rents and decide whether to renew contracts, Clarke placed a buffer between himself and the tenants in case of disputes. Likewise, when tenants contracted with neighborhood agents, they accepted a moral imperative to abide by those agreements to preserve community order.[9]

Clarke's ejectment suit against Nicholas Fero of Charleston, Montgomery County, in 1875 revealed the complex relationships engendered by customary leases. Fero rented a ninety-acre plot from Clarke for $67.50 per year. Clarke made the agreement orally and his agents granted subsequent contracts. Clarke did not recall when Fero first occupied the farm, but he thought that it was in 1872. When Fero paid his rent each year, he asked if he could have the farm on the same terms for the ensuing season. Each year, Clarke's agents asked for higher rent but acceded to Fero's wishes with little argument. Fero believed that he could hold the land as long as he wanted at the same rent.[10]

The arrangement between Fero and Clarke, though apparently simple, involved at least two other tenants and was tied to other customary practices. First, Clarke charged "by the acre" but had never surveyed the plot to discover its actual size. Clarke may not have known the precise acreage of many of his tenant farms. When he wished to increase Fero's rent to one dollar per acre, he did not make a survey but relied on the farm's approximate size to calculate the new rent. Fero accepted Clarke's word that the lot contained ninety acres and paid accordingly. The two other tenants with whom Fero shared the lot had no contract with Clarke, nor did a third tenant who occupied a small dwelling on the property. Fero grazed eighty sheep but shared the land with another man who needed pasturage. They split the available meadow roughly in half. Clarke, as was customary, honored these arrangements. Any single piece of lease property might therefore contain a number of secondary contracts that would be affected by any alteration of the primary lease. Clarke began issuing standard printed contracts in 1874 to alleviate the aggravation of lawsuits against these "customary" tenants.[11]

What happened on the Clarke estate in the 1870s exposed the deep rift opening between large landowners such as Clarke, who wished to expand commercial production, and tenants trying to preserve their customary rights. A protracted and violent struggle, this tenants' strike shows that small farmers were anything but complacent and satisfied. The strike had its roots in the resolution of the Anti-Renter's civil suit against Clarke in 1854. When New York state attorney general Levi S. Chatfield had challenged Clarke's title to the land at the behest of the Montgomery County Anti-Rent Association in 1849, he charged that one tract of Clarke's land, the Corry Patent, had been fraudulently granted and that its settlement requirements had not been met in the allotted time. If the court voided the patent, as he argued it should, the land would revert to the state, which, the Anti-Renters presumed, would sell the land to its occupants. Clarke produced letters patent for the tract and claimed that the statute of limitations had expired on the alleged fraud. But the crux of his argument was that he and his father had received rents on the land for forty years without challenge and that he therefore held the land by "custom." Clarke won the Montgomery County case and a similar one involving the Outhout Patent in Otsego County. He also survived each appeal, the last coming in 1854. Tenants alleged that Clarke had bribed the courts.[12]

Clarke's actions stood in stark contrast to many other landlords. Intent on changing the nature of the estate, Clarke converted life leases into annual cash or share terms and by 1865 had eliminated most life leases. Whereas life leases required a nominal rent of twenty-five dollars per year, the new leases (at least ideally) raised the rent to one dollar per acre on the ninety-acre farms while still requiring tenants to repair buildings and fences. Tenants who already believed that they were being deprived of additional rights were further angered when Clarke decided to raise rents to an average of two dollars per acre in 1865 to help pay his debts. As the Fero case indicated, many farmers had not yet been convinced to pay even one dollar per acre; two seemed onerous indeed. When tenants balked at the new terms, Clarke offered to reinvest 25 percent of the rent in improvements, but the tenants still refused to cooperate.[13]

Trouble for Clarke began on his land in Oneida County, where tenants set several fires on lease farms to protest the new rents. Otsego County farmers then demanded a renegotiation of their terms. When Clarke began eviction proceedings, they organized "tenants' mutual associations" to strike for the old rates, to remain on the land in defiance of Clarke, and to use terrorism and intimidation to discourage outsiders from leasing the land. By 1874, tenants successfully prevented Clarke from letting his farms under the new terms in both Montgomery and Otsego Counties.[14]

Some parallels existed between the tenant associations of the 1840s and 1870s. Each attacked the landlord's title to the land, formed associations, levied taxes to fund activities, and conducted covert intracommunity activities. But the tenant movement of the 1870s took place within a significantly different social and economic context and thus made notably different demands. First, Clarke's tenants operated in a far more commercialized world than did their fathers. Whereas Anti-Renters had believed that gaining title to the land would counteract the evils of the manor system and provide each man with a competence, the Montgomery County Mutual Tenants' Association sought security for farm families by extending the length of contracts and limiting rents. They blamed Clarke's cash leases for forcing them into dairy and hops production, accelerating the emigration of poor families, and undermining the fabric of the community in the process. Only affordable lease terms and, therefore, limited market production could ensure their survival. Unlike the Anti-Rent movement, tenants in the 1870s now viewed tenancy as a means of escaping economic and social disruption.[15]

The effects of commercialization on central New York agriculture can be seen in the differences between the two tenant organizations. The Anti-Rent movement had drawn upon traditional popular rituals common to a preindustrial milieu through which adherents expressed their common rights and needs. In the 1870s, tenants still felt bound to the landlord but also besieged by the commercial world. On the one hand, "Well Wisher" counseled Clarke in 1877 to adopt a conciliatory course toward those "trespassing on your commons." Citing "an interest in right and best for individuals and communities," he advised Clarke, "If you will treat the people of Charleston [R]oot and Glen so that they can afford to respect and treat you well there your trouble will end." In case Clarke missed his meaning, "Well Wisher" offered a biblical analogy for the tenant troubles: "Your manner of doing business reminds me of Rekeboan Solomon's Son his subjects asked him to lighten their burthens he counseled with old men they advised him to do so he then counseled with young men (who had place & interest at stake) they advised increasing the burthens which thing he did the result was ten tribes revolted and so remained he sowed to the wind and reapt the whirlwind."[16] Tenants still viewed themselves as tied to Clarke by custom (the use of his "commons"—abandoned tenant farms—to graze livestock was one example) and demanded that he protect them from outside dangers.[17]

On the other hand, tenants also likened their relationship with Clarke to one between capital and labor. Their first set of resolutions, issued February 19, 1874, declared that Clarke belonged to the class of amoral speculators that had caused the financial crash that now engulfed them. Tenant actions against Clarke represented as much a protest against that development as against Clarke's more traditional power as a manor lord. Many refused to raise his cattle and grew hay instead.[18] J. J. Miller summed up the anticommercial stance of the tenants in 1874: "Samuel Bronson an Englishman and tenant went to pay Clark his annual rent. Mr. Clark told him he must pay twelve shillings. Mr. [Bronson] said I cannot for in order to pay you a dollar per acre I am obliged to draw Milk to the Cheese Factory. Mr. Clark said, 'shut up I'll hear nothing from you' and Mr. Bronson shut up now Mr. Editor he can do this with an Hinglishman one of his own countrymen, but John Bull can't do it with a live Yankee."[19] The result was that tenant demands in the

1870s reflected these changing circumstances. They wished to maintain a loose customary relationship with Clarke and have him protect them from commercial farming. Tenant E. L. Potter urged Clarke to perform his public duty as a wealthy citizen and "raise up" his less fortunate fellow tenants rather than exact higher tribute. Deference to Clarke, if it did not endanger tenant survival, was vastly more palatable than exposure to the full force of the market economy.[20]

The Montgomery County Mutual Tenants' Association operated within this dual understanding. Prominent community members in the 1870s led local associations and negotiated with Clarke, corresponded with other associations, penned manifestos for publication, and directed the covert activities of the organization. But their movement failed to create a statewide political wing through which to secure their desires. Instead, violence perpetrated by shadowy characters, especially in the guise of "Captain Jack Commander of the Modocs," overtook the more "respectable" methods of protest. Like the calico Indians of the 1840s, "Captain Jack" embodied several important symbols. The real Captain Jack had led the Modoc assassination of General E. R. S. Canby, sent by the federal government to force the withdrawal of the Modocs from Siskiyou County, California, in April 1873. Montgomery tenants clearly equated the dispossession of the Modocs with their own plight, and using the name "Modocs" symbolized their powerlessness, indigence, desperation, and willingness to fight a superior foe. But "Captain Jack" also referred to fire, and Clarke and his tenants spoke of "Captain Jack" as the act of arson.[21]

A typical Captain Jack case occurred on the 100-acre farm in Charleston originally let to John J. Frank for seventy-five cents per acre in the early 1870s. When Clarke endeavored to raise the rent to one dollar per acre, tenants immediately burned the barn. After the land "practically lay to commons" for a season, Charles Larne took the farm on Clarke's terms and agreed to pay Clarke the arrears of the former tenant by giving Clarke a note. Larne died in 1877, leaving the note unpaid. Henry Lounsberry, who had half interest in another 100-acre plot used by Larne, failed when Larne died and assigned the plot to E. L. Potter. Larne's executors, in the meantime, sold their lease for the remaining 150 acres to James Ryan. Clarke knew nothing of these transactions.[22]

When Potter and Ryan paid rent to Clarke in 1877, he refused to grant them another lease on the property unless they paid Larne's note. They refused, and he served them with a notice to quit. Clarke then let the farm to Peter Ostrander of Carlisle, Schoharie County, for two dollars per acre plus taxes. Potter convinced Ostrander to withdraw his offer. Ryan subsequently let the farm to Martin Russell, who occupied it until November 1877. On the night of December 29, 1877, Captain Jack burned the farm buildings. Meanwhile, two "interlopers," William and James Link, purchased the hay standing on Potter's fifty acres. Tenants sent threatening letters to the brothers, causing James to renounce the deal. William, however, cut the hay and posted a guard to prevent any trouble. Ten "emissaries" confronted the sentry with "threats of a terrifying nature." When he refused to relent, the tenants set fire to the hay and tools and "totally demolished" Link's "valuable mowing machine." Arson was so widespread that Clarke recalled waiting for a train in 1874 at Fonda and seeing fires on his farms illuminating the night sky all around the village.[23]

Whereas the calico Indians had claimed the land as their own, the "Modocs" acquiesced in their disempowerment and merely wished to be left alone to scrape a living from the land. Even Captain Jack's condemnation of the landlord class carried none of the sense of entitlement to the soil prevalent in Anti-Rent missives. "I think that any man who holds more than two thousand acres of land in free America," a man using the pseudonym Captain Jack wrote, "is a nuisance and a dangerous citizen to the commonwealth." But this hardly constituted a call for sweeping land reform. This Captain Jack instead hoped the government would interfere to ensure reasonable rents. The Modoc's disinclination to gain title to the land allowed for their nihilistic solution to their immediate problem: they would strip Clarke's farms of the capital improvements necessary to raise commercial crops successfully. From 1873 to 1878, in fact, Clarke lost $75,000 worth of property in Montgomery County alone.[24]

Other Captain Jack strategies indicated that the tenants viewed themselves as something of an agricultural proletariat. Much like English agricultural rioters in the 1830s, they organized strikes and "mutual associations"—tactics that drew more heavily on those of industrial laborers than peasants. They also attacked machinery. When Captain Jack visited a farm, insurgents destroyed reapers, mowers, and other tools, even those that did not belong to Clarke. "A Looker On" believed that Clarke's complex lease agreements that required tenants to keep a number of his cattle or perform maintenance on farms made tenants into laborers, and "the tenants declare that Clark demands higher rental than they can pay and still have left a just remuneration for their labor."[25] Clandestine tenant activities thus primarily targeted Clarke's property and symbols of commercial agriculture.[26]

Clarke and his tenants fought to a draw by the end of the decade. Clarke prosecuted delinquent tenants and hired detectives to investigate the activities of the arsonists. The detectives unearthed sufficient evidence against one alleged leader to indict him in 1878, though Clarke eventually dropped the charges. Meanwhile, he could find few takers for his farms and, faced with spiraling insurance costs, slowly abandoned efforts to raise rents. Tenants, who hoped to retain rights to the land under the customary system and avoid higher rents and further immersion in the marketplace, ultimately contributed to Clarke's downfall at the hands of his creditors, who cornered him in 1886 and began rapidly selling his estate to repay the massive mortgages he had taken to finance his hops speculations. Tenant efforts inadvertently accelerated the breakdown of the last colonial estate in eastern New York, a process that exposed the tenants to displacement by new landlords, creditors, and western competition for their products. The process of rationalization of land tenures by banks and other creditors initiated the surveying of lands, written contracts, and rents that reflected the relative fertility of the land and its proximity to canal and rail transportation. Customary rights and privileges between the gentry and tenants ceased to be the bonds that held the agricultural community together in Montgomery and Otsego Counties. For tenant farmers, therefore, their late-nineteenth-century efforts to preserve a local way of life ended in definitive defeat.[27]

* * *

Central New York commercial farmers also found that new economic circumstances challenged their conceptions of community and economics. Even though farms rapidly increased in size over the last three decades of the century in the three counties, commercial farmers felt the decline in agricultural prices and organized to improve their position in the market economy. Large-scale hops and dairy farmers banded together to fight import duties, establish information pools on prices and methods, and rationalize marketing structures for their benefit—sometimes through cooperative buying and selling. Finally, they sought improved education and technology to increase production and establish efficient farm units that could compete in the national marketplace. They, like tenants, were not merely lashing out in frustration but were attempting to reassert control over their lives and communities.[28]

Agricultural price declines beginning in 1868 struck hard at large growers who had mortgaged their property while hops and dairy markets expanded. George Clarke's financial situation as one of the leading commercial farmers in the region sheds light on the concerns of that class of farmers. Clarke's need for cash had motivated his attempt to enforce payment of his leases and to eject tenants who held farms under customary leases from primary tenants, or to garner such rents for himself. His overseer, Tanner O. Clarke, suggested further steps. He counseled his employer to withdraw from cattle breeding and dairy production since tenants would not take mature cows on shares and that he should invest instead in less costly sheep or heifers. By the end of 1870, Clarke managed to pay off six of his outstanding Otsego County loans, but slipping agricultural prices created further setbacks.[29]

In 1871, Clarke could not pay his remaining Otsego County debts, and the sheriff sold a number of Clarke's sheep and cattle to pay judgments against him. Cooperstown bankers tightened their hold on Clarke in the fall, forcing him to remain outside the county in order to avoid having papers served by the court. Clarke instructed his agent, T. C. Smith of Milford, to sell all remaining butter and cheese before the market deteriorated further.[30] Intent on selling Montgomery County property, Clarke meanwhile announced to Smith, "I can smell money in the wind here and consequently am not anxious to lease at present." Combining luck, determination, and bravado, Clarke refinanced the loans and saved his estate, at least for the moment.[31]

Financial problems beset Clarke for the next five years. In 1872, he could not purchase sufficient cattle to stock his tenant farms. By 1877, despite rising prices for hops, his main source of cash, Clarke's expenses in Otsego County surpassed his income. He continued to buy property in the county to expand his operations, despite his earlier troubles. Yet Clarke found New York bankers unwilling to accept notes from agricultural sources. Then in 1878, a creditor received a $25,000 judgment against Clarke, and the North American Detective Police Agency presented him with a $4,000 bill for investigating the Montgomery County fires. In 1882, the Equitable Insurance Company demanded payment of arrears on Clarke's $100,000 loan, while Dutchess County bankers refused to grant Clarke forbearance despite pleas that his agricultural pursuits benefited the county as a whole. All the while, Clarke continued a pattern he had established in the 1850s with devastating results. Unwilling to sell his hops for the going rate, he waited too long and lost his opportunity, year after year, to pull himself out of debt with a windfall profit.[32]

As western competition began to affect the central New York hops market in the 1880s, prices rapidly declined. Clarke could no longer escape his creditors in 1885. In March, the Otsego County sheriff seized two hundred bales of hops to pay a judgment against Clarke. It was the first of a series of sales that brought an end to Clarke's estate in 1887, when his assignee began selling farms. Clarke's demise had serious consequences for central New York farming. First, the breakup of his estate marked the beginning of the decline of hops farming in the area. Clarke had essentially forced production of the crop despite indications that it would not pay in the long term to compete with western producers. Second, the sale of this last large estate removed the gentry from power in the community. Finally, the end of the Clarke estate freed bankers and other "middle-class" professionals from one of the remaining impediments to the development they sought for the area.[33]

Clarke had been the center of local commerce by virtue of his wealth. But Clarke's position was at the center of an economy based on barter and credit. He purchased supplies for his tenants and took payments in labor or goods. When tenants could not pay their rent, Clarke accepted personal notes. Clarke traded these personal notes, many of which were worthless, as a matter of course, even when he knew they probably would not be paid by the originator. Rather, he took them on honor and goodwill. Bankers disliked this chaotic system because it retarded investment and caused substantial losses for the last person to hold the notes. They were only too glad to be rid of Clarke in 1886 and greatly facilitated his fall. Clarke and his fellow landlords had also irritated speculators. While the gentry invested in local railroad projects, they did so only to service the agricultural economy, not to reap windfall profits. And Clarke had further annoyed Cooperstown villagers by refusing to sell them lands they wanted for commercial development.[34]

The local tension between Clarke and his creditors pointed to a wider disparity between agricultural interests and financial and industrial interests after the Civil War. Central New York Democrats, generally sympathetic to agriculturalists, deplored the activities of "shyster" financiers local and abroad who engaged in scandalous activities. Though Clarke's tenants lumped him with "speculators," Clarke believed that he oversaw a system that offered dignity to labor and an honest profit for himself as a "planter." Clarke remained a Democrat, clinging to the notion that he was a producer, not a capitalist. He eagerly joined with other hops growers in the 1850s and 1860s to stymie the growing influence of middlemen—brokers and consignment agents—over the local economy. In 1855, Clarke spearheaded the call for a hops growers convention in Cooperstown, hoping to bring large growers together to share information on farming methods, quality, and marketing. The short-lived organization sought to establish standards of trade to maximize their share of the profits by selling only to actual consumers at "reasonable" prices. Though it met only annually and stopped short of forming a cooperative organization and the hops trade remained chaotic, the convention indicated the widening rift between producers and marketers of agricultural products even before the Civil War.[35]

Events after the war convinced Clarke and other large producers that further action was needed to shore up a flagging agricultural sector. In 1865, T. O. Clarke advised his employer "to join the fraternity of Free Traders" because "the Country—more particularly the agricultural interest—cannot and will not stand the prices it has to pay for the aggrandizement of the New England Manufacturers." He argued that free trade benefited "the farmer and agricultural land owner" because it allowed him "to sell where he can get the most for his products and he wishes to buy where he can get the best article for the least price without restriction." Enough large-scale producers agreed to warrant inviting William Cullen Bryant and David Wells to address a free trade meeting at Cooperstown in 1872.[36] T. O. Clarke, like other free traders, viewed the motives of capitalists with suspicion: "I have little doubt in my own mind but that to create a large National Debt has been, a primary motive in the late War, by the manufacturing interests. Have Free Trade and specific direct taxation and they will whine under the Demon they have raised. There is no Tax more just or equitable in principle. But all they want is to amass wealth at the expense of agriculture. 'Tis a good and significant sign to see the Commercial interest siding with the rural community."[37]

The influence of entrepreneurs on hops production also placed economic pressure on older production systems. Hops growers James F. Clark and David Wilber approached growing and marketing hops differently; nevertheless, each employed methods that were more efficient than that of the gentry. Clark "industrialized" hops farming. He hired wage labor to grow, harvest, and cure his hops and adopted methods similar to those of the western growers who preferred the efficiency of large hops farms to the patchwork, nonstandardized methods that share farming encouraged in central New York. Oneonta banker and agricultural broker David Wilber instead viewed hops as a speculative investment and shed many of the customary ways of doing business—such as purchasing low-quality hops from loyal growers or paying more than they were worth. Wilber and George Clarke also viewed local railroads differently. The former invested to make a profit on stocks, while Clarke wanted railways built to move his goods more easily to market. Wilber had been one of several local businessmen who had sided with outsiders Jay Gould and Jim Fisk during the "Erie War" in 1869—a contest that pitted Joseph H. Ramsey and "local" directors against Erie Railroad financiers Gould and Fisk for control of the Albany & Susquehanna Railroad.[38] Both Clarke and Wilber had belonged to the coalition of large landholders and merchants who worked to build the A&S. But in the post–Civil War years, the alliance disintegrated, and the specter of Wilber siding with "interlopers" Gould and Fisk to seize the road seemed to many to confirm that some members of the community would willingly sacrifice local interests for pecuniary gain. "New" monied interests, led by Wilber, saw an opportunity to expand their financial dealings nationwide and openly sided against the "old" merchant interest of the three counties who wanted to retain local control of the road. Scandals involving the railroad commissioners of Westford, among others, who illegally speculated with town finances to aid Gould, increased popular indignation with financiers and corporations. Most of the railroad's supporters were Republicans, which fostered blanket denunciations of the party's corruption, but Republicans were also divided; Radicals (often former Democrats) impeded state support for the A&S from 1864 to 1868, and Stalwarts repeatedly pushed for approval of such projects.[39]

Splits also developed between hops growers and dealers as both large and small producers struggled for survival while dealers like Wilber prospered. This struggle became an integral element of local politics as dealers routinely opposed protective legislation, hoping to drive down prices in the countryside, while farmers looked for politicians who would vote for protective legislation. Thus, when Wilber ran for Congress as a Republican in 1876 and 1878, Democrats charged that he had done little to aid the cause of hops growers because protection conflicted with his interests as a dealer. He ultimately supported such legislation, but in 1884 he sided with brewers and dealers in their efforts to increase the packaging tare on each bale of hops. During a heated debate at the Otsego County Hop Growers' Association—an outgrowth of Clarke's earlier organization—Wilber defended his actions, saying, "*It was hard to strive against the brewing interest.*" Hops growers blanched, and Democrats made his stance a political issue again in the 1886 election.[40]

Still, other Republican growers, such as James F. Clark, distanced themselves from Wilber. By the 1888 congressional election, Clark and Wilber were embroiled in a protracted "railroad" fight. Clark and Cooperstown hops dealer E. M. Harris operated the railroad from Cooperstown to Colliersville, where it connected with the A&S. The dispute began when Wilber demanded that the Cooperstown company place new switches, telegraph lines, and fences in Milford to accommodate his expanding shipping business and to protect hops fields. Also angered at prejudicial freight rates the road charged him to ship hops, Wilber had launched a "West Shore" stage line to compete with the railroad.[41]

Such local political fights over economic control mirrored national trends that had spawned the Greenback Party. Before the Civil War, northeastern farmers, especially in New York State, had advocated pro-rata legislation to end discrimination against local producers. The western "anti-monopoly revolt" of 1865 and 1866 spread east with the National Anti-Monopoly Cheap Freight Railway League. Many large hops growers, including George Clarke, offended by the Republican Party's support of high tariffs on manufactured goods, the National Banking Act, and the Democrats' refusal to back inflationary monetary policies, defected to the Greenbackers in 1878. Here, more than in the factional disputes of the mainstream parties, lay the essence of elite debtor-farmers' critique of capital and the Republican Party. Overinvestment in real property and farm improvements had paralyzed large farmers such as Clarke nationwide after the high prices of the Civil War subsided. For such men, the Republican Party's steadfast defense of financial and industrial interests rather than agricultural producers and the Democratic Party's failure to advocate agricultural relief led them to embrace radical political ideas.[42]

Badly in debt and convinced of his "partnership" with labor—his tenants—Clarke championed the Greenback cause as the way to fight hops dealers and bankers, who he believed threatened his prosperity. Clarke had favored the repeal of the National Banking Act in 1868. Then he blamed it for the depression of the 1870s and told his wife in 1875, "Again the control which the National Banking law has given to the Banking interest is being met with marked disfavor and a feeling for its repeal is growing steadily throughout the land," though "it may take several years yet to strike it from the statutes." Clarke believed capitalists responsible for the country's woes and hoped the Ohio Platform—which demanded inflationary policies—would rid the country of "this pernicious law [that] has had more to do with our present ailment than has been until lately generally understood."[43] Clarke, who generally remained behind the scenes in Democratic politics, accepted the nomination of the Otsego County Greenback Party for state assembly in 1878.[44]

The regular parties quickly attacked the Greenbackers, fearing their potential to attract significant numbers of debtor voters. The Democratic Cooperstown *Freeman's Journal* accused Clarke and other large farmers of trying "to save themselves, and themselves only." Inflation might help debtors with surplus goods to temporarily reduce their burdens, but "excessive indulgence" would drive up the cost of farming and manufacturing, discourage exports, and diminish long-term growth.[45] Hoping to head off a Democratic-Greenback alliance, the Republican Franklin *Register* argued that the Democratic proposal to curtail government expenditures would contract the paper currency more than the National Banking Act or Republican policies, the exact opposite effect Greenbackers sought. The *Register* concluded that despite the protestations of the Greenbackers, no conflict existed between capital and labor. Only expanding farming and industry would ensure the needs of both, the editor argued, and Republican spending facilitated that outcome.[46]

Although the Greenback Party appeared to have significant strength early in the campaign, by election day the old parties had recaptured enough voters to hand the Greenbackers an embarrassing loss in Otsego and Schoharie Counties, despite their relative success statewide. The Republican candidate for Congress, David Wilber, polled 6,698 votes (50 percent), the Democrats received 5,346 (40 percent), and the Greenback Party tallied 978 votes (7 percent), insufficient to constitute the swing vote. Clarke fared even worse, earning only 103 votes in his district. Although, as Lee Benson pointed out, the Greenback vote may not have reflected adequately the extent of local sympathy for its platform, the results of the election demonstrated that the party had a marginal appeal at best. Once specie payments resumed in 1879 and prices again rose, the Greenback Party no longer appealed even to those voters. Yet it is incorrect to argue also, as Benson did, that Greenback voters were interested only in short-term economic relief rather than motivated by deeper concerns over the nature and effects of economic change in their communities.[47]

The large commercial farmers like Clarke who supported the Greenback Party in central New York also championed the National Anti-Monopoly Cheap Freight Railway League in the 1880s, a nonpartisan political body whose motto, "We advocate and will support and defend the rights of the many against privileges for the few. Corporations, the creation of the State, shall be controlled by the State. Labor and Capital—allies, not enemies; justice for both,"[48] defended older notions of the relationship between capital and labor. Ironically, considering Clarke's long history of difficulties with tenants, he believed that share tenancy offered an excellent opportunity for men to labor honorably and receive just compensation. He dealt with his tenants on a personal basis and considered their particular needs when making decisions, a situation in which each benefited. Bankers and industrialists, he believed, denied men that dignity and stifled organic relationships between men (which included hierarchies of wealth, education, and power).[49]

Still, the anti-monopoly movement did not significantly alter electoral alignments in central New York. The most important consequence of the political feuds of the decade between 1876 and 1886, rather, was the resulting redefinition of local political and social power. When George Clarke left the Democratic Party to champion Greenbackism, he lost the political and economic support of many of his old associates. His personality had won him few friends over the years, and when hops prices again declined from 1882 to 1886, Otsego County bankers and hops merchants called in debts Clarke could not possibly repay. Among those who brought Clarke down was David Wilber, who secured a sizable judgment against Clarke for back debts. Wilber's role in Clarke's bankruptcy was fitting, if not central, for it symbolized the triumph of the "new" economic elite over the gentry in central New York. Creditors feasted on the bankrupt estate, leading one family friend to observe of a particularly zealous businessman, "There is no honor about him and we must do the best we can" to preserve a remnant of the estate for the family.[50] With Clarke gone, these men would shape the economic development of the region for the rest of the century.[51]

Hops growers' efforts to advance their group interests generally ignored the needs of small producers. After George Clarke's death in 1889, his son Hyde presided over the remains of the estate, now reduced to nine of Clarke's "home" share farms and those of Hyde's wife, the granddaughter of William H. Averell. Hyde had taken great pains to study agricultural methods and established good working relationships with his tenants while acting as his father's overseer in Otsego County after 1884. Hyde embraced his father's political ideals, became an active Democrat, and took a leading role in the New York State Hop Growers' Association. Hyde opposed the reduction of hops duties in 1892 and worked closely with Republican James F. Clark to advocate the pure beer bill in Congress, a joint effort that submerged partisanship to promote the hops growers' economic agenda. Hyde differed from his father in advocating protection of farm products instead of free trade, but his actions demonstrated his commitment to a policy of cooperation between hops farmers that his father had been unable to achieve.[52]

The crisis for central New York hops growers deepened in the 1890s. The McKinley tariff of 1890 raised duties on hops, sacking, and twine (neither of the latter was manufactured domestically), and growers lost money on baling materials. They also feared that England and Germany would retaliate and damage the export market. Dealers thought the tariff reduced their trade in foreign hops without substantially aiding domestic producers, but dealers could pass their costs on to producers or consumers. The New York State Hop Growers' Association called for "a Hop Growers' Exchange" to stabilize prices and increase the power of farmers in the marketing of the crop. Hyde Clarke chaired the committee appointed to explore such an institution. He asserted that the exchange was necessary because New York City firms controlled the hops trade in central New York. He was essentially correct; only two local merchants remained in the business. All of the other dealers in Cooperstown acted on behalf of New York brokers. He also believed that in this situation, country growers had to depend on misinformation passed to them by agents of New York merchant houses, thus giving farmers little reliable knowledge of international prices and crop conditions.[53]

The main purpose of a hops growers' exchange, however, would be to market hops on a cooperative basis so as to reduce the power of New York merchant houses. According to Clarke, "Middlemen are in most cases a blessing to all producers of country products, but when these middlemen become the mere tools of what is almost a monopoly then they are more a curse than a blessing." The exchange would eliminate that evil: "I would suggest that a Hop storehouse and samples room be established at every convenient railroad station in the hop growing districts. Every member of the Growers' Exchange should draw his Hops to this storehouse and place them on deposit with the officers of the Exchange. The bales would then be sampled by an expert who would grade and classify each lot, according to quality. This would be a great benefit to brewers and dealers, for here they would find large quantities accumulated, of whatever quality they desired to purchase. It would also be a great advantage to the Grower, for here would be the real opportunity for him who raised the best Hop to receive the highest price." Clarke further argued that the system would prevent poor producers from selling hops during periods of low prices for the same or higher rates given to better farmers. Indeed, he contended, "The price, under the proposed system, would soon regulate itself by the amount of business done, according to the actual demand and supply, which is the real fundamental principle of all trade. It would greatly diminish the speculation in Hops, which has been the curse of Hop-raising for years." The exchange would allow growers to borrow five cents per pound of hops deposited at the exchange to pay their pickers and the cost of preparing the crop for market. They could thus circumvent the banking interests that plagued farmers.[54]

The Hop Growers' Exchange marked the first concerted effort to unify the industry as a statewide cooperative. Each county and town in hops farming areas would have a branch that would, Clarke recommended, include "all real Hop Growers who raise Hops for sale, and in order to secure a bona fide membership I would suggest that each member should either own or operate a Hop kiln." Finally, he suggested that the state organization elect officers annually to coordinate the efforts of the separate exchanges. The state exchange would either appoint local officers or allow town and county branches to elect them. Clarke's plan removed a tremendous amount of personal control from the farmer and put it in the hands of officials entrusted with the sale of the hops. Exchange agents would sell the crop as soon as a buyer appeared, and thus "each Grower should have a fair annual profit." Clarke's plan excluded small producers or tenants who took their hops to their landlord's kilns for drying. The proposal was never put into action because many growers did not wish to relinquish control over the marketing of their crop. Yet it was a dramatic attempt to temper the effects of the free market on hops farming. Instead, hops farming in central New York continued its move toward extinction at the hands of western growers and natural disasters.[55]

* * *

The slow evolution toward cooperation in hops farming lagged significantly behind similar developments in the dairy industry. Dairy farmers demonstrated a willingness to act in concert earlier and more extensively than hops farmers. Part of the reason lay in the lower profit margins earned by dairy farmers. Even large herds of fifty or more head produced gross incomes under $1,500 per year in the late 1880s. Since this unity took place among farmers who, at least early in the period, could rely on other crops if dairy prices fell, dairy farming encouraged more cooperation among less competitive farmers.[56]

State- and local-level cooperation functioned from the 1870s onward. The New York State Dairymen's Association, with county branches, acted as a clearinghouse for information on new methods and marketing strategies for large producers. The association petitioned the government for patent protection, favorable tariffs, and restrictions on the marketing of oleomargarine. But local dairy associations also functioned as community efforts, holding fairs and picnics in addition to their regular meetings to widen their appeal. Dairy farmers also organized independent clubs to process and disseminate information or to act as semicooperative marketing organizations.[57]

Organizations such as the Deposit Farmer's Club of Delaware County, begun by a number of leading local dairymen, revealed the motivation of large-scale producers for joining cooperative ventures. The club had twenty-five members when it was chartered on April 10, 1884, with George D. Wheeler, a member of one of Delaware County's prominent families and a colonel in the militia during the suppression of the Anti-Rent movement, serving as its first president. Wheeler had been elected town supervisor of Tompkins in 1873 before leading the movement to form a separate town, Deposit, in 1880. He was the new town's first supervisor and postmaster and was elected from his district to the New York state assembly in 1875. Wheeler had operated his Laurel Bank Farm on the line of the New York & Erie Railroad since 1858 with two different partners. His brother, Nelson K. Wheeler, one of the presiding judges at the Anti-Rent trials, owned a neighboring farm worth $20,000. In 1887, with his son, George W., a full partner, Laurel Bank Farm boasted twenty-two milch cows (sixty-five total) that produced $1,380 worth of milk. The farm's 150 tillable acres produced 70 tons of hay, 450 bushels of oats, 400 bushels of corn, 325 bushels of beets, 330 bushels of turnips, 110 bushels of carrots, 200 bushels of potatoes, and 5 acres of fodder corn to feed the cows. Wheeler also bred cattle and rented several adjoining farms for cash or shares. Wheeler sold much of his milk to a local creamery that shipped it to New York City.[58]

Along with Wheeler, the club's other members, by and large, also saw dairy farming as a business venture. The Axtell brothers owned a farm that also contained a cheese and butter factory patronized by local farmers. The farm consisted of 250 acres, forty-one cows, fifteen sheep, and fourteen colts. The Axtells owned four neighboring farms that they rented for ninety pounds of butter per cow—essentially a share agreement that gave the Axtells two-thirds of the dairy products of each farm. One tenant farm contained 400 acres and boasted thirty-six milking cows; two others had 150 acres and thirteen cows and 112 acres and twenty cows, respectively. Each of the five farms the Axtells operated produced hay and little else. They streamlined all of their activities to suit the needs of milk production and avoided crops that detracted from that effort. Even the sheep they owned served the needs of the dairy cattle: the Axtells used them to clean the pastures of small shoots and weeds; any profit from wool or lambs merely supplemented the dairy business. They built a butter factory in 1884 and received milk from 300 local cows. They had thirty-five patrons with 500 cows in 1888. Envisioning dairy farming as a business, they quickly adopted innovations that would increase their net profits and production whenever possible.[59]

The organizational structure of the Deposit Farmer's Club reflected a similar emphasis on the rationalization of agricultural production and marketing. Unlike the Grange, which sought similar goals, the club restricted itself to activities and debates that touched directly on "the improvement of its members in all branches of agriculture." The club excluded women, eschewed social issues connected with the development of commercial agriculture, and limited membership to farmers engaged in or intending to enter large-scale production.[60]

The club at first limited itself to discussing fertilizers, the use of root crops to feed cattle, and proper breeding methods. Members placed topics on the agenda such as "Shall we Breed for a Specific Purpose & How shall we do it?" and "How to Feed Stock in Winter" that met the concerns of men confronting the initial imperatives to specialize. The *Delaware Dairyman* noted that "the butter of the future is to be the product of the creamery. No dairyman making butter in a small way and by the old methods at home can compete with the creamery. He cannot make as good an article; he cannot make it at so small a cost, and he cannot begin to sell it for as much."[61] By the end of 1885, however, the club had to consider the root causes of the changes that had prompted them to organize in the first place. For the next four years, members weighed the merits of shipping milk to New York City instead of making butter. The debate sprang from the first direct challenge offered to the dairy industry nationwide by manufacturing interests: the development of a cheap substitute for butter, oleomargarine. Delaware County butter producers joined with state and county dairy associations to lobby Congress to restrict the new product. The Deposit Farmer's Club sent the following resolutions to their representatives on January 2, 1886: "That it is the sense of this meeting that the Legislature of this State and the Congress of the United States should adopt measures that will put an end to the fraudulent traffic in Artificial Butter, and that our Senators, Members of Congress, and Members of Assembly be requested to do all in their power for the protection of the Agricultural interests." They also asked their state legislators to support Senator Warner Miller for chairman of the Agricultural Committee of the New York state senate because of his support for prohibitive legislation.[62]

While dealing with this issue, the club considered the benefits of cooperative marketing. In January 1886, the club invited two members of the Orange and Sullivan County Milk Association—a private dealer—to speak to members about the association's purchase of local milk for shipment to New York. The two men explained their purchasing practices, quality standards, and payment procedures, as well as their handling methods. One year later, after many had contracted to ship milk with the association, the club again debated whether to sell milk or make butter. Members spent much of the spring listening to each other's experiences, usually finding that those who made butter reaped higher profits. Armed with these findings, a committee called on the manager of the association, Mr. Kling, and demanded that he raise the price for milk.[63]

Several members clearly hoped to steer the club toward cooperative marketing, but most opposed such a step. Kling offered to purchase milk delivered from April to September for one-half cent less than the latest price on the New York Milk Exchange and for one-quarter cent less during the winter months. The farmers asked to be paid a preset price for milk. The agent instead informed producers that those who could not produce the same amount of milk in the winter as in the summer would have to take one-quarter cent less than the New York price for nine months of the year. They broke off negotiations, and several farmers suggested that they patronize a creamery being built at Deposit whose owners would purchase at a preset price. The Axtell brothers, who ran their factory on a share basis in 1886 (a semicooperative arrangement in which the factory manager made butter or cheese and paid proportional shares of the profit to patrons), interjected that they might purchase milk outright during the coming year and would pay ninety cents per 100 pounds of milk or the rate paid by the Deposit creamery. During the debates, however, members "repeatedly and plainly stated that there was no intention on the part of the farmers to combine and hold their milk and compel Mr. Kling to pay more than he has already offered, but it seemed the general purpose of each to consider for himself whether he could best afford to sell his milk to Mr. Kling, or make up his milk at home or patronize a butter or cheese factory."[64]

The refusal of members of the Deposit Farmer's Club to enter cooperative marketing in 1887 marked the limits of large-scale market farmers' willingness to step outside of independent decisions based on the needs of their own business. Few wanted to place their fortunes in the hands of the many or see their profits decline in order to help neighbors who produced less or lower quality milk. Like large hops farmers, they believed that incentives to remain independent outweighed signs that cooperation provided their best hope of success in a competitive marketplace.[65]

The debates in 1888 revealed significant rifts within the club over cooperation that led to its downfall. During the March 3 meeting, club members devoted their second session to discussing "Are Dairy Products the most Profitable for this Region, and how shall we Dispose of them?" M. D. Whitaker suggested that while dairying had always been considered the most profitable business for farmers in Delaware County, sheep farming would likely pay as much, partly because of competition with margarine producers and partly because milk returned a better profit, unless one had a large enough family to produce butter.[66]

But milk prices did not meet the cost of production. Another member suggested that "the farmers here could form a co-operative association, build a creamery and handle their own products to good advantage." Farmers on the line of the New York, Ontario & Western Railroad, he said, successfully adopted this course and gained flexibility to make butter or milk, depending on the market. A. G. Loomis reported that the Erie Railroad would ship milk from a cooperative creamery for the same price as commercial shippers. Loomis cautioned that "he would rather sell as we do now if present buyers would pay what milk is worth, but is in favor of co-operation when they want all the profits." C. Van Shoyck and George D. Wheeler argued that dealers underpaid farmers and that cooperative action would yield higher profits. But while they carried the debate over to the next session, the farmers faced making their annual contracts with milk dealers in April and merely agreed that they received too little for their product. Unable to free themselves from the demands of the market and unwilling to take a chance on a cooperative venture, the club slowly declined after 1889.[67]

The Deposit Farmer's Club demonstrated that large-scale dairy producers, like their counterparts in the hops industry, remained wary of cooperative solutions to their marketing problems in 1890. It also showed that large-scale dairy farmers viewed themselves as different from other farmers who produced smaller dairy surpluses for sale by forming a separate club. The club exhibited little propensity to challenge the fundamental assumptions of the market economy and its social effects. These farmers believed that only systematized, scientific methods and hard work would produce a "paying" farm and looked down on those who spent time in the village socializing or who contented themselves with mere subsistence. The club thus functioned as a forum for disseminating information among male businessmen primarily interested in dairy farming as a profit-making venture.[68]

As the government assumed the role of arbiter between mercantile and farm interests at the federal and state levels, often regulating quality and price, many states created county farm bureaus to disseminate information at the local level. New York State, for example, funded agricultural societies and experimental stations and worked to establish favorable toll rates on canals. Farmer's Institutes, established in 1886, formed the keystone of New York's early efforts. The program, directed from the state-funded agricultural college at Cornell University, sent lecturers to various parts of the state to educate local farmers in all phases of agriculture—including soil preservation, direct marketing, and citizenship. The intent of its founders was to eliminate destructive practices and folk superstitions that surrounded the planting, harvesting, and preparation of farm products. Then, it was hoped, New York State farmers could better compete with western growers.[69]

The founders of the Farmer's Institutes intended to reach all classes of farmers, but in practice the program tended to work from the top down. Lecturers generally spoke on specific subjects before clubs or Grange chapters. Usually, a single meeting constituted a visit to an area. Many of the solutions that lecturers offered on soil improvement, proper breeding, orchard care, and farm maintenance may have been scientifically sound but could be implemented only by farmers with ample means. Tenants and poor freeholders who did not join farm organizations or who believed that farming for profit was dishonest or corrosive remained outside of the circle of beneficiaries of the institutes. In short, the institutes probably helped a number of middle- and upper-class farmers compete more effectively in the marketplace while doing little to end the weeding out of nonmarket farmers or the poverty attendant to tenancy and marginal farming.[70]

The benefits of the educational programs at Cornell University and the Farmer's Institutes did have an important effect on dairy farming in central New York, however, and helped establish the industry as the main source of income for Otsego, Delaware, and Schoharie County farmers to the present day. Jared Van Wagenen Jr. of Lawyersville, Schoharie County, epitomized the type of farmer who prospered despite the demanding competition of the late nineteenth century. Born to an elite local family that owned the 244–acre Hillside Farm, Van Wagenen attended nearby Cobleskill Free Academy and graduated in 1887. Just sixteen years old, Jared entered the Cornell Agricultural School and graduated in 1891. Cornell offered him a lectureship in 1895, which he accepted, and he remained there four years before spending a year teaching at Rutgers. He then began a long association with the Farmer's Institutes, holding lectures throughout the state into the 1920s, and wrote several articles for farm journals. Van Wagenen also served in the New York state assembly in 1920.[71]

Van Wagenen steadfastly championed rural values as a bulwark against the corrupting influences of industrial society, but he also personified the alterations in agricultural rhythms and strategies that upstate New Yorkers adopted in response to market demands. One observer wrote of him in a 1919 book that celebrated the success of several farmers: "If you are looking for a high-class combination of safe and sane yet progressive agriculture, for comfortable success in farming independent of props and outside sources of revenue, for theoretical knowledge wedded to practical everyday routine, for a man who thinks more of his family than of his best livestock and who puts his duty as a citizen up near the top of the list—then it is worth considering the outfit and person here described."[72] Van Wagenen employed a three- or four-crop rotation in his fields, relying primarily on corn, oats, and hay with an occasional crop of wheat. He planted alfalfa, clover, timothy, and alsike in his fallow fields and bluegrass and white clover on his pastureland. He harrowed his fields and carefully selected clean seed to avoid troublesome weeds, in addition to spraying. Van Wagenen's father had begun the slow evolution toward dairy production in 1880, and by 1919, Jared owned fifty head of purebred grade Guernseys, including a registered bull and thirty-five milch cows. Although he could make more money by selling whole milk, Van Wagenen sold only cream, shipping it to Albany on the A&S Railroad. He owned a milk separator housed in his barn, built in 1895 with a 150-ton hay capacity and two silos to hold feed. Van Wagenen fed the skim milk to his pigs and calves, which he believed netted a greater return in the long run. He also boasted forty-five grade Shropshire ewes and sold the lambs and wool. In 1918, he became the first farmer in his area to purchase a tractor—at a cost of $1,100. By then, the value of the farm, implements, and livestock neared $25,000. It was a model central New York dairy farm.[73]

The Farmer's Institutes partially succeeded in preparing farmers already established in the marketplace for the continuing changes in the rural economy. Larger, more efficient producers stood the best chance of survival. Wealthier farmers like Van Wagenen benefited the most from the educational opportunities offered by the state and could look forward to a reasonable chance of success. The members of the Deposit Farmer's Club likewise remained competitive because of their heavy investments in labor, purebred livestock, and their location near New York City, despite price fluctuations that endangered their livelihoods.

Hops and dairy farmers found little in common during the last years of the nineteenth century, even though they had similar incentives to specialize and consolidate. But where dairy farmers had readily available markets in northeastern cities, hops farmers found that western competitors fared well in international and domestic markets centered in the Midwest. They fought as much for survival as for profit. At the same time, the market, as the Montgomery tenant troubles demonstrated in the 1870s, placed tremendous pressure on small producers and threatened to drive them out of farming altogether. People joined tenant associations, farmers' clubs, and the Farmer's Institutes to find relief from their difficulties and hope for the future. But the mass of farmers in central New York operated in between the extremes of tenancy and large-scale production. They experienced the problems that beset those above and below them, making their adjustments the most striking in many ways. The path they chose would define the nature of rural society in central New York as it entered the twentieth century.

8. The Grange Movement, 1874–1900

The Grange, or the Patrons of Husbandry, was founded in 1868 by Oliver H. Kelley, an employee of the United States Post Office, as a forum for education and political discourse. Kelley wanted farmers nationwide to band together to preserve and promote farming as the cornerstone of the American Republic. To succeed in this effort in post–Civil War America, Kelley believed that farmers needed to understand and then manipulate the marketplace. Kelley formed the first Subordinate Grange at Fredonia, Chautauqua County, New York, on April 16, 1868; the New York State Grange was organized in November 1873; and by the end of 1874, there were nearly three hundred subordinate Granges in the state, thirteen in Schoharie and Delaware Counties.[1]

Despite these large numbers and its considerable effect on the lives of rural Americans, the Grange rarely has been studied for more than its role in attacks on corporate monopolies in the Midwest and its educational and moral impact on family farming. The social and economic background of members, why they joined, or the discussions and rituals at meetings remain largely unexplored. The secrecy of the Grange partly explains this, but because scholars assumed that Grange members were the more commercialized, wealthier farmers, they have drawn dubious conclusions about the order in New York. Historian Solon J. Buck asserted that northeastern farmers ignored the movement because it advocated favorable railroad rates for western farmers, while Lee Benson and others later argued that Grange members came from the upper echelons of rural society. But the Grange expanded in New York State after it declined in other sections of the country, and it had a large membership that made it one of the most important Grange states in 1900. A closer scrutiny of the membership and activities of the Grange in New York can help us see why this was the case.[2]

Moreover, in mistaking the membership and duration of the Grange in New York, historians have underestimated the organization's broader social, political, and economic activism. In fact, despite protestations to the contrary, the Grange had a deeply political agenda from the outset, though it steadfastly refused to enter the electoral arena. When the financial scandals of 1873 revealed secret deals between finance capitalists and politicians, Grange members were outraged by such betrayal of the public interest in what appeared to them to be a political culture that caused the economic crash. Locally, the embezzlement of township funds by railroad commissioners and a series of insurance scams aimed at farmers added to the sense that corporate interests were destroying public trust. At the same time, strongly influenced by its female members, the Grange began to agitate for woman suffrage, state administration of highways, and temperance to check the corrupting influences of members associated with urban industrial society. To members, the Grange was more than a farmers' guild; it was the vehicle for the regeneration of the sickly American Republic. Thus, the Cooperstown *Freeman's Journal* would sing the praises of the Grange in 1874 as a vehicle for moral reform. Many central New Yorkers, especially dairy farmers, would take up the call by 1900.[3]

* * *

The Grange attracted large numbers of middling farmers in central New York—the group most likely to succumb to the pressures facing agriculture. As farms decreased in size and number, hops prices fell, butter-making gave way to the milk industry, and migrant workers flooded the countryside each summer, many freeholders feared losing their land and the destruction of the family farm as the cornerstone of rural society. Caught between tenants wanting insulation from the market and large commercial farmers battling agricultural brokers for economic and political dominance, middling farmers turned inward, hoping that hard work and Protestant morality would stem the tide of change. Central New Yorkers thus rejected the radicalism of Populism and gave the Grange widespread support because it simultaneously addressed local concerns over corporate capitalism and upheld a modified paternalist household, Protestant values, and a producer ideology strikingly similar to that once voiced by Anti-Renters and opponents of the A&S. Shorn of electoral insurgency as a strategy for effecting change, the Grange nonetheless sought to reduce government spending, retain local control of political institutions, and cultivate republican virtue.[4]

Central New York Grangers overwhelmingly were Protestant men and women who practiced mixed farming and approximated their neighbors in real wealth and commercial production. The backgrounds of the charter members of six subordinate Granges in Delaware, Otsego, and Schoharie Counties in 1880 suggest that Grangers came overwhelmingly from landholding families from the stable farm population and were well into their adult lives. The average age of those who formed Granges was forty-one—men generally were older (forty-two) than women (thirty-nine); 92 percent were married and joined the Grange together; and 92 percent were native New Yorkers. Among males, 58 percent listed their occupation as farmer, and three-quarters lived in households that engaged in farming. Craftsmen and laborers formed the second largest occupational group, with 17 percent of the total. Among women, 83 percent were listed as housekeepers in the federal census; the 9 percent with no occupation formed the next largest group.[5]

The exact economic status and crop production of individual Grange members varied within and among chapters but in total had similar economic profiles to average farmers in their counties. For example, of the eleven Seward Grange #344 members listed as farm operators in 1880, 55 percent owned their land. They cultivated an average of 124 acres valued at $5,851, owned equipment worth $416 and livestock worth $574, and produced $1,517 in goods in 1879. The eleven Huntersland Grange #375 farmers all owned their land, cultivating 104 acres valued at $2,700. They invested $118 in implements and $530 in livestock and grew $991 in farm products. Seward farmers owned on average 7 milch cows that produced 809 pounds of butter, raised 11 sheep, and devoted 3.8 acres to hops that produced 1,696 pounds for market. Huntersland Grangers owned 5 milch cows, made 605 pounds of butter, had a herd of 25 sheep, and produced 979 pounds of hops on 1.4 acres. The average Schoharie County farmer owned a 102-acre farm worth $3,633, invested $192 in implements and $434 in livestock, owned 5.4 milch cows and 6.5 sheep, produced 701 pounds of butter, and had 1.5 acres that yielded 766 pounds of hops for a total output of $827. Despite some variation, therefore, Seward and Huntersland Grange members averaged slightly greater acreage than their fellow Schoharie farmers, practiced diversified farming, and produced average or above-average yields of market products like butter, wool, and hops.[6]

Figures from Otsego and Delaware Counties, with some variation, confirmed that Grange members represented a "middle class" of farmers. The average farmer in Otsego County in 1880 owned a 105-acre farm valued at $3,790; held $168 in equipment and $513 in livestock; owned 8 milch cows and 7 sheep; and produced 751 pounds of butter, 729 pounds of hops on 1.5 acres, and 156 bushels of oats on 5.8 acres, with total output valued at $867.

Members of Elk Creek Grange #506, the county's first chapter, and Westville Grange #540 approximated these numbers.[7] In Delaware County in 1880, the average farm consisted of 140 acres worth $3,212, with $150 in equipment, $560 of livestock, and $718 of output; housed 11 milch cows and 6 sheep; produced 1,469 pounds of butter; and grew 5 acres of oats that yielded 148 bushels. In comparison, the farms of Rock Valley Grange #470 and China Grange #475 members tended to be smaller in size and production. All eleven Rock Valley Grangers owned the farms they operated, cultivated 133 acres valued at $1,300, owned $44 of equipment and $285 of livestock, and grew $369 of products. They owned 4 milch cows and 9 sheep, made 415 pounds of butter, and grew 50 bushels of oats on 2.2 acres. Of China Grangers, 85 percent owned their land but had holdings of 190 acres. They valued their farms at $3,314, with $108 of tools, $649 of livestock, and $781 of commodities. They owned 12 milch cows, made 1,533 pounds of butter, owned 21 sheep, and produced 146 bushels of oats on 4.5 acres. As in Schoharie and Otsego, Delaware Grange members came from a wide economic range of farm operators but were not among the largest in the county. Rather, the economic and social background of Grange members in all three counties indicates that they approximately represented the successful farmers of their townships, even when their output deviated from county averages.[8]

The extant membership lists of the China Grange and the Deposit Farmer's Club make possible a direct comparison of members from the same township. The lists reveal that Grangers engaged in market activity but held significantly different notions about the purpose and meaning of that activity than did farmers who considered agriculture to be a business venture. Moreover, the Deposit Farmer's Club did not include women, a reflection of the structural difference in the organization of production on their farms. The average age of men in the China Grange (forty-nine) resembled that of the Deposit Farmer's Club (forty-eight), but the latter had a 27 percent merchant membership, while none of the thirteen Grangers claimed to be merchants. All China Grange members operated farms. Club members cultivated an average of 132 of their 256 acres valued at $3,271 and had heavier capital investments in farm implements ($209) and livestock ($726), though their overall farm operations produced less cash ($745) than the China Grangers. They grew 75 percent more corn and oats than their neighbors. They had the same number of milch cows (12) as the Grangers and produced less butter (1,288 pounds) and had fewer sheep (2) per farm, an indication of their specialization. Club records show that sales of fresh milk to New York City probably offset the low butter production figures, even though the census enumerator did not list such trade. Four Deposit Farmer's Club members listed farming as their occupation but did not appear on the census as farm operators. Three of these men were over sixty, and all four had extended families or employees living in their households. Most likely they had retired from active farming and had others working their property. Two leather manufacturers, a railroad agent, and the editor of the Deposit *Courier* were on the club's rolls, confirming its more commercial nature.[9]

Thus, while both groups were prosperous and competed in the marketplace, China Grangers organized their lives around a distinct set of social relations based on assumptions and values that resonated with Grange rhetoric and ritual. Indeed, the Grange and its message drew members in the three counties for the rest of the century; seventy-eight subordinate chapters opened in Delaware, Otsego, and Schoharie Counties between the years 1874 to 1900, with over 1,600 charter members and perhaps 8,000 or more who joined later. Out of a total of 15,000 farms in the three counties in 1890, the Grange claimed a sizable following among farmers. The Grange thus was the primary agrarian voice in central New York in the late nineteenth century.[10]

Farmers joined the Grange because it gave voice to their growing apprehension about industrial society. The Grange uniformly promoted the commonly shared ethos in central New York of hard work, temperance, self-reliance, and conservative farm strategies as the solution to rural problems. Embracing what Lee Benson labeled "agricultural fundamentalism," the Grange argued that all members of republican society had to understand the legal and social restraints that fostered "democratic" government and protected civil liberties. Grange leaders stressed that local, decentralized government restrained politicians by making them directly accountable to their constituency, unlike administrative government, which separated voters from policy makers. Political parties compounded the unresponsiveness of government because they dictated policy rather than responded to public sentiment. By uniting farmers, the caretakers of the Republic, the Grange hoped to educate the masses in proper republican government and eliminate the influence of political bosses, corporate rings, poverty, and decay that marked industrial, urban America.[11]

Central New Yorkers were receptive to the Grange's denunciations of cities, the home of the landless, unskilled, non-Protestant laborers who farmers believed threatened the social order. Grange agitation had commenced in the three counties in 1874 amid an economic depression that simultaneously threatened many small producers with ruin while flooding the countryside with "hoboes" who brought drunkenness, disorder, and petty crime into the area. Throughout the 1870s and 1880s, many farmers and villagers fought to curtail or closely regulate migrant farm workers who were becoming an increasingly larger portion of the available workforce, especially during the hops harvest, and agitated for temperance reform. The first Grange chapters in each county had formed in former Anti-Rent areas that supported the Working-Men and Know-Nothings and maintained dairy products, not hops, as their primary cash crop.[12]

This deep concern about the effect of "nonproducers" and intemperate landless laborers on society caused the Grange to include women as members. The Grange recognized women's labor as crucial to the welfare of the family farm and believed that women should enjoy the public rewards of hard work, and like temperance societies, the Grange considered women central to protecting the home. A Grange could not be formed without women holding at least four offices—Ceres, Pomona, Flora, and lady assistant steward—and women enjoyed an equal vote in Grange affairs. Subordinate Granges frequently elected them to the offices of lecturer, treasurer, secretary, and chaplain, which Kelley had reserved originally for men (along with master, overseer, steward, assistant steward, and gatekeeper). As John W. McArthur, the master of one of the first Granges organized in Delaware County, noted, "Until this Order came into existence," the average farm woman endured "the toils of life" but did not enjoy its "honors and privileges." The Grange "takes her by the hand and recognizes her as a friend, companion, educator, and equal of man; opening the door of opportunity, and bidding her to make the most of the opportunities of life."[13]

Most Grangers hoped that including women would help counter the corrupting influence of immigrant wage laborers on the rural social order, perhaps adapting New York State Agricultural Society president Marsena R. Patrick's idea that a cottage system rather than wage labor would demonstrate the virtues of the republican family and teach landless male immigrants how to become proper American citizens.[14] Unlike day laborers, women had a stake in the social order and could be better trusted to help uphold it. The Grange thus asked men to give women a greater role not merely in labor—for they already contributed heavily to the family economy—but also in the strategies of the household and public sphere, making concrete in public life the gendered mutuality of rural households. According to McArthur, each woman was the moral guardian of her family and could transfer that role into the wider community: "It is part of her mission to make all places where her feet oft tread, home-like, and she is making each Subordinate Grange like a home." In the Grange, he continued, "man invites her to stand by his side, on the same level with himself, his constant companion, his counselor, and his co-worker, and opens to her all the joys of extended social intercourse, and all the avenues to knowledge and usefulness." The Grange consistently agitated for woman suffrage in the 1880s and 1890s and encouraged whole families to join (the minimum age being fourteen) so that all members would learn their proper roles in society. By such united efforts, Grangers thus hoped to reinforce the family farm and be an uplifting, stabilizing influence in the community.[15]

For the Grange, the cleansing of American society was the task of temperate white Protestant husbands and wives who valued hard labor and viewed the traditional family farm as the key ingredient of republican stability. As such, the organization was openly nativist. In 1886, McArthur declared: "A new world is awakening to life in the realms of the Anglo Saxon people. This world is the farming class, and it is organizing as the Grange. It is gentle, conservative and peaceful. And, perhaps, it is the harbinger of the golden future, which, ever and anon through the ages, has risen before the vision of the bards and prophets, of whom, verily, the world has not been worthy. This future is not a mere Utopia. The time of peace and happiness, symbolized by beating swords and spears into plowshares and pruning-hooks, will yet dawn upon this strife-cursed and blood-stained world."[16] Such sentiments were broadly shared. Thomas S. Goodwin, writing under the pseudonym "Gracchus Americanus," conceded that blacks and immigrants might learn to be virtuous citizens eventually, but Anglo-Saxon farmers intuitively understood the requirements and needed to reform the nation. It was the "irrepressible racial development"—perfection achieved through the adoption of Anglo-Saxon culture and sensibilities. Roman Catholics in particular would have a difficult transition, he wrote, because of the Church's aversion to "non-monarchical" society and their loyalty to the Vatican. Membership in the movement, at least initially, would exclude most people.[17]

The Grange's *Declaration of Purposes* embodied this almost evangelical spirit of agrarian republicanism:

> To develop a better and higher manhood and womanhood among ourselves. To enhance the comforts and attractions of our homes and to strengthen our attachment to our pursuits. To foster mutual understanding and co-operation. To maintain, inviolate, our laws and to emulate each other in labor; to hasten the good time coming. To reduce our expenses, both individual and corporate. To buy less and produce more, in order to make our farms self-sustaining. To diversify our crops and to crop no more than we can cultivate. To systematise our work and calculate intelligently on probabilities. To discountenance the credit system, the mortgage system, and every other system tending to prodigality and bankruptcy.[18]

The Grange thus tapped into the middling farm men and women's understanding of work, politics, and society. It criticized the market economy's negative influence on farmers—mortgages, credit, debt, and the commercial agriculture that generated them; insisted that the fruits of a family's toil should remain in its own hands, not devolve to middlemen or "nonproducers"; and asserted that frugality, knowledge, and mutual aid would ensure independence and security. As the document put it, "We shall earnestly endeavor to suppress personal, local, sectional and national prejudices, all unhealthy rivalry, all selfish ambition."[19]

Local Grangers might accept these principles, but they did not wholeheartedly embrace the political agenda of the state or national Grange, instead acting independently. Local chapter meetings tended to dispatch with "business" fairly quickly—purchasing policy, agricultural methods, correspondence from the state Grange—devoting time, instead, to instructing members in the order's rituals. Grange rituals and songs blended agrarian antagonisms toward corporate capitalism with a millennialist Protestant sensibility that called on members to dampen personal ambition and work for moral uplift in their communities. Though secular, the Grange taught that Protestant values, not the rules of the market, should govern people's lives. In emphasizing ritual content, the Grange in central New York attempted to tie its members securely to a past ethos of independence and mutuality to better resist the new economy. The rituals themselves thus are important to understand.[20]

Oliver H. Kelley devised the initiation rituals with four degrees for men and women that would corresponded to the seasons, to the family, and to a person's place within the Republic: "In early spring, Laborers are received to prepare the ground for seeding, and Maids are welcomed to the work of the mansion and the dairy. As summer approaches, these Laborers become Cultivators, and the Maids are promoted to the care of the flocks. When the crops ripen, Cultivators are advanced to be Harvesters, and the Shepherdess becomes the Gleaner. And when the year is crowned with goodness, and the season for social intercourse and mental improvement is at hand, Harvesters and Gleaners are exalted to the dignity of Husbandmen and Matrons, to oversee farm and household, and to dispense, as well as to enjoy, the blessings of Providence."[21] The four virtues of faith, hope, charity, and fidelity could be seen in each season; Grange membership would spread these virtues from the farm household to society as a whole.[22]

The Grange chaplain offered a prospective Laborer, the first degree for men, his first and most important instruction: "He that will not plow by reason of cold shall beg in the harvest and have nothing. He that tilleth his land shall be satisfied with bread; but he that followeth vain persons is void of understanding. The hand of the diligent shall bear rule, but the slothful shall be unto tribute. Happy is the man that findeth wisdom, and the man that getteth understanding, for the merchandise of it is better than the merchandise of silver, the gain thereof than fine gold."[23] The inductee learned that "promises of progress and improvement are delusive" and that farming "is full of obstacles, rough and uneven, environed with dangers, and leads, you know not where." Only hard work could cure the ills of the farmer, and the Grange would provide "ease and comfort" along the way.[24]

The Laborer received gifts that signified his duties as a farmer and as a member of the community: a plow emblazoned on a pouch reminded the Laborer that, as with plowing a field, his conduct must be "straight" and he must cultivate a "deep thinking and active" mind to overcome hardship; an axe, a plow, a harrow, and a spade symbolized the process of clearing, planting, cultivating, and improving a fruitful farm. Female officers addressed the new member. Ceres gave an ear of corn, admonishing him to eat some but to save the rest to plant and to make sure his neighbors had enough to eat and sow. Flora gave him a garland emphasizing the importance of the forest, orchard, and flowers to the function and beauty of the farm, and Pomona offered fruit as a symbol of refined agriculture and health.[25]

The other three degrees involved less lengthy instruction but likewise were designed to promote proper economic and social behavior. The second degree, Cultivator, taught farmers that to labor diligently, learn new methods, and be a virtuous citizen, his mind—much like a farm—had to be cultivated "with the virtues of Love, Truth, and Charity." The Grange master gave the Cultivator a pruning knife and a hoe that would remind him of "that proper degree of restraint so necessary to improve yourself, to keep your passions within due bounds, and prevent your fancy from leading you astray after the vanities and vices of the world."[26]

The Grange usually conferred the remaining degrees of Harvester and Husbandman simultaneously. Harvesters were told that "a man's life consisteth not in the *abundance* of things he *possesseth,* but in the *right use* of God's blessings" and that he must not denigrate those below him or attempt to rise falsely above his neighbors but must join with his neighbors in their labor. The Harvester received a sickle to represent the highest and most important goals of the order—to nurture both the soil and the community. Husbandman represented the last stage of life and the seasons. The Laborer represented spring and the commencement of working life; the Cultivator the summer of a man's mature years in which he planted the crop he hoped to harvest; the Harvester symbolized the successful attainment of that goal in the fall; and the Husbandman took care to safeguard the fruits of one's toil for the next season during winter. The Husbandman also was charged with setting a good example for the young: "We may tell them of the pleasures and independence of the farmer's life, but if their daily intercourse with us shows it to be tedious, irksome labor, without any recreation of body or mind, they will soon lose all interest in it, and seek fascinations elsewhere. Therefore strive to make your home more pleasing . . . adorn your family circle with the noble traits of man—a kind disposition; govern them with affection; teach them to love and not fear you, for love is power." The Husbandman received a piece of agate to symbolize fidelity, character, and friendship. With bestowal of this degree, a man reached full maturity.[27]

As the market economy forced families apart and weakened men's ability to fulfill the role of provider, Grange songs celebrated both the importance of family and neighborhood to a successful farmer and individual, spiritual bonds with other members of society. In "The Hand That Holds the Bread," Grangers were reminded that farmers had a duty to feed the world but also a right to

A fair reward for toil,
A free and open field,
An honest share for wife and home,
Of what your harvests yield.[28]

"Farmer's March" told men that their duty was to preserve their families' rights to the land and "snatch the earth" from "idlers" who placed a "stigma" on labor that "mocks your manhood." "Be ye men today," the song ran, and "fell oppression in your way."[29]

The songs thus invoked traditional notions of gender by casting men as the caretakers of their families' political and economic welfare; they also demanded that men alter their behavior for the good of their families and the community. They must not, for example, succumb to partisan politics but adopt moral suasion—which had worked well for women's voluntary societies. "Hither Come" provided a good example:

. . .May kind heav'n the glad day hasten,
When, in one fraternal band,
We may number, in our Order,
All who till the land,
As a mighty host with banners,
Peaceful vict'ries we will gain;
Moved by Right's resistless purpose,
Held by Love's electric chain. . . .

Serfs and vassals, then, no longer,
Chain'd to ceaseless labor's oar,

> Deaf to heaven's highest teaching,
> Blind to nature's grandest lore;
> But with minds that honor freedom,
> Strong in strength that shields the weak,
> And, with freedom's peaceful weapons,
> We'll enforce the rights we seek. . . .[30]

Although men's labor defined manhood and secured their right to participate in politics, only a wider, morally based effort to transform society would protect them from oppression. For central New York Grange members, at least, the political party was not "a natural lens through which to view the world."[31] Rather, they believed in an organic political economy of independent producers as an alternative to electoral politics as a means of effecting change.[32]

This idea caused the Grange to view women as partners with their husbands in the labor and decisions of the household. The four women's degrees Kelley established followed the same general pattern as men's. Women entered the order as Maids, symbolizing the initial stage of a farm woman's life. The Grange master proffered a bouquet of different grasses, explaining that while it possessed "little beauty and less of interest to the careless observer," it was vital to the cycle of life. The ritual stressed that women should act as helpmeet, rear children, and safeguard the moral underpinnings of the home. "Woman," the officer said, "is the educator of youth and our co-student through life, and to be this she must acquire knowledge and wisdom. . . . The interests, the social relations, and the destiny of man and woman are identical. She was intended by the Creator to be the helpmeet, companion, and equal of man; each shares the glory or the shame of the other." Women were told, however, not solely to aid in that duty but to properly manage a farm if accident or ill health left her "without a protector" so that "she may not be entirely dependent on the bounty of others." And she must eschew beauty, fashion, and vain ambition: "Let the *modesty* and usefulness of the humble grass be to *you* an object of imitation, as a sister in our Order."[33]

Successive degrees laid out the Grange's ideal of womanhood more fully. The second degree, Shepherdess, symbolized woman's role in guiding the family through peril by "practising all the social virtues." She was told to study and learn the processes of nature to better understand the "greatness of God" and always to extend charity to those in need, regardless of their social station. The degree of Gleaner carried little ceremony, instructing only that life often brought great hardship but that she must see the beauty of even nature's unhappiest times, be faithful, guard against anger, and exercise leniency and forgiveness toward the faults of others—to "glean" good from the world.[34] The highest degree for women, Matron, embodied all the virtues of motherhood, and, like the Husbandman, reminded women that they must pass their knowledge and love on to the next generation: "As matrons in our Order, remember that the mother writes her own history on the imperishable mind of her child. That history will remain indelible. On the tablet of the mind you write for everlasting good or ill, which storms cannot wash out, nor the moving ages of eternity obliterate. Be careful, then, to engraft those truths which shall guide and teach when your voice shall be silent, and you have passed from this to another world."[35]

The rituals never clearly delineated women's work and its importance to the household economy, assuming it would fall under the husband's authority as the head of the household. But in practice, Grangers expanded women's role to fit better with the partnership between men and women that functioned on central New York dairy farms. Believing that women and their labor needed to be recognized and rewarded with civic responsibility in order to perpetuate the family farm, the Grange supported increased sharing of decision-making in the household, cooperation between men and women in their labors, and participation in the promotion of the agrarian way of life.[36] The lack of additional materials makes it difficult to hypothesize much further about how central New York women themselves visualized and manipulated their role in the Grange. That the rituals stemmed from an organic vision of agrarian society predicated on mutuality, however, provides important insight into the nature of the organization, even if in the end this vision only partly transcended late-nineteenth-century notions of proper gender roles.[37]

* * *

By the end of the century, few central New Yorkers could claim to exist outside the marketplace, and Grange members, coming from the middling class of farmers, were no exception. This did not prevent Grange members from voicing a sincere critique of the effects of the market on their lives. John W. McArthur, a prosperous dairy farmer, identified the cash economy as the primary culprit in the decline of rural society. The urban middle class, he observed, denigrated traditional ways of conducting business and social intercourse in the country: "Corrupting influence comes from city and large town. These places are hotbeds of vice, and nurseries of extravagance and foolish and harmful fashions. The trading city and village are centers of extravagance. There Fashion has her altars, and there her votaries worship. And through the city and village social circle she holds sway over the surrounding country, threatening social ostracism for non-compliance with her prescribed ritual of service."[38] Quoting Rev. Thomas K. Beecher's speech before the National Grange in 1879, McArthur voiced his anger with "the commercial idea": "It is an undeclared, smouldering universal war. The armies of industry yield their wounded, dead and missing after every great commercial crisis and panic in greater numbers than any army that has followed banners. . . . The machinery of manufacture, the resistless tides of commerce, the great law of demand and supply, inspired from end to end with selfish fears and selfish ambitions, grind out their superb results of gaudy glory. But the machinery itself is lubricated with blood; its bearings are heated with human hate; the grist is bone-dust of crushed men."[39] Rather than experience such isolation and failure, Grange gatherings promoted rural values and the social interaction of farm families while reducing the influence of merchants and commercial mores.[40]

Members also believed that government and society were structurally unprepared to cope with the processes of capitalist development. Political parties had been grossly negligent in addressing the concerns of producers, so the only avenue for farmers was to alter society from within. The Grange thus embarked on a number of political and economic reforms during the nineteenth and early twentieth centuries that reflected the needs of farmers embedded in but unhappy with the market economy.

Continuing railroad scandals provided fertile ground for exploiting such unhappiness. When Grange organizers reached Delaware County in 1874, they found farmers disenchanted with local and national railroads. Stamford farmers, in fact, had founded the first county Grange chapter soon after the New York, Ontario & Western refused to lay track in several towns that had extended municipal bonds to the company. The malfeasance of many town railroad commissioners along the route of the A&S during the Erie Wars of the late 1860s, the shadow cast by Jay Cooke's financial adventures, and the refusal of the Delaware & Hudson Railroad (which had leased the A&S in 1869) to pay a proportionate share of local taxes deepened farmer anger. Even as late as 1897, the Sidney Grange asserted that canals and rivers provided more useful transportation routes, even though the community had access only to the A&S.[41]

Railroad corruption, most Grangers believed, threatened every aspect of American life, a fear expressed by McArthur in 1886: "We desire a proper equality, equity, and fairness; protection for the weak; restraint upon the strong; in short, justly distributed burdens, and justly distributed power. These are American ideas, the very essence of American independence, and to advocate the contrary is unworthy of the sons and daughters of an American republic."[42] Railroad pools that boosted prices trampled the rights of individual citizens and gave railroads exorbitant profits to bribe the people's representatives in government. The state's duty to protect its citizens thus gave it the authority to regulate railroads to prevent abuses.[43]

McArthur also argued that the founding fathers had sacrificed their individual interests to create a republic that foresaw equal rights regardless of social status. "They abolished the law of primogeniture and entail," he wrote, "and supposed they had thrown sufficient safeguards around the people to protect the most humble citizen in all his rights." But technological advances and rapid social and economic expansion created new, corrupting forces within the system that could not be anticipated in the days when competing carriers hauled goods on canals and turnpikes "and their business was no more lucrative than that of their neighbors who were engaged in other pursuits." "Steam and electricity," McArthur noted, "have opened new avenues by which men have acquired vast wealth in a short space of time, and thus become the oppressors of the people." Railroad corporations monopolized the movement of goods to market into the hands of a few individuals. Defunct railroads left communities that invested in them obliged to repay bonds, while operating roads sought special state and local tax exemptions. The economic and political sway of railroads had robbed producers of the fruits of their toil.[44]

Since scholars of the railroad "question" generally focus on who supported government regulation and why it emerged when it did, they have not investigated how agitation over railroads exposed other ideological rifts in areas such as central New York.[45] In Delaware, Otsego, and Schoharie Counties, farmers blamed political parties for railroad corruption and for a general decline in community morality and created receptive audiences for Grange charges that the two-party system was causing the Republic to decay. Grange writer Thomas S. Goodwin had argued that the Civil War signaled a sickness in the body politic symbolized by the agitation of contrived sectional issues by both parties. Without strong competition since the war, the Republican Party degenerated into a cabal of hungry, unprincipled office seekers. The shattered Democratic Party likewise had begun to scramble for pecuniary rewards. Goodwin viewed the splintering of parties over programs, not principle, as confirmation of the system's demise. He concluded that party failure to call for reform of the system proved its unworthiness.[46]

Internal differences made if difficult for the Grange to translate such sentiments into a program to combat "monopolies." A number of wealthy western New York farmers, New York State Grange leaders, and farmers' club members founded the Farmers' Alliance at Rochester on March 21, 1877, to provide a political voice for the Grange. Membership was open to all farm organizations, but most middling farmers found that the Alliance did little to protect their interests, and tenant farmers were completely excluded. The Alliance, which ultimately represented at best 10 percent of the "organized" farmers in the state, advocated pro-rata legislation, tax reform, mutual insurance, and free coinage of silver. Overtures made to New York City merchants in their battle for railroad reform likely alienated radicals who lumped merchants into the same class of antagonists as railroad managers. Unable to cultivate the support of rank-and-file Grangers, the Alliance failed to mount a successful political insurgency, receding rapidly after 1881, despite helping to create the New York State Board of Railroad Commissioners and the National Farmers' Alliance.[47]

According to some historians, the Grange and agrarianism failed in the Northeast because the Farmers' Alliance did not significantly influence elections in the region. The Populist Party's inability to make inroads in rural areas like Delaware, Otsego, and Schoharie Counties has been cited as further evidence that farmers accepted the political status quo. The refusal of central New York farmers to support the Farmers' Alliance, the Greenback Party, the Anti-Monopoly League, and the People's Party suggested that pro-rata legislation, inflationary policies, and free silver coinage did not generate sufficient interest in the region to warrant breaking party loyalties.[48] Paula Baker argues, however, that rural New Yorkers rejected partisan politics because they held the parties responsible for much of the corruption and unjust legislation. The legacy of the Anti-Rent movement and the failure to block construction of the A&S in the 1850s, as shown in previous chapters, support her argument. Agrarian producers by the late nineteenth century had ample history from which to conclude that their program would be subsumed by other political considerations. Adopting localized nonpartisan, issue-oriented political tactics, they believed, would accomplish much more than an electoral insurgency.[49]

In this regard, the influence of rural women was significant. Women formed the vanguard of temperance agitation and activities in central New York. In the Grange, the Woman's Christian Temperance Union, and other voluntary associations, women emphasized suasion, not partisanship, since women could not participate in elections. Grange women mobilized the entire order for the cause and made the Grange's record on social legislation far more impressive than its economic reforms. Women spurred temperance activity at the first meeting of the State Grange in 1874; twelve years later the Grange called for an amendment to the state constitution banning the importation, manufacture, and sale of liquor, again at the behest of a large number of female delegates. Male Grangers generally supported such reform. Partly spurred by the agitation of the Grange, the Republicans passed the Raines Law in 1896, which created a state board of excise to replace local licensing. Whether the Grange directly influenced the rise of the Republican Party to dominance in Otsego and Delaware by 1896 is unclear. But social reform found a more sympathetic ear with Republicans.[50]

At the same time, both parties proceeded cautiously on prohibition as they competed for the votes of hops growers and dairy farmers in Otsego and Schoharie Counties. Otsego Democrats won closely contested elections until 1886, when the hops crash pushed small and larger producers alike out of the market.[51] Whereas in 1872, the party's gubernatorial candidate had carried the county by 61 votes, in 1886 the total Democratic vote fell by 454 and the Republicans won by a comfortable 1,300 ballots, while the prohibition candidate tallied 700. Many Otsego farmers thereafter entered mixed-dairy farming. By 1896, there were twenty-five Grange chapters in the county, which had become a solid Republican district, with the party enjoying a hefty 2,000-vote margin in the gubernatorial race. Schoharie, on the other hand, increased hops production throughout the century. Only nineteen Granges formed in the county in the nineteenth century, and Schoharie remained steadfastly Democratic, despite the pro-temperance stance of many party members. Hops production was small in Delaware, where the Republican Party held power from 1872 onward. That year, the Republicans gained 56 percent of the vote for governor, and by 1896, that total increased to 63 percent. Prohibition was popular in 1886 in the county, drawing 8 percent of the vote; added to the Republican total, the two parties controlled 52 percent of the vote in the dairy county. Significantly, the county boasted thirty-one Granges during that period. Importantly, none of the counties responded to Populism. In 1896, for example, the People's Party mustered only 159 votes in Otsego, its strongest showing.[52] In short, though central New York dairy farmers expressed deep concerns over the solidification of a capitalist political economy nationwide, they found the moderation of the Grange, not the radicalism of the Populists, more suited to their reformist agenda.

Most Grangers believed by the end of the century that woman suffrage would help weed out corruption in political parties and counter radical elements in the electorate. At the State Grange's first meeting in 1874, Master Hinckley had appointed a committee of eleven women to recommend actions that would improve the order's usefulness to women. They did not then call for the vote, but suffrage leader Eliza C. Gifford introduced resolutions advocating the vote before the State Grange during Hinckley's term.

When her husband, Walter C. Gifford, became state master in 1877, he successfully lobbied the state legislature to enact a law enabling women to vote for school commissioners, an important early victory for the movement.[53] The National Grange declared at its 1885 meeting: "One of the fundamental principles of the Patrons of Husbandry, as set forth in its official declaration of purposes regulating membership, recognizes the equality of the two sexes. We are therefore prepared to hail with delight, any advancement in the legal status of women, which may give to her the full right of the ballot-box, and an equal condition of citizenship."[54]

Locally, male members evinced ambivalence about women's roles and often referred state circulars regarding women to women's committees. Even so, extant records from Otsego, Delaware, and Schoharie Counties show them uniformly approving woman suffrage measures in the 1890s. The Sidney Grange decided that men and women contributed equally to society and that both should earn the same pay for similar tasks. The Rock Valley Grange turned its September 13, 1892, meeting over to female members in a lighthearted effort to alter gender roles in the chapter, if only for a day. Women performed their duties as officers of the order at public functions—including Grange speeches—and there is evidence that members attempted to bring Grange understandings of gender to other community institutions like the church.[55]

Grange support for temperance, woman suffrage, and political reform did not help it develop a strategy for attacking its most formidable foe, corporations. Granges remained divided into groups that impeded such an effort. The most active were located in prospering agricultural townships; many more were in stable towns; and a few served declining ones. Poorer Granges proved reluctant to decide contentious issues like railroad regulation, and large farmers were unlikely to call for the breakup of railway corporations. As a result, the Grange staked out an uneasy middle ground between agrarian activism and fears of provoking further social unrest, as John McArthur wrote: "We wage no aggressive warfare against any other interests whatever. On the contrary, all our acts, and all our efforts, so far as business is concerned, are not only for the benefit of the producer and consumer, but also for all other interests that tend to bring these parties into speedy and economical contact. Hence we hold that transportation companies of every kind are necessary to our success, that their interests are intimately connected with our interests, and harmonious action is mutually advantageous."[56] Without challenging the right of capitalists to invest in and operate railroads, Grange members could not mount a sustained attack against them.[57]

As had been evident during the A&S struggle in the 1850s, it remained difficult to know where to draw the line between individual rights to profit and gouging the public. "In our noble order there is no communism, no agrarianism," the *Declaration of Purposes* stated. "We are opposed to such spirit and management of any corporation or enterprise as tends to oppress the people and rob them of their just profits. We are not enemies to capital, but we oppose the tyranny of monopolies. We long to see the antagonism between capital and labor removed by common consent and by an enlightened statesmanship. . . . We are opposed to excessive salaries, high rates of interest, and exorbitant per cent profits in trade. They greatly increase our burdens, and do not bear a proper proportion to the profits of producers."[58] Yet corporations responded to the needs of the marketplace, not ties of kinship and neighborhood that defined rural communities by demanding accountability to those affected by one's actions. Historian Robert Wiebe described the bind in which the Grange found itself: "The Grange movement hinted at problems that one day would assume giant proportions, but neither formulated them clearly nor held its followers tightly to the cause."[59]

So Grange members turned to railroad regulation as their only hope for redress. Regulation, they believed, would reform a legislative process that had allowed rail corporations to unduly influence government. Pointing to the success of the post office, McArthur argued that government surely could not operate railways more corruptly than capitalists, especially if the system included local cooperative boards consisting of the producers and buyers who used the lines. Organized like the Grange at local, state, and federal levels, railroads then would become a service rather than a business and provide citizens equal service.[60]

The Grange, the Farmers' Alliance, and the Anti-Monopoly League did not possess sufficient power to secure railroad regulation without cooperating with New York City merchants who favored pro-rata legislation. Their combined efforts resulted in the formation of the Hepburn Committee in 1879 to investigate railroad practices. The committee's discovery of widespread price fixing and corruption among railroad managers prompted national efforts to investigate railway affairs. Even so, regulation did not come until 1882. When the Republican Party split and Tammany Hall championed Anti-Monopoly in the Democratic Party, the legislature created the New York State Board of Railroad Commissioners. It had only limited powers, but nonetheless its existence removed a source of farmer unrest. Moreover, the Anti-Monopoly League and the Farmers' Alliance faded from the political scene after supporting free tolls on the Erie Canal, a measure that would have harmed New York farmers competing against western growers. Grange calls for further state regulation never materialized.[61]

These examples show how the Grange in fact did function as a political interest group, albeit a nonpartisan one that cooperated with other agricultural organizations in New York. Appalled by the growing disparity of wealth, which they attributed to unrestrained capitalist development, Grange members supported other political measures. They urged an income tax and reform of the current tax assessment laws in the 1880s and 1890s, debated and approved the use of the Sherman Anti-Trust Act against the Northern Securities Company and other corporations, and took steps to protect farm interests in the legislature. The State Grange helped secure the passage of federal regulation of the manufacture of oleomargarine in 1886 and advocated the appointment of a state dairy commissioner to enforce similar state legislation. Individual chapters frequently endorsed candidates for state and federal agricultural posts who supported Grange programs.[62]

Grange economic policies emphasized many of the same principles as their political programs, especially contesting the influence of "monopolies." Cooperative action formed the cornerstone of local Grange activity. Subordinate Granges corresponded on political and social issues; formed cooperative marketing agencies; visited each other for debates, lectures, and social activities; and voluntarily sent money to help other chapters build new halls. The state grange master called for more effective cooperative measures at the first State Grange meeting in March 1874 and sought the creation of county councils (later called Pomona Granges) to facilitate the effort "to secure a better reward for well directed toil and investment, by a system of purchasing direct from the manufacturer and selling as nearly as practicable to the consumer, saving to each a portion of the profits paid to middlemen or agents."[63]

But cooperative buying on a large scale generally succeeded only in the most prosperous farming areas. The Union Grange Trade Association (UGTA), for example, founded in Monroe County in western New York in 1874, became the strongest Grange cooperative. Initially operated independently, State Grange leaders established a link with the UGTA in 1884, arranging for approved firms to pay a percentage of their total Grange trade to the State Grange to defray operating costs. By 1891, the UGTA reported over $218,641 worth of business. This remained the primary form of cooperative trade until the Grange Exchange supplanted it in 1919. The UGTA enjoyed an impressive record, though like other cooperatives, it functioned better as a purchasing agency than as a joint sales organization. Smaller, less active Granges often could not agree on purchasing policies, refused to patronize approved merchants, or generated little cooperative buying or selling.[64]

Most local Granges elected a male member purchasing agent each year and engaged in some measure of cooperation. The purchasing agent had little real authority because few members wanted others to control their purchases, and some disliked the idea of forming agricultural "trusts." Members negotiated the price they would accept for products or pay for goods and which merchants and brands to patronize, then left the agent to carry out their wishes as best he could. Debate over "the flour & feed business" was often heated and took several meetings to resolve. Still, cooperative trade gave Grangers some bargaining power, and local merchants made an effort to court their custom. To maintain cooperation and minimize abuses, many Granges required members to purchase with cash. The Rock Valley Grange bought a sizable amount of bulk goods each year—flour, seed, feed—reporting $1,738 worth of business in 1892, $1,862 in 1893, $2,044 in 1894, and $1,568 in 1895. Sidney Grange #729, also in Delaware County, followed State Grange policy and traded with "Grange houses" for large orders, such as four plows bought in 1896. Local merchants filled most orders for seed, groceries, insecticides, fertilizer, and flour.[65]

Market considerations by 1886 forced most farmers to keep either farm or crop insurance to protect heavy investments in equipment, livestock, or speculative commodities like hops. But they purchased insurance from large carriers based in cities and claimed they were charged high rates to cover claims made by urban policyholders, even though farmers had lower risks. James Leal of Delaware County thus advocated cooperative insurance companies as the farmers' solution. By sharing risk, neighbors would be more apt to safeguard each other's property, while none of the proceeds would "enrich a soulless corporation, or . . . swell the pile of the millionaire."[66] After insurance scandals rocked the nation in the 1870s, the State Grange at its first meeting in 1874 advocated legalizing mutual insurance companies. By 1878, local Granges supported twenty-five unincorporated mutual insurance companies statewide.[67]

Strict state requirements for capitalization that were designed to protect consumers forced many of the small companies to operate on the "honor system" rather than incorporate. To ward off further farm activism, the state legislature legalized mutual insurance companies in 1879. By 1882, forty-seven of the sixty mutual fire insurance companies connected with the Grange had been formally organized under the statute. The State Grange created the New York State Central Organization of Co-operative Fire Insurance to oversee the companies. Ten years later, the Grange secured legislation allowing mutual insurance companies to operate outside of the standard policy strictures of the state's general insurance law. Under this system, 121 companies organized statewide by 1901. Like other Grange programs, mutual insurance companies suffered from dissension. Early companies, for example, allowed non-Grange members to join and had little power to discipline delinquents. But even when the companies failed, members took solace in knowing the individuals responsible. The success at promoting fire insurance companies led to efforts in the twentieth century to extend life and liability coverage to members.[68]

* * *

The New York Grange movement was a tentative and often conservative movement, insofar as it did not launch a systematic attack on the social order or property, nor did it mount an agrarian electoral insurgency. Yet it voiced the prevailing critique of industrial society advanced by middling central New York farmers, one in which echoes of the past resonated within a program of agricultural advancement. Eschewing radical political parties like the Populists, these farmers worked within their communities for social, economic, and political change. Mirroring earlier New York farmer movements, the Grange combined economic and social activities with educational strategies that strove to inculcate agrarian values. Family events such as picnics, lectures, readings, plays, and ice cream socials were designed to expose farm children to the recreational aspects of rural life, which in turn would teach that ties to family, neighborhood, and township formed the cornerstone of active, enlightened citizenship. Such events were the late-nineteenth-century equivalents of pre–Civil War logging bees, barn raisings, threshing cooperatives, and other community labor-sharing activities.[69]

Central New York Grangers sought answers to social and economic uncertainties, thus, by looking inward rather than out toward the larger society. They ardently believed that they could resolve the complex issues of the day by turning to each other at the local level, as when the Sidney Grange asked permission from the State Grange to debate the important political issues of the day within their hall in 1892, or when the Rock Valley Grange attempted to answer in 1893 why the Grange was not more united, or when West Exeter Grange members wondered in 1904, "What is the matter with the farmers and the farms?" The answer to the latter, members most often decided, was that they needed to work harder individually and in their communities to overcome divisions that prevented their towns from enjoying economic and social harmony.[70]

The Grange was the forum for considering crop strategies and farm improvements because members believed that they could overcome the challenges of commercial farming if they prepared properly. Most weekly meetings featured a presentation of a farming problem, after which members expressed their opinions. Topics might include what types of feed would increase the amount of butter fat in milk, whether to milk in the winter, milk testing, what crops generated the best returns, when to plant crops, and the best materials for constructing outbuildings. Local Granges hosted lecturers from the State Grange and the Farmer's Institutes who instructed them in agricultural improvements, often inviting the public to the sessions.[71]

Yet these New York Grangers also realized that to protect their way of life, they had to extend respect for rural life and republican virtue beyond the immediate community. Agricultural colleges became an important vehicle toward this goal. The New York State Grange had early charged Cornell University with focusing too little on the needs of smaller producers and the importance of maintaining the family farm. The Grange reconciled its differences with Cornell in the 1880s, and by the 1900s local Granges throughout the state actively supported state funding for the university. The State Grange also established a number of scholarships to Cornell to enable qualified farm children to attend the school. In addition, the State Grange founded the Juvenile Grange in 1904 to supplement the work of the parent organization and bring children into contact with the advantages of rural life before they reached adulthood.[72]

Rather than dismissing the Grange as conservative and politically ineffectual because it spurned radical strategies, the case of central New York suggests a more complex way to understand the organization. Members were attracted to the Grange's message of self-discipline, community uplift, and agrarian republicanism, all of which they believed were the foundation of America's social and political well being. This led them to try to recover these virtues as the only means to ensure their own prosperity as well as that of the nation. In this sense, they were conservative. They were drawing upon past traditions and historical insurgencies in their region. This too often led them to advocate nativist policies. Belief in older virtues also meant that chapters operated as a confederation that did not accept state or national dictates. This situation indeed kept the Grange from being a force on the new national political party scene. On the other hand, it made local Granges more radical on woman suffrage, anti-railroad legislation, and higher education than was the state organization. Finally, when the Grange advocated using structures of government to regulate business and protect the interests of small producers, whether by curbing corporate monopolies or providing access to public education, the order can be perceived as comparatively radical in relation to other prevailing doctrines of social Darwinism and laissez-faire.

When taken out of the local context and forced into national economic, social, and political contexts, an understanding of the virtues and vices of central New York farmers becomes distorted. Whether they acted for good or ill, these were real people facing national changes that were causing them serious distress and dislocation. They looked inward for a solution while staking out new political positions that challenged the control industrial capitalism exerted over their lives. They lost most of their battles, but their actions and ideas were an integral part of the struggle over modernization in the United States.

Conclusion

The white Protestant farmers who championed the Grange shared a social outlook that seemed increasingly under attack by corporations, immigrant labor, and political corruption at the end of the nineteenth century; it was an outlook that was very much a product of the historical experience of farmers in central New York. Deeply involved with the market and equally troubled by its implications, Grangers did their best to strip away the worst features of the capitalist system and secure space within it for the family farm to succeed as the economic and social foundation of rural life. Like their predecessors, they attempted to create a political economy that would at once open avenues for producers to prosper while protecting their interests through a vigorous democratic political system. Much of the responsibility for this, they held, rested on their own shoulders, for they believed unflinchingly that becoming better family farmers and citizens would protect them from the forces that they identified as the most dangerous to their livelihoods—credit, dependence on a single crop, and western competition for their goods.[1]

In many ways, the Grange has been asked to bear an unjustified historical burden by scholars who define radicalism solely in class terms. As Thomas A. Woods has noted, the Grange invoked a more simplified understanding of class than Marx, stopping more or less at the point that the Jacksonians did: that the few (nonproducers) tended to monopolize wealth and power at the expense of the many, the working classes. But Grange members did not wish to initiate anything like a class war, for it would undermine the fundamental rights, liberties, and institutions that lay at the foundation of the Republic. They recognized, indeed, that the United States provided an unusually fertile ground for establishing and sustaining family-based agricultural systems—land was relatively cheap, taxes low, markets nearby, and political participation widely enjoyed. Their radicalism came from their ardent devotion to an agrarian republicanism that entailed an attack on capital and a xenophobic mistrust of unskilled laborers and other "outsiders."[2]

The same commitment to democracy animated the other agrarian groups analyzed in this book. Anti-Renters, for the most part, agreed that the agrarian message voiced by the Bucktail Republicans in 1821 could be advanced only through democratic channels; that is, it was difficult for farmers to "syndicate" and was therefore imperative that mechanisms be put in place to ensure that their voices could rise above the din of more readily organized groups. Rebuilding the state government in a way that shifted power to localities was an ingenious and effective means of promoting Jacksonian agrarianism.[3] But the panic of 1837, the emergence of slavery as a national issue, and the entropy caused by devotion to localism in politics planted the seeds of failure for the Jacksonian Democrats. Anti-Renters did not discard democracy, agrarianism, or localism—not to mention support for working men and antagonism toward corporations—but the rent issue put them at odds with Jacksonian leaders, mostly Barnburners, at a critical juncture; hence, Whigs and Hunkers were able, at least in the years 1840–46, to launch a coordinated attack on the constitution of 1821, the party system it spawned, and the social relations it encouraged by using "anti-monopoly" against its very authors. The purge of the Anti-Rent radicals that took place after the Steele murder, combined with the framing of a new, decidedly Whiggish constitution in 1846, has made it tempting to conclude that tenants were attempting to situate themselves individually to compete more advantageously in the market. Yet this study demonstrates that those decisions were political, not ideological: tenants reunited with Barnburner Democrats and managed a stunning victory for the Free Soil Party in 1848 in Delaware County, and the Jacksonian message would animate the brief Working-Men's Party in the early 1850s and attacks on the Albany & Susquehanna Railroad.

The Jacksonians' overarching contention—that the centralization of economic and political power was a direct threat to local self-determination—achieved greater clarity during the debate over the A&S. When A&S promoters chartered the road in 1851, few central New Yorkers rallied to its cause, and even fewer spoke out against the project. But as soon as evidence emerged that the company had misled or defrauded stockholders, widely held concerns about the danger corporations posed to republican society immediately surfaced and, it is important to note, came first from the commercially oriented sector of rural society that had invested in the road. The town bonding acts widened the breach, for now all citizens of these agricultural counties were faced, first, with paying for a road they did not demand and, second, entrusting public funds to a corporation, which in this case had already demonstrated a failure to follow honest business practices. Opponents responded much as they had done during Anti-Rent: they organized meetings at the township level, assuming that collective action would be sufficient to head off the designs of A&S company officials. What they had misunderstood was that the constitution of 1846 and the General Railroad Act of 1850 had all but stripped township and county governments, or for that matter the state legislature, of power over corporations. The only recourse for antirailroad forces was the courts, and here judges upheld the letter of the law, specifically ruling that appeals to the "spirit of the law" or community ideas of injustice were naive and inapplicable. In a legal sense, then, the kinds of democratic insurgency imagined by farmers in the Jacksonian era were effectively dead, and subsequent movements would have to formulate new strategies to promote an agrarian message.

What is striking about both the Anti-Rent wars and the controversy over the A&S is that insurgents could not count on political parties to respond to their needs. Though scholars have christened the period 1828–56 the "second party system," this study suggests that what characterized the period was instability generated by factionalism, local jealousies, and personal ambition with a strong dose of divisive national issues. And this incoherence persisted through the first six years of the "third party system" as well. At one level, the state constitution of 1821 and its dispersal of power was to blame. The unintended consequence of decentralization was the encouragement of factionalism within the parties and a constant jockeying for power between townships at the county level—so endemic that frustrated townships or portions of townships frequently sought a redrawing of civil boundaries in the years 1837–50. The dominant party, the Democrats, endured the worst uncertainty. Though Democrats constituted a majority of citizens right up to the Civil War, various factions bolted to the other side on an election-by-election basis. Hence, William C. Bouck, the Hunker boss of Schoharie County, worked both sides of the political street in the 1840s in his successful effort to maintain control of the party machinery. In 1840, he funded a pro-Whig newspaper to bring down Martin Van Buren, even while he was running on the Democratic ticket for governor. During the Anti-Rent crisis, he likewise played Whigs, Barnburners, and Anti-Renters against each other on local issues and managed to pass the county on to his son-in-law, Lyman

Sanford, more or less restored to its Hunker ways. Part of the failure of Anti-Rent came from its trust of township leaders to follow evolving Anti-Rent principles rather than long-tested political alliances. And, of course, Bouck, William H. Seward, and other state-level leaders were fighting a bigger battle, positioning themselves within their own parties to take advantage of swirling national debates over slavery, free soil, economic development, and other reforms. And it was this situation that made a local issue like Anti-Rent viable, even if only for a short time.

The coming of the third party system generally is attributed to the increasing dominance of national issues over local issues. But this study shows that this was only part of the story. At the national and state levels, party leaders in fact were engaged in the process of recrafting coalitions in direct reference to sectional conflicts, and this trickled down to county politics. Yet the debate over the A&S also indicates that Republican success in the counties was predicated on Democratic divisions over political economy. Sanford's political maneuverings in Schoharie County were again indicative of the lack of party loyalty. Both Democrats and Republicans of various stripes joined informally together to defeat slates of candidates who did not support their notion of proper economic development or political ideology. Hence Sanford and Republican Joseph H. Ramsey traded votes repeatedly in order to get friends of the A&S elected. In Delaware, Otsego, and Schoharie Counties, Democrats therefore remained a numerical majority yet could not win elections because they either ran separate tickets or suffered local defections. For A&S opponents, this spelled doom, for Soft Shell Democrats, who espoused a similar notion of political economy, could never muster the full strength of the party to turn back the aggressive champions of the railroad. With both the courts and the parties seemingly aligned with the forces of centralization, it is not surprising that farmers in the region entered the Civil War disaffected with the political system under which they lived.

Given their experiences during the antebellum years outlined here, it follows that central New York farm families turned away from electoral politics after the Civil War. While previous studies have attributed this phenomenon either to naïveté or complacency, the weight of evidence shows that it was a logical response to new political and economic circumstances. In the 1850s, democratic insurgency failed to achieve agrarian goals within a political economy that disillusioned New York Democrats rightly believed was dominated by "Sewardism," the increasing centralization of all aspects of life that had been encoded in the New York state constitution of 1846 and the General Railroad Act of 1850. The Civil War saw the consolidation of these trends at the national level.[4]

Thus by 1870, with an economic depression looming, farmers had little choice but to adopt nonelectoral strategies to attempt to effect change. But the rapidly advancing national economy was creating sharp divisions among farmers that led to the emergence of fragmented farmer movements that represented different classes or crop specialties, not farmers as a whole. Despite widespread rural agitation against aspects of the new order, therefore, farmer groups rarely cooperated together to advance a single agrarian message. Tenants, for example, sought paternalist protection from the disruptions of the market, while commercial farmers worked to forge organizations that would advance the interests of large-scale producers at the state and national level. Frequently they advocated policies that would push poor farmers out of business and off the land. Grange members, who occupied the middle strata of the farm community, sought solutions that would maintain rural society and the competitiveness of family farmers through self-discipline, education, temperance, and mixed-farming strategies that limited risk. Their position frequently put them at odds with tenants, who they believed were poor farmers, and large-scale producers, who they believed cared little for the rural community at large. Previous scholarship has not delineated these divisions carefully and thus has concluded that farm movements as a whole were contradictory, confused, or reactionary. But parsing the messages of the separate organizations demonstrates just how persistent agrarian opposition to the postwar political economy was and that it came in multiple forms that were difficult to reconcile under the pressures facing central New York farmers.

This opposition gets back to the original intent of this book: to determine the character of agrarian movements in central New York in the nineteenth century in order to rethink previous interpretations of the extent to which northern farmers embraced or resisted the emergence of a capitalist political economy. This study offers two conclusions. First, agrarian agitation formed a constant element of political discourse in the nineteenth century in central New York, but alterations in the fundamental law of the state, not to mention political culture, made pursuit of unified agrarian action increasingly difficult to achieve. Second, the radicalism of central New York farmers took on increasingly conservative tones as the century progressed because the manifestations of capitalist development in their midst—social disorder, political corruption, economic uncertainty—threatened the agrarian world they thought necessary to the future of the Republic. Their xenophobia must be recognized for what it was, but so too should their commitment to the democratic principle that they believed acted as a counterweight to the exaggerated strength of financial wealth; it was, after all, rural representatives like

Jared Van Wagenen Jr. who would stand up in the state legislature in 1920 and demand the seating of socialists in the midst of the Red Scare.[5]

It is tempting to assume that after 1900, the kinds of agrarian reforms embraced by the Grange and other farmers' groups were unsustainable, but in fact, as Elizabeth Sanders has pointed out, the Progressive era was marked by the passage into law of many of their most fundamental demands: an income tax, regulation of business, woman suffrage, a national agricultural policy, direct election of senators, and so forth. And whereas recent studies have argued that the spread of consumer culture all but finished off agrarianism in places like central New York, it was the women and men of the dairy country, including Grangers, who staged milk strikes during the Great Depression in defense of a still-thriving family system of farming. And, at the same time, one should be careful not to consider central New York farmers to be exclusively business-oriented. In her study of folklore in Schoharie County in the early twentieth century, for example, Emelyn E. Gardner discovered that many hill farmers clung to folk beliefs, including notions that successful farmers practiced witchcraft to elevate butter yields or cast spells on neighbors' milch cows that lowered production. As well, there were still men and women who chose to farm more as a way of life than as a business. Fred Lape recalled that in 1902, his father, Herman, who had been trained as a watchmaker, commenced farming in Esperance, Schoharie County. When Herman chose a farm, he picked the "Oak Nose" farm for its view atop the high banks of Schoharie Creek, not for its soil. In fact, the soil was poor and difficult to work. In less than a decade, the toll of hard work and divisions within the family caused by the small returns on such labor led them to sell the farm. But the fact that men and women like Herman and Emma Lape would try to make a go of it should remind us that historically, farming in the United States has symbolized something more than an economic exercise. Only with this in mind can the agrarian movements of the nineteenth century be understood on their own terms.[6]

Notes

List of Abbreviations

AIHA	Albany Institute of History and Art
CUL	Division of Manuscripts and University Archives, Cornell University Libraries
DCHA	Delaware County Historical Association, Delhi, New York
JAH	*Journal of American History*
NMAH	National Museum of American History, Smithsonian Institution
NYSAD	*New York State Assembly Documents*
NYSHA	New York State Historical Association, Cooperstown, New York
NYSL	New York State Library, Albany, New York
SUL	Arents Special Collections, Syracuse University Library

Introduction

1. Carl Carmer, *Dark Trees to the Wind: A Cycle of York State Years* (New York: William Sloane Associates, 1949), 319.

2. Ibid., 321.

3. Examining both social and political life over the long duree of the nineteenth century has enabled me to see trends easily hidden by studying one or the other or a smaller chronological period. That approach distinguishes this book from the superb books on New York that have appeared in recent years, including Nancy Grey Osterud, *Bonds of Community: The Lives of Farm Women in Nineteenth-Century New York* (Ithaca: Cornell University Press, 1991); Donald H. Parkerson, *The Agricultural Transition in New York State: Markets and Migration in Mid-Nineteenth-Century America* (Ames: Iowa State University Press, 1995); Sally McMurry, *Transforming Rural Life: Dairying Families and Agricultural Change, 1820–1885* (Baltimore: Johns Hopkins University Press, 1995); Charles E. Brooks, *Frontier Settlement and Market Revolution: The Holland Land Purchase* (Itha-

ca: Cornell University Press, 1996); and Martin Bruegel, *Farm, Shop, and Landing: The Rise of Market Society in the Hudson Valley, 1780–1860* (Durham, N.C.: Duke University Press, 2002). Politics has received less attention, but three studies are particularly significant: see Paula Baker, *The Moral Frameworks of Public Life: Gender, Politics, and the State in Rural New York, 1870–1930* (New York: Oxford University Press, 1991); Reeve Huston, *Land and Freedom: Rural Society, Popular Protest, and Party Politics in Antebellum New York* (New York: Oxford University Press, 2000); and Charles W. McCurdy, *The Anti-Rent Era in New York Law and Politics, 1839–1865* (Chapel Hill: University of North Carolina Press, 2001). For the early national period, see Alan Taylor, *William Cooper's Town: Power and Persuasion on the Frontier of the Early American Republic* (New York: Vintage, 1995).

4. I use the term "political economy" in the same way as Drew McCoy, who defined it as a broad concept that "signified the necessary existence of a close relationship between government, or the polity, and the social and economic order" that assumed the interdependence of each. See Drew R. McCoy, *The Elusive Republic: Political Economy in Jeffersonian America* (Chapel Hill: University of North Carolina Press, 1980), 6.

5. A notable exception is Catherine M. Stock, *Rural Radicals: From Bacon's Rebellion to the Oklahoma City Bombing* (Ithaca: Cornell University Press, 1996), which traces the influence of common themes on rural movements and concludes that they have often been conservative. My local focus enables me to trace in more detail the reasons for this conservatism. For examples of more traditional approaches to understanding rural unrest, see Solon J. Buck, *The Granger Movement: A Study of Agricultural Organization and Its Political, Economic, and Social Manifestations 1870–1880* (Cambridge: Harvard University Press, 1933); Fred A. Shannon, *The Farmer's Last Frontier: Agriculture, 1860–1897* (New York: Holt, Rinehart and Winston, 1945); Theodore Saloutos and John D. Hicks, *Agricultural Discontent in the Middle West, 1900–1939* (Madison: University of Wisconsin Press, 1951); John D. Hicks, *The Populist Revolt: A History of the Farmers' Alliance and the People's Party* (Lincoln: University of Nebraska Press, 1961); and Robert H. Wiebe, *The Search for Order, 1877–1920* (New York: Hill and Wang, 1967).

6. For a summary of scholarship that discusses the ways in which rural Americans viewed themselves and their place in the Republic, see David B. Danbom, *Born in the Country: A History of Rural America* (Baltimore: Johns Hopkins University Press, 1995). Examples of works that discuss the nonpartisan political strategies of farmers are Hal S. Barron, *Mixed Harvest: The Second Great Transformation in the Rural North, 1870–1930* (Chapel Hill: University of North Carolina Press, 1997), and Lawrence Goodwyn, *Democratic Promise: The Populist Moment in America* (New York: Oxford University Press, 1976). The details of New York State's constitutional transformation at midcentury can be found in McCurdy, *Anti-Rent Era,* and L. Ray Gunn, *The Decline of Authority: Public Economic Policy and Political Development in New York State, 1800–1860* (Ithaca: Cornell University Press, 1988).

7. Carmer, *Dark Trees to the Wind,* 324.

Chapter 1: Republican Fathers

1. Irving Mark, *Agrarian Conflicts in Colonial New York* (New York: Columbia University Press, 1940), 19–49; Jay Gould, *History of Delaware County* (Roxbury, N.Y.: Keeny and

Gould, 1856), 6–12, 242–56; *History of Delaware County, N.Y.* (New York: W. W. Munsell, 1880), 45–48; Francis W. Halsey, *The Old New York Frontier* (New York: Charles Scribner's Sons, 1901), 87–105; *History of Otsego County, New York* (Philadelphia: Everts and Fariss, 1878), 12–14; Ruth L. Higgins, *Expansion in New York* (Columbus: Ohio State University Press, 1931), 47–55, 70–82; Edith M. Fox, *Land Speculation in the Mohawk Country* (Ithaca: Cornell University Press, 1949), x, 1–9; Edward P. Cheyney, *The Anti-Rent Agitation in the State of New York* (Philadelphia: University of Pennsylvania Press, 1887), 5–10; David M. Ellis, *Landlords and Farmers in the Hudson-Mohawk Region, 1790–1850* (Ithaca: Cornell University Press, 1946), 9, 59–62 (epigraph quote, 56).

2. Cheyney, *Anti-Rent Agitation,* 10–11; Halsey, *Old New York Frontier,* 106–11; Gould, *History of Delaware County,* 247–48; Ellis, *Landlords and Farmers,* 45; James A. Frost, *Life on the Upper Susquehanna Valley, 1783–1860* (New York: King's Crown Press, 1951), 7.

3. William E. Roscoe, *History of Schoharie County, New York* (Syracuse: D. Mason and Co., 1882), 134; Edwin Williams, ed., *New-York Annual Register, 1836* (New York: Edwin Williams, 1836), 99, 135; Edwin Williams, ed., *New York Annual Register, 1843* (New York: J. Disturnell, 1843), 35–36; *History of Otsego County,* 23–24; John H. French, *Gazetteer of the State of New York* (Syracuse: R. P. Smith, 1860), 257, 532, 600; Henry U. Swinnerton, "The Story of Cherry Valley," *Proceedings of the New York State Historical Association,* vol. 7 (Cooperstown: NYSHA, 1907), 744–98; T. Wood Clarke, *Émigrés in the Wilderness* (New York: Macmillan, 1941), 98–108; Frost, *Life on the Upper Susquehanna Valley,* 7–16.

4. Alan Taylor, *William Cooper's Town: Power and Persuasion on the Frontier of the Early American Republic* (New York: Vintage, 1995), 44–85; Harry B. Yoshpe, *The Disposition of Loyalist Estates in the Southern District of the State of New York* (New York: Columbia University Press, 1939), 113–19, 166–86; Halsey, *Old New York Frontier,* 106–15.

5. Mark, *Agrarian Conflicts,* 21–22; Yoshpe, *Disposition of Loyalist Estates,* 113–19, 166–86; William Cooper, *A Guide to the Wilderness* (Cooperstown: *Freeman's Journal,* 1836), 7; *History of Delaware County, N.Y.,* 46; *History of Otsego County,* 257; Halsey, *Old New York Frontier,* 111–15; Gould, *History of Delaware County,* 140–44; Taylor, *William Cooper's Town,* 63–75.

6. Indenture: G. Banyar to John Livingston for Lands in Scott's Patent, November 3, 1798, Box 2, Goldsbrow Banyar Papers, DU10703, NYSL; John Kiersted to Henry Overing, November 30, 1830, October 9, 1848, Kiersted Family Papers, 17820, NYSL; Albany *Anti-Renter,* September 13, 1845; Halsey, *Old New York Frontier,* 111–15; *History of Delaware County, N.Y.,* 45–48; Gould, *History of Delaware County,* 247.

7. Halsey, *Old New York Frontier,* 104, 365–67; Frost, *Life on the Upper Susquehanna Valley,* 19; Deposition of George Clarke, *George Clarke v. Charles Montanye,* Supreme Court of the State of New York, 1874, MSS, Box 59, George Hyde Clarke Papers, #2800, CUL; Fox, *Land Speculation,* xi, 49–50; E. B. O'Callaghan, *Voyage of George Clarke, Esq. to America* (Albany: J. Munsell, 1867), 29–109.

8. G. William Beardslee, "An Otsego Frontier Experience: The Gratzburg Tract, 1770–1795," *New York History* 79 (July 1998): 233–54; Halsey, *Old New York Frontier,* 99–115; Gould, *History of Delaware County,* 150–51.

9. *History of Delaware County, N.Y.,* 47; French, *Gazetteer of the State of New York,* 262, 600–608; Cheyney, *Anti-Rent Agitation,* 10; Ellis, *Landlords and Farmers,* 7–65.

10. Josiah Priest, *Stories of the Early Settlers in the Wilderness* (Albany: J. Munsell, 1837), 10; Levi Beardsley, *Reminiscences* (New York: Charles Vinten, 1852), 18–19; Frost, *Life on*

the Upper Susquehanna Valley, 17, 20; Gould, *History of Delaware County,* 152–53; Taylor, *William Cooper's Town,* 87–101.

11. Frost, *Life on the Upper Susquehanna Valley,* 18–20, 122; Gould, *History of Delaware County,* 149.

12. Cooper, *Guide to the Wilderness,* 39; Frost, *Life on the Upper Susquehanna Valley,* 122; Gould, *History of Delaware County,* 254; Taylor, *William Cooper's Town,* 86–114; Paul W. Gates, *Landlords and Tenants on the Prairie Frontier* (Ithaca: Cornell University Press, 1973), 303–25.

13. Francis W. Halsey, ed., *A Tour of Four Great Rivers: The Hudson, Mohawk, Susquehanna and Delaware in 1769* (New York: Charles Scribner's Sons, 1906), 29–80; Beardslee, "Otsego Frontier Experience," 233–54. Small speculators might also purchase chunks of land from a great proprietor, then sell them to settlers. See Warrantee Deed: Eliza Nichols, Helen Bateman, Pixlee and Catherine Judson, William and Sarah Deforest, Bridgeport, Conn., William and Rebecca Prince, Catherine, N.Y., Isaac and Anna Nichols and Lewis and Amy Nichols of Kirkland to Charles Harley, 1828, Misc. Manuscripts, Box H, NYSHA. Harley paid $60 for 156.6 acres in Roxbury, Delaware County.

14. S. A. Law's Improved Estate at Meredith, Correspondence, 1796–1817, Box 1, Samuel A. Law Papers, CX10277, NYSL; French, *Gazetteer of the State of New York,* 263; Ellis, *Landlords and Farmers,* 62, 64.

15. "Report of the Committee on so Much of the Governor's Message as Relates to the Difficulties between the Landlord and Tenants of the Manor of Rensselaerwyck," *NYSAD,* no. 271 (1840); Mark, *Agrarian Conflicts,* 50–84; *History of Delaware County, N.Y.,* 45–48, 65–66, 238; Cheyney, *Anti-Rent Agitation,* 9, 13–19; French, *Gazetteer of the State of New York,* 257, 266, 530–39, 600–608; Ellis, *Landlords and Farmers,* 64, 88; Henry Christman, *Tin Horns and Calico: An Episode in the Emergence of American Democracy* (New York: Henry Holt, 1945), 32, 60–61. Perpetual leases were sometimes made between small landholders and tenants. See Lease: Charles Harley to Felix Sears, 1826, Misc. Manuscripts, Box H, NYSHA. Harley let 100 acres to Sears for 20,000 years at $24 per year. Sears paid all taxes and could redeem the lease for the annual rent plus $500.

16. C. Pepper Jr., *Manor of Rensselaerwyck* (Albany: Albany and Rensselaer Anti-Rent Associations, n.d.), 3–34; Anson Bingham and Andrew J. Colvin, *A Treatise on Rents, Real and Personal Conveyances and Conditions* (Albany: W. C. Little, 1851), 1–4, 9–44, 55–59, 100–113; Charles W. McCurdy, *The Anti-Rent Era in New York Law and Politics, 1839–1865* (Chapel Hill: University of North Carolina Press, 2001), 1–31; *History of Delaware County, N.Y.,* 45–48, 65–66; Cheyney, *Anti-Rent Agitation,* 13–19; Charles E. Brooks, *Frontier Settlement and Market Revolution: The Holland Land Purchase* (Ithaca: Cornell University Press, 1996), 56–60.

17. "Report of the Committee on . . . the Governor's Message"; Pepper, *Manor of Rensselaerwyck,* 3–34; Bingham and Colvin, *Treatise on Rents,* 1–4, 9–44, 55–59, 100–113; Mark, *Agrarian Conflicts,* 50–84; McCurdy, *Anti-Rent Era,* 10–31; Martin Bruegel, "Unrest: Manorial Society and the Market in the Hudson Valley," *JAH* 82 (1996): 1393–424.

18. Schedule of Banyar Lands, 1836, Box 3, Folder 7, Banyar Papers; Beardsley, *Reminiscences,* 19; T. H. Wheeler to Richard Cooper, January 20, 1841, Clarke Papers; Cheyney, *Anti-Rent Agitation,* 13–19; Gould, *History of Delaware County,* 247–53; Ellis, *Landlords and Farmers,* 64.

19. H. G. Munger, "Reminiscences of the Anti-Rent Rebellion," MSS, NYSHA; *History of Delaware County, N.Y.,* 258–59; Gould, *History of Delaware County,* 247–53; French, *Gazetteer of the State of New York,* 258; Christman, *Tin Horns and Calico,* 13–21; Ralph Birdsall, *The Story of Cooperstown* (Cooperstown: Augur's Book Store, 1948), 233–37.

20. Otsego County Supreme Court, George Clarke vs. ———, ca. 1830, MSS, Box 2, Otsego County Court Records, #281, NYSHA; Share Lease: S. A. Law to Gaius Coon, July 31, 1840, Box 14, Law Papers; Share Lease: Hiram Every to Lewis Hammond, March 27, 1846, Misc. Delaware County Papers, #1–88+, NYSHA. Every and Hammond, both of Kortright, Delaware County, agreed that Hammond would take the farm in exchange for the wool shorn from 100 sheep. Every reserved all timber rights except lumber for fences and held the produce as security. The lease ran one year, and Every retained possession of the house and lot around it.

21. Frost, *Life on the Upper Susquehanna Valley,* 19. Examples include Lease: Philip Sawyer to Joseph C. Gates, 1834, Lake Family Papers, #64, NYSHA; Lease: Thomas and John Clarke to Joshua Clarke, March 24, 1843, Box 1, Clarke Family Papers, #115, NYSHA; Schedule of Banyar Lands, 1836, Box 14, Banyar Papers.

22. Albert C. Mayham, *The Anti-Rent War on Blenheim Hill: An Episode of the 40's* (Jefferson, N.Y.: Frederick L. Frazee, 1906), 28; Gould, *History of Delaware County,* 153; Ellis, *Landlords and Farmers,* 70; Samuel Ostrander to Robert Campbell, February 21, 1836, Box 3, Robert Campbell Papers, #100, NYSHA.

23. Mark, *Agrarian Conflicts,* 85–106; Gould, *History of Delaware County,* 253–59; Frost, *Life on the Upper Susquehanna Valley,* 27, 122; Cooper, *Guide to the Wilderness,* 7; Halsey, *Old New York Frontier,* 357–64; Bruegel, "Unrest," 1393–424; Taylor, *William Cooper's Town,* 97–101; Brooks, *Frontier Settlement and Market Revolution,* 18–19, 82–105.

24. Alan Taylor, "The Art of 'Hook and Snivey,'" *JAH* 79 (March 1993): 1371–96.

25. Cooper, *Guide to the Wilderness,* 19; Mayham, *Anti-Rent War,* 28; Ellis, *Landlords and Farmers,* 16–117; Cheyney, *Anti-Rent Agitation,* 21–22; Frost, *Life on the Upper Susquehanna Valley,* 20.

26. Priest, *Stories of the Early Settlers,* 20–28; Beardsley, *Reminiscences,* 18–48; Ellis, *Landlords and Farmers,* 68–76; Frost, *Life on the Upper Susquehanna Valley,* 19–30; Gould, *History of Delaware County,* 139, 147, 153.

27. Priest, *Stories of the Early Settlers,* 20–24; Beardsley, *Reminiscences,* 18–48; Gould, *History of Delaware County,* 153; *History of Delaware County, N.Y.,* 228–29, 238; Frost, *Life on the Upper Susquehanna Valley,* 20–21.

28. Ellis, *Landlords and Farmers,* 114, 124–27; Frost, *Life on the Upper Susquehanna Valley,* 122; Taylor, *William Cooper's Town,* 91, 119–37; Gould, *History of Delaware County,* 195–96.

29. Halsey, *Old New York Frontier,* 379–439; John D. Monroe, *Chapters in the History of Delaware County, N.Y.* (Delhi, N.Y.: DCHA, 1949), 98–118; Ellis, *Landlords and Farmers,* 159–83; Frost, *Life on the Upper Susquehanna Valley,* 40, 59–64.

30. Roscoe, *History of Schoharie County,* 138; Gould, *History of Delaware County,* 192; Ellis, *Landlords and Farmers,* 151–58.

31. William Armitage to G. Banyar, March 20, 1803, Box 3, Banyar Papers; Widow Saunders to Stephen Van Rensselaer, March 15, 1840, Henry Christman Papers, #10, NYSHA; Gould, *History of Delaware County,* 247–49; Frost, *Life on the Upper Susquehanna Valley,* 34, 40.

32. Henry Conklin, *Through "Poverty's Vale": A Hardscrabble Boyhood in Upstate New York, 1832–1862* (Syracuse: Syracuse University Press, 1974), 29–30; Sally McMurry, *Transforming Rural Life: Dairying Families and Agricultural Change, 1820–1885* (Baltimore: Johns Hopkins University Press, 1995), 6–99; Joan M. Jensen, *Loosening the Bonds: Mid-Atlantic Farm Women, 1750–1850* (New Haven: Yale University Press, 1986), 3–76; Benjamin D. Gilbert, "The Cheesemaking Industry of the State of New York," *U.S. Bureau of Animal Industry Bulletin,* no. 15 (Washington, D.C.: USDA, 1896), 7–44; Frost, *Life on the Upper Susquehanna Valley,* 24, 73.

33. Beardsley, *Reminiscences,* 18–48; Louis C. Jones, ed., *Growing Up in Cooper Country: Boyhood Recollections of the New York Frontier* (Syracuse: Syracuse University Press, 1965), 80–81; McMurry, *Transforming Rural Life,* 51–61.

34. Jones, *Growing Up in Cooper Country,* 80–82, 155–56; Conklin, *Through "Poverty's Vale,"* 77; *History of Otsego County,* 22–23; Beardsley, *Reminiscences,* 29–48, 85–87; Francis W. Halsey, ed., *The Pioneers of Unadilla Village, 1784–1840 and Reminiscences of Village Life and of Panamá and California from 1840 to 1850* (Unadilla, N.Y.: Vestry of St. Matthews Church, 1902), 162. The importance of mutuality coincided in many ways with artisan republicanism of the 1820s described in Sean Wilentz, *Chants Democratic: New York City and the Rise of the American Working Class, 1788–1850* (New York: Oxford University Press, 1984), 95.

35. Jane Barton to S. A. Law, June 25, 1844, Law Papers; Thomas F. Gordon, *Gazetteer of the State of New York* (Philadelphia: Thomas F. Gordon, 1836), 160–63; Frost, *Life on the Upper Susquehanna Valley,* 23. For an excellent framework for understanding "plebeian" conceptions of "customary rights," see E. P. Thompson, *Whigs and Hunters: The Origin of the Black Acts* (New York: Pantheon, 1975), 219–44.

36. James Fenimore Cooper, *The Chainbearer* (Boston: Houghton Mifflin, n.d.), 322.

37. Ibid. Squatters elsewhere argued that their toil gave them title to the land and "that laws interfering with this right were unjust and of non-effect." See George M. Stephenson, *The Political History of the Public Lands from 1840 to 1862* (New York: Russell and Russell, 1967), 21–23.

38. Bruegel, "Unrest," 1393–424.

39. Beardsley, *Reminiscences,* 50; Halsey, *Old New York Frontier,* 114; Frost, *Life on the Upper Susquehanna Valley,* 34; Christman, *Tin Horns and Calico,* 35, 38–39, 50; J. Starkweather to G. Banyar, undated (ca. 1800), Petition from residents of Stewart's Patent, Otsego County to G. Banyar, July 12, 1802, Agents Reports, July 1837, Stewart's Patent, Otsego County, Memorandum of Letters Mailed to Tenants of Stewart's Patent, Otsego County, January 13, 1838, Box 2, Folder 16, Banyar Papers. For a discussion of backcountry rebellions on the early national frontier, see Thomas P. Slaughter, *The Whiskey Rebellion: Frontier Epilogue to the American Revolution* (New York: Oxford University Press, 1986); Alan Taylor, *Liberty Men and Great Proprietors: The Revolutionary Settlement on the Maine Frontier, 1760–1820* (Chapel Hill: University of North Carolina Press, 1990); and Rachel N. Klein, *The Unification of a Slave State: The Rise of the Planter Class in the South Carolina Backcountry, 1760–1808* (Chapel Hill: University of North Carolina Press, 1990).

40. William Armitage to G. Banyar, March 20, 1803, Box 3, Banyar Papers; William Knight to John Kiersted, December 18, 1846, Box 14, Joseph Scott to John Kiersted, April 1, 1851, Box 1, Folder 1, Kiersted Family Papers; Bruegel, "Unrest," 1393–424; Schoharie *Patriot,* November 29, 1844; William Beardsley to John Hancock, December 2, 1844, Box

1, Hancock Family Papers, #169, NYSHA; Christman, *Tin Horns and Calico,* 117–18; Ellis, *Landlords and Farmers,* 62, 263–64.

41. Roscoe, *History of Schoharie County,* 134.

42. Priest, *Stories of the Early Settlers,* 34–35; Beardsley, *Reminiscences,* 191–99, 471–94, 511–17; Julia A. Perkins, *Early Times on the Susquehanna* (Binghamton: Malette and Reid, 1870), 179–81, 206–10; Gould, *History of Delaware County,* 182–93; Frost, *Life on the Upper Susquehanna Valley,* 82; Alan Taylor, "The Great Change Begins: Settling the Forest of Central New York," *New York History* 76 (July 1995): 265–90.

43. Taylor, *William Cooper's Town,* 317–45, 363–405; Ellis, *Landlords and Farmers,* 118–58, 187.

44. Taylor, *William Cooper's Town,* 317–45, 363–405; Ellis, *Landlords and Farmers,* 118–58, 187.

45. Fox, *Land Speculation,* xi; Beardsley, *Reminiscences,* 78–442, 445, 450; Jabez D. Hammond, *The History of the Political Parties of the State of New-York* (Syracuse: Hall, Miles, and Co., 1852), 1:490; Frost, *Life on the Upper Susquehanna Valley,* 21, 73–74; *History of Delaware County, N.Y.,* 77–78; Ellis, *Landlords and Farmers,* 151–58.

46. Frost, *Life on the Upper Susquehanna Valley,* 75–77, 105.

47. Richard P. McCormick, *The Second American Party System: Party Formation in the Jacksonian Era* (Chapel Hill: University of North Carolina Press, 1966), 104–14; Craig Hanyan and Mary Hanyan, *DeWitt Clinton and the Rise of the People's Men* (Montreal: McGill University Press, 1996), 5–14; Carol Sheriff, *The Artificial River: The Erie Canal and the Paradox of Progress, 1817–1862* (New York: Hill and Wang, 1996), 25–51, 79–137.

48. Hanyan and Hanyan, *DeWitt Clinton,* 5–14; Isaac H. Tiffany, *Address to the Committee of Schoharie to the Convention of Delegates of the State of New York* (Albany: E. and S. Hosford, 1817), 1–16; Roscoe, *History of Schoharie County,* 138; Halsey, *Pioneers of Unadilla Village,* 167.

49. Hanyan and Hanyan, *DeWitt Clinton,* 5–14; Tiffany, *Address to the Committee of Schoharie,* 1–16; Roscoe, *History of Schoharie County,* 138; Halsey, *Pioneers of Unadilla Village,* 167; J. Hammond, *History of the Political Parties,* 1:488–89, 2:1–81; Beardsley, *Reminiscences,* 216–17; Wells S. Hammond, *Oration Delivered at Cherry Valley, on the Fourth Day of July, 1839* (Albany: J. Munsell, 1839), 14–15; John Ashworth, *"Agrarians" and "Aristocrats": Party Political Ideology in the United States, 1837–1846* (New York: Cambridge University Press, 1987), 21–51.

50. Hanyan and Hanyan, *DeWitt Clinton,* 5–14; Tiffany, *Address to the Committee of Schoharie,* 1–16; Roscoe, *History of Schoharie County,* 138; Halsey, *Pioneers of Unadilla Village,* 167; J. Hammond, *History of the Political Parties,* 1:488–89, 2:1–81; Beardsley, *Reminiscences,* 216–17; W. S. Hammond, *Oration Delivered at Cherry Valley,* 14–15; Ashworth, *"Agrarians" and "Aristocrats,"* 21–51.

51. Ellis, *Landlords and Farmers,* 46, 298; Frost, *Life on the Upper Susquehanna Valley,* 39, 104–5, 125; Taylor, *William Cooper's Town,* 372–405.

Chapter 2: Democratic Children

1. Stuart B. Blakely, *A History of Otego* (Cooperstown: Christ, Scott, and Parshall, 1907), 46.

2. Horatio G. Spafford, *Gazetteer of the State of New York* (Albany: Packard and Van Benthuysen, 1824), 224.

3. Ibid., 564.

4. Ibid., 140.

5. Ibid., 452.

6. Ibid., 122, 145, 214, 315–16, 477–78, 487; David M. Ellis, *Landlords and Farmers in the Hudson-Mohawk Region, 1790–1850* (Ithaca: Cornell University Press, 1946), 184–224; James A. Frost, *Life on the Upper Susquehanna Valley, 1783–1860* (New York: King's Crown Press, 1951), 78–79.

7. William E. Roscoe, *History of Schoharie County, New York* (Syracuse: D. Mason and Co., 1882), 60; *History of Otsego County, New York* (Philadelphia: Everts and Fariss, 1878), 39.

8. Roger Hayden, ed., *Upstate Travels: British Views of Nineteenth-Century New York* (Syracuse: Syracuse University Press, 1982), 88–91.

9. *In the Court of Appeals. The People of the State of New York v. George Clarke. Opinion of the Court* (Albany: Munsell, 1854); *History of Otsego County,* 322; Levi Beardsley, *Reminiscences* (New York: Charles Vinten, 1852), 446; Edith M. Fox, *Land Speculation in the Mohawk Country* (Ithaca: Cornell University Press, 1949), 28–50.

10. Spafford, *Gazetteer of the State of New York,* 554.

11. Ibid., 405, 438–40, 502, 554–55.

12. Daniel D. Barnard, *A Discourse on the Life, Services and Character of Stephen Van Rensselaer* (Albany: Hoffman and White, 1839), 30–95.

13. James Fenimore Cooper, *The American Democrat* (Indianapolis: Liberty Press, 1981), 180–81.

14. Charles Smith to William H. Averell, September 28, 1836, Box 9, Averell Family Papers, #82, NYSHA; S. A. Law to W. Gaylord and L. Tucker, editors of the *Cultivator,* January 17, 1842, Box 7, Samuel A. Law Papers, CX10277, NYSL; *History of Otsego County,* 43; Frost, *Life on the Upper Susquehanna Valley,* 67–68, 81.

15. T. H. Wheeler to Richard Cooper, January 20, 1841, Richard Cooper Journal, 1854–55, MSS, Box 100, George Hyde Clarke Papers, #2800, CUL; Albany *Freeholder,* March 13, December 18, 1850, March 26, 1851; F. W. Beers, *History of Montgomery and Fulton Counties, New York* (New York: F. W. Beers and Co., 1878), 281, 335; Washington Frothingham, *History of Montgomery County, New York* (Syracuse: D. Mason and Co., 1892), 281, 335, 345; Ralph Birdsall, *The Story of Cooperstown* (Cooperstown: Augur's Book Store, 1948), 233–37; Ellis, *Landlords and Farmers,* 298, 308–9. Agricultural economists in the 1920s recommended the abandonment of share farming in dairying because returns were too low. Clarke may well have recognized this same problem and thus wished to combine hops and dairying together, especially since the manure from dairy cattle could be used to fertilize the demanding hop plant. See E. G. Misner, "Farm Management Problems in the Northeastern Dairy Belt," *Journal of Farm Economics* 7 (April 1925): 251–73.

16. John Kiersted Jr. to Wynkoop Kiersted, October 1, 1838, Box 1, Kiersted Family Papers, 17820, NYSL. Joseph D. Reid Jr. also noted that share farming systems are characteristic of regions in which risk and the need for entrepreneurship are low. This certainly would describe central New York in the 1830s. Joseph D. Reid Jr., "Sharecropping in History and Theory," *Agricultural History* 49 (April 1975): 426–40.

17. Ellis, *Landlords and Farmers,* 147; Albert C. Mayham, *The Anti-Rent War on Blenheim Hill: An Episode of the 40's* (Jefferson, N.Y.: Frederick L. Frazee, 1906), 20.

18. Agent's report, Stewart's Patent, Otsego County, July 1837, Box 2, Goldsbrow Banyar Papers, DU10703, NYSL; Blakely, *History of Otego*, 46.

19. Ellis, *Landlords and Farmers*, 232; Labor account: S. A. Law and Adam Hoffman, March 15, 1844, Law Papers.

20. Harry L. Watson, *Liberty and Power: The Politics of Jacksonian America* (New York: Hill and Wang, 1990), 42–72.

21. Conceiving of Jacksonian democracy in New York as rooted in a localist democratic political economy may help reconcile the debate between those who argue that the Jacksonians spoke for the working class or "men on the make." See Arthur M. Schlesinger Jr., *The Age of Jackson* (Boston: Little, Brown and Co., 1945); Marvin Meyers, *The Jacksonian Persuasion: Politics and Belief* (Stanford: Stanford University Press, 1957); Walter Hugins, *Jacksonian Democracy and the Working Class: A Study of the New York Workingmen's Movement, 1829–1837* (Stanford: Stanford University Press, 1960); and John Ashworth, *"Agrarians" and "Aristocrats": Party Political Ideology in the United States, 1837–1846* (New York: Oxford University Press, 1987).

22. Major L. Wilson, "What Whigs and Jacksonian Democrats Meant by Freedom," in *The Many-Faceted Jacksonian Era*, ed. Edward Pessen (Westport, Conn.: Greenwood Press, 1977), 192–211; Lee Benson, *The Concept of Jacksonian Democracy: New York as a Test Case* (Princeton: Princeton University Press, 1961), 86–109, 288–328; Ashworth, *"Agrarians" and "Aristocrats,"* 17–73; Daniel Walker Howe, *The Political Culture of American Whigs* (Chicago: University of Chicago Press, 1979), 8–9, 96–122; Douglas T. Miller, *Jacksonian Aristocracy: Class and Democracy in New York, 1830–1860* (New York: Oxford University Press, 1967), 1–80.

23. Edwin Williams, ed., *New York Annual Register, 1840* (New York: Edwin Williams, 1840), 368–69; Edwin Williams, ed., *New York Annual Register, 1843* (New York: J. Disturnell, 1843); Delaware *Gazette*, November 23, 1836; James R. Sharp, *The Jacksonians versus the Banks: Politics in the States after the Panic of 1837* (New York: Columbia University Press, 1970).

24. Paul E. Johnson, *A Shopkeeper's Millennium: Society and Revivals in Rochester, New York, 1815–1837* (New York: Hill and Wang, 1978), 79–94, 116–41; Mary P. Ryan, *The Cradle of the Middle Class: The Family in Oneida County, New York, 1790–1865* (New York: Cambridge University Press, 1981), 60–144; Watson, *Liberty and Power*, 179, 188–93.

25. Financial Statement: William C. Bouck, April 8, 1835, Robert McClellan to William C. Bouck, January 6, 1839, John I. Mumford to William C. Bouck, August 31, 1842, Vincent Quackenbush to William C. Bouck, October 25, 1842, Box 1, William C. Bouck Papers, #2206, CUL; Mitchell Sanford to Lyman Sanford, August 27, 1838, Box 1, Folder 12, Lyman Sanford Papers, BY462, AIHA; Thomas F. Gordon, *Gazetteer of the State of New York* (Philadelphia: Thomas F. Gordon, 1836), 156–68; Roscoe, *History of Schoharie County*, 62, 67–68, 82, 104, 108, 110; Jabez D. Hammond, *The History of the Political Parties of the State of New-York* (Syracuse: Hall, Miles, and Co., 1852), 1:563–64; Frost, *Life on the Upper Susquehanna Valley*, 42–57.

26. A sense of this culture can be gained from Glenn C. Altschuler and Stuart M. Blumin, "Limits of Political Engagement in Antebellum America: A New Look at the Golden Age of Participatory Democracy," *JAH* 84 (1997): 855–85. My findings suggest even greater instability in the political system than they discovered.

27. Eldridge H. Pendleton, "The New York Anti-Rent Controversy, 1830–1860" (Ph.D. diss., University of Virginia, 1974), 164–67; Charles Hathaway to ———, July 12, 1844, MSS #13702, NYSL; John D. Monroe, *The Anti-Rent War in Delaware County, New York: The Revolt Against the Rent System* (Delhi, N.Y.: John D. Monroe, 1940), 17–23; Henry Christman, *Tin Horns and Calico: An Episode in the Emergence of American Democracy* (New York: Henry Holt, 1945), 155; Ellis, *Landlords and Farmers,* 147–48; S. A. Law to Erastus Root, January 30, 1840, R. W. Morgan to S. A. Law, February 4, February 5, 1840, S. A. Law to Orson M. Allaben and Erastus Root, February 7, 1840, S. A. Law to Thomas Clymer, March 3, 1840, E. G. Barnes to S. A. Law, September 2, 1840, Repudiation Notice of Stockholders of the Meredith Turnpike Company, ca. 1843, Box 6, S. A. Law to Erastus Root, February 14, 1842, Ira Emmons to S. A. Law, December 28, 1842, Plaister Trippen to S. A. Law, February 23, 1843, Box 7, Law Papers.

28. Recollections of Hon. George Anson Starkweather, MSS, Morgan Family Papers, #38, NYSHA; Levi C. Turner Diary, July 20, 1839, typescript, NM25.55, NYSHA.

29. William C. Bouck to Anonymous, April 2, 1840, Box 1, Bouck Papers.

30. D. M. Hard to William H. Averell, July 23, 1838, Box 6, Averell Family Papers.

31. Sumner Ely to Joseph Peck, March 14, 1840, Box 1, Peck Family Papers, #230, NYSHA; Recollections of Hon. George Anson Starkweather.

32. John Sawyer, *History of Cherry Valley from 1741 to 1898* (Cherry Valley: Gazette Printing, 1898), 98–99; *History of Otsego County,* 124–25; Beardsley, *Reminiscences,* 445–50; Frost, *Life on the Upper Susquehanna Valley,* 105–7.

33. *The Centennial Celebration at Cherry Valley, Otsego Co., N.Y., July 4th, 1840* (New York: Taylor and Clement, 1840), 6–31; *History of Otsego County,* 280; Benson, *Concept of Jacksonian Democracy,* 165–85; J. Hammond, *History of the Political Parties,* 2:29.

34. Williams, *New York Annual Register, 1840,* 368–69; Frost, *Life on the Upper Susquehanna Valley,* 91, 96, 107–9; Eugene D. Genovese, *Roll, Jordan, Roll: The World the Slaves Made* (New York: Vintage, 1973), 5–6, 27–28. Williams's *New York Annual Register* indicated that the three counties lagged substantially behind other areas of the state economically. While the average value of an acre of land in the state in 1839 was $18.80, the Otsego average was $8.16, Schoharie's $5.32, and Delaware's $3.52. This created a paradoxical situation. On the one hand, taxes were low enough so that marginal farmers could survive. On the other, however, the low values made it difficult to obtain credit to improve farms, and once a farmer got into debt, he might find himself unable to liquidate the debt by selling out his farm. As many observers noted, this dampened enterprise. This was a situation that most farmers seemed willing to accept, but merchants voiced increasing frustration with the constant underdevelopment of the region.

35. Edwin Williams, ed., *New-York Annual Register, 1836* (New York: Edwin Williams, 1836), 86–89; John D. Monroe, *Chapters in the History of Delaware County, New York* (Delhi, N.Y.: DCHA, 1949), 98–118; Beardsley, *Reminiscences,* 224–29; Benson, *Concept of Jacksonian Democracy,* 64, 288–328; Watson, *Liberty and Power,* 231–53; Pendleton, "New York Anti-Rent Controversy," 164–67; Monroe, *Anti-Rent War in Delaware County,* 17–23; Christman, *Tin Horns and Calico,* 155.

36. Robert Eldredge to William C. Bouck, March 24, 1834, Box 1, Bouck Papers.

37. Edwin Williams, ed., *The New York Annual Register for the Year of Our Lord 1837* (New York: G. and C. Carvill and Co., 1837), 40–44.

38. Benson, *Concept of Jacksonian Democracy,* 150–54; Cheyney, *Anti-Rent Agitation,* 23–24; James Fenimore Cooper, *The Redskins* (Boston: Houghton and Mifflin, n.d.), 30–31; Timothy Corbin Jr. to William C. Bouck, August 15, 1844, Box 1, Bouck Papers; Mayham, *Anti-Rent War,* 43.

39. Thomas Skidmore, *The Rights of Man to Property* (reprint, New York: Burt Franklin, 1976), 3–77; Ashworth, *"Agrarians" and "Aristocrats,"* 21–51; Sean Wilentz, *Chants Democratic: New York City and the Rise of the American Working Class, 1788–1850* (New York: Oxford University Press, 1984), 102–3, 330.

40. Marcus Clarke to Mrs. William Clarke, September 9, 1840, Daniel A. Park to Norton T. Collins, May 16, 1841, Daniel A. Park to Clarissa Park, May 16, 1841, Lease: Thomas and John Clarke to Joshua Clarke, March 24, 1843, William Frove to Joseph Frove, December 4, 1845, Box 1, Clarke Family Papers, #115, NYSHA.

41. Mayham, *Anti-Rent War,* 12, 15, 18, 21, 23–32.

42. George Potts to William B. Campbell, September 2, 1836, Box 1, Campbell Family Papers, #313, NYSHA.

43. One of many examples is Thomas Wing to William C. Bouck, February 1, 1840, Box 1, Bouck Papers. This was a double-edged sword, and farmers sometimes found themselves foreclosed upon by their benefactors. Peter Shaver to Freeman Shaver, February 6, 1840, Shaver Family Papers, Box S, NYSHA.

44. David Morse to Lucas Elmendorf, March 14, 1836, Schoharie County Legal Papers, NYSHA.

45. Henry Conklin, *Through "Poverty's Vale": A Hardscrabble Boyhood in Upstate New York, 1832–1862* (Syracuse: Syracuse University Press, 1974), 27.

46. Albany *Freeholder,* January 5, 1848; John A. Garraty, *Silas Wright* (New York: Columbia University Press, 1949), 319; Christman, *Tin Horns and Calico,* 3–14, 21.

47. Williams, *New York Annual Register . . . 1837,* 366; O. L. Holley, ed., *The New-York State Register for 1844* (New York: J. Disturnell, 1844); O. L. Holley, ed., *The New-York State Register for 1845* (New York: J. Disturnell, 1845), 60–68; Conklin, *Through "Poverty's Vale,"* xii–xvi, 77, 103; Ryan, *Cradle of the Middle Class,* 18–59; Joan M. Jensen, *Loosening the Bonds: Mid-Atlantic Farm Women, 1750–1850* (New Haven: Yale University Press, 1986), 36–128; Ellis, *Landlords and Farmers,* 106, 147–50, 160–65, 184–224; Benson, *Concept of Jacksonian Democracy,* 158–59. Craftsmen constituted the second highest percentage in each county—12 percent in Schoharie, 15 percent in Delaware, 19 percent in Otsego. The three counties had 128 tanneries and 746 sawmills in 1835. But by 1840, 21 tanneries and 142 sawmills closed, leaving many men without a vital source of income.

48. Conklin, *Through "Poverty's Vale,"* xii–xvi, 77; Ellis, *Landlords and Farmers,* 184–224; Ryan, *Cradle of the Middle Class,* 18–59; Jensen, *Loosening the Bonds,* xiii–xv. The children earned a fee, such as $100, at the close of the term, which was about four years. Four of Mary's sons and daughters worked under this system in 1845.

49. Williams, *New York Annual Register . . . 1837,* 366; Rolla Tryon, *Household Manufactures in the United States, 1640–1860* (Chicago: University of Chicago Press, 1917), 288–311; Thomas Dublin, "Women and Outwork in a Nineteenth-Century New England Town: Fitzwilliam, New Hampshire, 1830–1850," in *The Countryside in the Age of Capitalist Transformation: Essays in the Social History of Rural America,* ed. Steven Hahn and

Jonathan Prude (Chapel Hill: University of North Carolina Press, 1985), 51–69; Jensen, *Loosening the Bonds,* 79–128.

50. Phoenix Cotton Manufactory Financial Statement, undated (ca. 1832), Box 9, Averell Family Papers.

51. Union Cotton Manufactory Financial Statement, March 27, 1832, Phoenix Cotton Manufactory Financial Statement, undated (ca. 1832), Charles Smith to William H. Averell, January 25, 1836, Charles Smith to William H. Averell, September 28, 1836, Box 9, Averell Family Papers; By-laws of the Laurens Bag Mill, May 7, 1849, Box 1, Strong Family Papers, #44, NYSHA; Williams, *New-York Annual Register, 1836,* 359; Thomas Dublin, *Women at Work: The Transformation of Work and Community in Lowell, Massachusetts, 1826–1860* (New York: Columbia University Press, 1979), 1–74. Otsego County had seven cotton factories located south of Otsego Lake in Middlefield and Hartwick, two in Butternuts, and one in Pittsfield.

52. James M. Wilson to Freeman Shaver, April 15, 1840, Shaver Family Papers; John Mack Faragher, *Sugar Creek: Life on the Illinois Prairie* (New Haven: Yale University Press, 1986), chap. 8. Central New York and southern Illinois bore a striking resemblance to each other in this regard.

53. Levi C. Turner Diary, April 3, 1839, MSS, NM25.55, NYSHA.

54. Morris Bishop, ed., "The Journeys of Samuel J. Parker," *New York History* 45 (April 1964): 145.

55. Peter Shaver to Freeman Shaver, February 6, 1840, Misc. Manuscripts, Shaver Family Papers. Peter Shaver tried to get his son, Freeman, to plant more wheat on his Ohio farm to help pay the costs of Peter's mortgage to Mr. Williams. He also considered having his brother-in-law take a half interest in the farm to keep Williams from seizing it.

56. Conklin, *Through "Poverty's Vale,"* 103.

57. Labor contract: S. A. Law and John Crayton, June 19, 1840, Labor contract: S. A. Law to John McKeever, March 18, 1843, Labor contract: S. A. Law and William Craig, March 25, 1844, Labor account: S. A. Law and Edward Butler, August 14, 1843, Labor account: S. A. Law and J. Ingraham, May 16, 1843, Receipt: S. A. Law Estate to G. P. and William Snyder, February 13, 1845, Law Papers; Frost, *Life on the Upper Susquehanna Valley,* 86; David E. Schob, *Hired Hands and Plowboys: Farm Labor in the Midwest, 1815–60* (Urbana: University of Illinois Press, 1975), 209–33.

58. Labor contract: S. A. Law and Ornan Crane, July 10, 1840, Labor contract: S. A. Law and Ornan Crane, July 16, 1841, Law Papers.

59. S. A. Law to Francis R. and Charles Wharton, April 10, 1841, Anonymous to Wharton Family, December 14, 1844, S. A. Law to Francis and William Wharton, n.d., Box 7, Law Papers. Quote is in first letter cited. Gates noted that timber-hooking from absentee landlords in the west was also common. Owners found that negotiation produced much better results than recourse to the law, which invariably met stiff resistance. Paul W. Gates, *Landlords and Tenants on the Prairie Frontier* (Ithaca: Cornell University Press, 1973), 240, 248. Arguments over timber were extremely common among all classes. In a case between relative equals, Ira Emmons sued Jared Goodyear Jr., who later acquired extensive landholdings in Otsego County, and others for cutting timber on his Milford property. Timber, water, and fishing rights were jealously guarded, and the courts brimmed with suits over these and other property issues. Otsego County Court of Common Pleas, Complaint,

Ira Emmons v. Jared Goodyear Jr., William V. White, Christopher Teal, and Richard Every, December 14, 1835, Box 3, Robert Campbell Papers, #100, NYSHA.

60. Meyers, *Jacksonian Persuasion*, 60–95; J. Hammond, *History of the Political Parties*, 2:492–94, 500–502.

61. L. Ray Gunn, *The Decline of Authority: Public Economic Policy and Political Development in New York State, 1800–1860* (Ithaca: Cornell University Press, 1988), 144–69.

62. Ibid.

63. Francis W. Halsey, ed., *The Pioneers of Unadilla Village, 1784–1840 and Reminiscences of Village Life and of Panama and California from 1840 to 1850* (Unadilla, N.Y.: Vestry of St. Martin's Church, 1902), 167.

64. Ellis, *Landlords and Farmers*, 147–48; S. A. Law to Erastus Root, January 30, 1840, R. W. Morgan to S. A. Law, February 4, February 5, 1840, S. A. Law to Orson M. Allaben and Erastus Root, February 7, 1840, S. A. Law to Thomas Clymer, March 3, 1840, E. G. Barnes to S. A. Law, September 2, 1840, Repudiation Notice of Stockholders of the Meredith Turnpike Company, ca. 1843, Box 6, S. A. Law to Erastus Root, February 14, 1842, Ira Emmons to S. A. Law, December 28, 1842, Plaister Trippen to S. A. Law, February 23, 1843, Box 7, Law Papers. For examples of other local elites' involvement in similar projects, see Charles Smith to William H. Averell, February 16, 1836, Box 6, Averell Family Papers; Grant B. Palmer to Robert Campbell, September 17, 1839, Box 3, Robert Campbell Papers; Amos H. Brown to William B. Campbell, May 1, 1841, David Gilbert to William B. Campbell, May 10, 1841, Box 1, Campbell Family Papers, #313; Albany *Freeholder*, March 6, 1850; and Caroline Evelyn More and Irma Mae Griffin, *The History of the Town of Roxbury* (Walton, N.Y.: The Reporter Co., 1953), 128–29.

65. E. G. Barnes to S. A. Law, February 10, 1841, S. A. Law to Coleman and Samuel F. Fischer, August 1, 1844, Box 7, Law Papers.

66. G. Burr to Erastus Root, January 5, 1843, ibid.

67. S. A. Law to Francis Wharton, July 30, 1844, ibid.

68. *Report of the Railroads Committee upon Several Petitions for Legislative Aid to the Canajoharie and Catskill Railroad, 1838* (reprint, Cornwallville, N.Y.: Hope Farm Press, 1973), 6–16; Oneonta *Herald*, September 5, 1895; William C. Bouck to the directors of the Canajoharie & Catskill Railroad Company, March 13, 1841, Peter Osterhout to William C. Bouck, April 9, 1841, William C. Bouck to Peter Osterhout, April 10, 1841, Bouck Papers; Mitchell Sanford to Lyman Sanford, August 15, 1836, Box 1, Folder 6, Mitchell Sanford to Lyman Sanford, April 5, 1838, Box 1, Folder 11, Sanford Papers.

69. Gabriel Bouck to Lyman Sanford, April 17, 1848, Box 4, Folder 4, Sanford Papers.

70. Frost, *Life on the Upper Susquehanna Valley*, 76.

71. Beardsley, *Reminiscences*, 427–28.

72. James M. Wilson to Freeman Shaver, April 15, 1840, Shaver Family Papers.

73. Walter B. Smith and Arthur H. Cole, *Fluctuations in American Business, 1790–1860* (Cambridge: Harvard University Press, 1935), 41–49, 87–90; Sawyer, *History of Cherry Valley*, 112.

74. George W. Reynolds Diary, December 20, 1842, MSS, NYSHA.

75. Revivalism struck the area later than the Burned-Over District, stretching between 1835 and 1845, with Methodism drawing strongly in future Anti-Rent districts. See Anonymous to Friend Fripp, September 9, 1835, Minutes of Methodist Episcopal Church Trial

of Elisha Brown of Summit, Schoharie County, March 25, 1842, Box 1, Lake Family Papers, #64, NYSHA; G. A. Litner Diary, January 15, 1843, MSS, Litner Family Papers, KI13168, NYSL; Williams, *New-York Annual Register, 1836,* 449; Williams, *New York Annual Register . . . 1837,* 466; George W. Reynolds Diary, January 7, March 12, 1837; Conklin, *Through "Poverty's Vale,"* 12–13, 39, 62; Mayham, *Anti-Rent War,* 46–47, 63.

76. J. Hammond, *History of the Political Parties,* 2:428–537; Benson, *Concept of Jacksonian Democracy,* 96–97.

77. Gunn, *Decline of Authority,* 171–72.

78. D. M. Hard to William H. Averell, October 16, 1837, Henry Ogden to William H. Averell, October 7, 1837, Box 5, Averell Family Papers.

79. Mitchell Sanford to William C. Bouck, October 26, 1837, Box 1, Bouck Papers.

80. Mitchell Sanford to Lyman Sanford, November 8, 1838, Box 1, Folder 13, Sanford Papers.

81. Anonymous to William C. Bouck, February 19, 1840, Box 1, Bouck Papers.

82. Henry Ogden to William H. Averell, October 7, 1837, Box 6, Averell Family Papers.

83. New Lisbon Town Resolution, March 7, 1837, Harvey Strong to Joseph Peck, March 13, 1837, C. Jones to Joseph Peck, March 19, 1839, B. H. Marks to Joseph Peck, August 19, 1840, Special New Lisbon Town Meeting Resolutions, January 28, 1841, J. S. Sprague to Joseph Peck, September 20, 1843, G. K. Noble to Joseph Peck, October 14, 1845, Box 1, Peck Family Papers; "Report of the Surveyor General on the Petition of Sundry Inhabitants of the Town of Oneonta, &c.," *NYSAD,* no. 107 (1840), "Communication from the Surveyor General, O. L. Holley, on the Petition of Sundry Inhabitants of the Town of Oneonta, &c.," *NYSAD,* no. 124 (1840), "Report of the Committee on Towns and Counties Relative to Certain Lots in Wallace's Patent," *NYSAD,* no. 184 (1840), "Report of the Committee of the Erection and Division of Towns and Counties on the Petition of Inhabitants of Otsego, Delaware, and Otsego Counties," *NYSAD,* no. 242 (1840); Roscoe, *History of Schoharie County,* 56–58.

84. Special New Lisbon Town Meeting Resolutions, January 28, 1841, Harvey Hunt to Joseph Peck, November 4, 1843, Box 1, Peck Family Papers.

85. S. Crippen to Seth Doubleday, February 8, 1841, Box 6, Averell Family Papers.

86. Halsey Spencer to Joseph Peck, September 23, 1845, Box 1, Peck Family Papers.

87. Levi C. Turner Diary, June 6, September 11, September 14, September 15, September 16, September 17, September 20, October 28, November 4, November 5, November 6, 1839.

Chapter 3: The Anti-Renters

1. Epigraph, "Tar and Feather Letter," Anonymous to G. H. Edgerton, July 18, 1844, MSS 1979-114, Delaware County Historical Association. These interpretations appear respectively in Henry Christman, *Tin Horns and Calico: An Episode in the Emergence of American Democracy* (New York: Henry Holt, 1945); David M. Ellis, *Landlords and Farmers in the Hudson-Mohawk Region, 1790–1850* (Ithaca: Cornell University Press, 1946); Reeve Huston, *Land and Freedom: Rural Society, Popular Protest, and Party Politics in Antebellum New York* (New York: Oxford University Press, 2000); and Charles W. McCurdy, *The Anti-Rent Era in New York Law and Politics, 1839–1865* (Chapel Hill: University of North Carolina Press, 2001).

2. Unlabeled newspaper clipping, December 1839, Box 5, Folder 14, Henry Christman Papers, #10, NYSHA; Andrew J. Colvin, "Anti-Rentism in Albany County," in *History of*

Albany County, N.Y., from 1609 to 1886, ed. George R. Howell (New York: W. W. Munsell, 1886), 278; Edward P. Cheyney, *The Anti-Rent Agitation in the State of New York* (Philadelphia: University of Pennsylvania Press, 1887), 25–27; Christman, *Tin Horns and Calico,* 35, 38–39, 50; Ellis, *Landlords and Farmers,* 234; James Fenimore Cooper, *The Redskins* (Boston: Houghton and Mifflin, n.d.), vii, 28. For an excellent discussion of the phenomenon of "white Indians" in American history, see Philip J. DeLoria, *Playing Indian* (New Haven: Yale University Press, 1998).

3. Christman, *Tin Horns and Calico,* 30; Albany *Freeholder,* October 8, 1845; Colvin, "Anti-Rentism," 278; Eldridge H. Pendleton, "The New York Anti-Rent Controversy, 1830–1860" (Ph.D. diss., University of Virginia, 1974), 15, 27–28, 83–84.

4. "Anti-Renters' Declaration of Independence, July 4, 1839," in *We the Other People: Alternative Declarations of Independence by Labor Groups, Farmers, Woman's Rights Advocates, Socialists, and Blacks, 1829–1976*, ed. Philip S. Foner (Urbana: University of Illinois Press, 1976), 60–62.

5. Quote from Christman, *Tin Horns and Calico,* 42. The Anti-Renters showed important parallels to the "Windsor Blacks." Thompson said, "Blacking arose in response to the attempted reactivation of a relaxed forest authority." A similar case could be made for Van Rensselaer's "reactivation" of rent payments. See E. P. Thompson, *Whigs and Hunters: The Origin of the Black Acts* (New York: Pantheon, 1975), 55. For a concise discussion of the "Helderberg War" of December 1839, as the confrontation between militiamen and Anti-Renters was called, see Huston, *Land and Freedom,* 92–94.

6. Albert C. Mayham, *The Anti-Rent War on Blenheim Hill: An Episode of the 40's* (Jefferson, N.Y.: Frederick L. Frazee, 1906), 73.

7. Manuscript United States Census, 1850, #432, Schedule I, Delaware County, Rolls 494, 495, Otsego County, Rolls 579, 580, Schoharie County, Rolls 595, 596, National Archives, Washington, D.C.; Mayham, *Anti-Rent War,* 75–76; Statement of lands in Margaretville NLY 1/2 of Great Lot No. 39, Hardenburgh Patent Belonging to Margaret Livingston, February 7, 1848, MSS.1984-340.21c, DCHA; George Clarke Account with Richard Cooper, 1850, MSS, George Clarke Rent Rolls, Cherry Valley and Long Patents, 1851, MSS, Box 45, George Hyde Clarke Papers, #2800, CUL. For an analysis of the relationship between family size and farming strategies, see Lee A. Craig, *To Sow One Acre More: Childbearing and Farm Productivity in the Antebellum North* (Baltimore: Johns Hopkins University Press, 1993), 53–73.

8. Manuscript United States Census, 1850, #432, Schedule I, Delaware County, Rolls 494, 495, Otsego County, Rolls 579, 580, Schoharie County, Rolls 595, 596, National Archives; Mayham, *Anti-Rent War,* 75–76; Statement of lands in Margaretville NLY 1/2 of Great Lot No. 39, Hardenburgh Patent Belonging to Margaret Livingston, February 7, 1848, MSS.1984-340.21c, DCHA; George Clarke Account with Richard Cooper, 1850, MSS, George Clarke Rent Rolls, Cherry Valley and Long Patents, 1851, MSS, Box 45, Clarke Papers.

9. Manuscript United States Census, 1850, #432, Schedule I, Schoharie County, Rolls 595, 596, National Archives; Mayham, *Anti-Rent War,* 12, 15, 18, 21–23, 75–76; William E. Roscoe, *History of Schoharie County, New York* (Syracuse: D. Mason and Co., 1882), 116–17, 262; Cheyney, *Anti-Rent Agitation,* 38–39; Thomas S. Peaslee Methodist Episcopal Certificate, #3-58, NYSHA. Among the calico Indian chiefs, Henry A. Cleaveland was thirty-

nine years old in 1850, was married, had two children, and owned $700 worth of real property. Christopher Decker, fifty-four years old, listed his occupation as farmer in 1850, had no real property, and lived with his married brother Lewis, who had four children and no real estate. John McIntyre was thirty-eight years old, was married, had his eighty-eight-year-old father, Jacob McIntyre, and six children living at home, and listed himself as a farmer with $1,100 worth of property. All were born in New York.

10. Pendleton, "New York Anti-Rent Controversy," 177–212, 248–85.

11. Albany *Freeholder,* January 7, 1846; Cheyney, *Anti-Rent Agitation,* 38; Mayham, *Anti-Rent War,* 15, 18, 28–33; Christman, *Tin Horns and Calico,* 91–93.

12. Mayham, *Anti-Rent War,* 33; Christman, *Tin Horns and Calico,* 91–93; Albany *Freeholder,* January 7, 1846. I discuss in detail the meaning and origins of calico Indian symbolism in Thomas Summerhill, "The Farmers' Republic: Agrarian Protest and the Capitalist Transformation of Upstate New York, 1840–1900" (Ph.D. diss., University of California, San Diego, 1993), 38–70.

13. Cheyney, *Anti-Rent Agitation,* 9–40; Pendleton, "New York Anti-Rent Controversy," 81–82; Deloria, *Playing Indian,* 38–70.

14. Message from Governor William H. Seward to the Senate and Assembly of the State of New York, January 7, 1840, *NYSAD,* no. 2 (1840).

15. S. A. Law to Erastus Root, May 3, 1840, Box 7, Samuel A. Law Papers, CX10277, NYSL.

16. Levi C. Turner Diary, September 28, 1842, MSS, NM25.55, NYSHA; Peter Shaver to Freeman Shaver, February 6, 1840, Alfred Shaver to Freeman Shaver, January 1, 1841, Shaver Family Papers, Box S, NYSHA; Levi Beardsley to William H. Averell, February 2, 1841, Box 5, Averell Family Papers, #82, NYSHA; McCurdy, *Anti-Rent Era,* 54–55.

17. Ellis, *Landlords and Farmers,* 225–67.

18. Edward A. Danscourt to William C. Bouck, December 25, 1843, Box 1, William C. Bouck Papers, #2206, CUL; Thomas A. Devyr, *Odd Book of the Nineteenth Century, or, "Chivalry" in Modern Days, a Personal Record of Reform—Chiefly Land Reform, for the Last Fifty Years* (Greenpoint, N.Y.: Thomas A. Devyr, 1882), 42; Christman, *Tin Horns and Calico,* 72–74; Ellis, *Landlords and Farmers,* 251.

19. McCurdy, *Anti-Rent Era,* 80–95; Robert D. Dorr to William C. Bouck, November 7, 1842, Box 1, Bouck Papers; Lee Benson, *The Concept of Jacksonian Democracy: New York as a Test Case* (Princeton: Princeton University Press, 1961), 86–109; L. Ray Gunn, *The Decline of Authority: Public Economic Policy and Political Development in New York State, 1800–1860* (Ithaca: Cornell University Press, 1988), 137, 144–69.

20. Christman, *Tin Horns and Calico,* 64, 75–76; Ellis, *Landlords and Farmers,* 251; Pendleton, "New York Anti-Rent Controversy," 45; Presentation of the *Helderberg Advocate* before the Schoharie County Grand Jury, May 13, 1842, #16852, NYSL; Roscoe, *History of Schoharie County,* 82.

21. J. W. Edmonds to William C. Bouck, August 22, 1842, Box 1, Bouck Papers; Christman, *Tin Horns and Calico,* 215–16.

22. Christman, *Tin Horns and Calico,* 64, 75–76; Ellis, *Landlords and Farmers,* 251; Pendleton, "New York Anti-Rent Controversy," 45; Presentation of the *Helderberg Advocate.*

23. Christman, *Tin Horns and Calico,* 64, 75–76; Ellis, *Landlords and Farmers,* 251; Pendleton, "New York Anti-Rent Controversy," 45; Presentation of the *Helderberg Advocate.*

24. Devyr, *Odd Book,* 42–43.
25. H. G. Munger, "Reminiscences of the Anti-Rent Rebellion," MSS, NYSHA.
26. Albany *Freeholder,* May 14, 1845.
27. Albany *Anti-Renter,* January 31, 1846.
28. Ibid., February 21, 1846.
29. J. Houck Jr. to William C. Bouck, May 2, 1842, Box 1, Bouck Papers. Chatfield opposed the control of Cooperstown politicians of both parties over county politics. One Laurens Whig asked William H. Averell to support his 1841 candidacy to run for assembly against Chatfield, who had become, he charged, self-important and unmindful of Cooperstown's interests. D. D. Comstock to William H. Averell, September 26, 1841, Box 6, Averell Family Papers.
30. George A. Starkweather to William C. Bouck, November 25, 1842, Box 1, Bouck Papers.
31. John B. Steele to William C. Bouck, November 2, 1842, ibid.
32. Henry Ogden to William H. Averell, October 7, 1837, D. M. Hard to William H. Averell, October 16, 1837, William W. Campbell to William H. Averell, November 14, 1837, Dewitt C. Bates to William H. Averell, January 17, 1840, Thurlow Weed to William H. Averell, September 27, 1840, Jacob Livingston to William H. Averell, October 13, 1840, Whig State Central Committee to William H. Averell, October 29, 1842, Box 6, Averell Family Papers.
33. William C. Bouck to Lyman Sanford, August 2, 1844, Box 8, Folder 14, Lyman Sanford Papers, BY462, AIHA.
34. Christman, *Tin Horns and Calico,* 81–82, 91; Ellis, *Landlords and Farmers,* 244–46.
35. Mayham, *Anti-Rent War,* 29–30; Smith A. Boughton, "Autobiography," typescript, Christman Papers; Ellis, *Landlords and Farmers,* 252.
36. Ellis, *Landlords and Farmers,* 244–47, 252–56; Charles Hathaway to ———, July 12, 1844, MSS #13702, NYSL; Christman, *Tin Horns and Calico,* 83–94, 97; Matthew Griffin Diary, July 25, 1845, MSS, DCHA; Jay Gould, *History of Delaware County* (Roxbury, N.Y.: Keeney and Gould, 1856), 260–61.
37. Both quoted in Christman, *Tin Horns and Calico,* 88–89.
38. Ibid., 90.
39. Munger, "Reminiscences of the Anti-Rent Rebellion"; Christman, *Tin Horns and Calico,* 102; Ellis, *Landlords and Farmers,* 247–48.
40. S. A. Law to Francis R. Wharton, July 30, 1844, Box 7, Law Papers.
41. Timothy Corbin Jr. to William C. Bouck, July 12, 1844, Box 1, Bouck Papers.
42. Ibid., July 13, August 7, August 15, 1844.
43. Munger, "Reminiscences of the Anti-Rent Rebellion."
44. Devyr, *Odd Book,* xv–xvi.
45. Ibid., xvii–xviii, 146; Ellis, *Landlords and Farmers,* 252–54.
46. Seymour Boughton, Abraham Van Tuyl, Samuel Baldwin, and I. W. Baird to William C. Bouck, August 15, 1844, Box 1, Bouck Papers; Albany *Freeholder,* October 1, 1845.
47. Demosthenes Lawyer to William C. Bouck, July 8, 1844, Box 1, Bouck Papers; Robert Salton, "The Anti-Rent War in Delaware County," *Delaware County History* 4 (Fall 1970): 26–27; Benson, *Concept of Jacksonian Democracy,* 125, 137, 267.
48. John A. Garraty, *Silas Wright* (New York: Columbia University Press, 1949), 319–22,

357; Christman, *Tin Horns and Calico,* 111, 117–47; Ellis, *Landlords and Farmers,* 260–61; Benson, *Concept of Jacksonian Democracy,* 137, 234–35.

49. Schoharie *Patriot,* February 1, February 22, 1845; Albany *Argus,* January 31, 1845; Christman, *Tin Horns and Calico,* 145–47.

50. Charles Hathaway to James Dexter, March 12, 1845, #12385, NYSL; Munger, "Reminiscences of the Anti-Rent Rebellion"; Gould, *History of Delaware County,* 267–72; John D. Monroe, *The Anti-Rent War in Delaware County, New York: The Revolt Against the Rent System* (Delhi, N.Y.: John D. Monroe, 1940), 20; Mayham, *Anti-Rent War,* 49; Ellis, *Landlords and Farmers,* 265; Christman, *Tin Horns and Calico,* 149, 154–60; Pendleton, "New York Anti-Rent Controversy," 145–47.

51. "Excerpts from the Diary of Mary Stockton St. John," in *The Story of Walton,* ed. Helen Lane (Walton, N.Y.: Walton Historical Society, 1975), 40; Monroe, *Anti-Rent War in Delaware County,* 20–21; Mayham, *Anti-Rent War,* 44–47; Christman, *Tin Horns and Calico,* 154–57; Matthew Griffin Diary, July 25, 1845; Pendleton, "New York Anti-Rent Controversy," 147–48.

52. G. A. Litner Journal, March 30, 1845, MSS, NYSL; Christman, *Tin Horns and Calico,* 152–54; Ellis, *Landlords and Farmers,* 264; Mayham, *Anti-Rent War,* 34–37, 80–83.

53. Mayham, *Anti-Rent War,* 37.

54. Quoted in Christman, *Tin Horns and Calico,* 171.

55. Ibid., 167–72, 180–81.

56. Albany *Freeholder,* October 8, 1845; Gould, *History of Delaware County,* 280–82; Report of Judge Amasa J. Parker Jr. to the New York State Senate on Penalties of Anti-Renters, February 16, 1846, #10907, NYSL; Matthew Griffin Diary, November 19, 1845; Monroe, *Anti-Rent War in Delaware County,* 43–51.

57. G. H. Noble to Joseph Peck, October 14, 1845, Box 1, Peck Family Papers, #230, NYSHA.

58. Albany *Freeholder,* October 15, October 22, 1845; Schoharie *Republican,* September 23, 1845.

59. Matthew Griffin Diary, September 20, November 9, 1845. The Law and Order faction nominated Griffin for town supervisor in September after the resignation of the Anti-Rent incumbent. He thought the party enjoyed a majority of 70 or more votes. If Griffin was correct, it suggests that in mid-September, a number of Anti-Rent supporters would have voted Law and Order, since a 70-plus majority was much different from the 358 to 107 drubbing the Anti-Renters handed Law and Order in November.

60. "THIRTEEN GOOD REASONS Why no honest WHIG should vote the 'Equal Rights' or Anti Rent ticket" (Broadside 1845), Box 28, Folder 1, Law Papers.

61. Thomas Lawyer to William C. Bouck, October 24, 1845, Box 1, Bouck Papers.

62. Albany *Freeholder,* October 22, 1845.

63. Ibid., May 21, 1845.

64. Ibid., October 22, 1845.

65. Ibid.

66. Cooperstown *Freeman's Journal,* October 18, October 25, October 25, November 1, November 8, November 22, 1845; Halsey Spencer to Joseph Peck, September 23, 1845, Box 1, Peck Family Papers.

67. Albany *Freeholder,* November 19, December 3, 1845.

68. Ibid., November 12, November 19, 1945; Demosthenes Lawyer to William C. Bouck, December 23, 1845, Box 1, Bouck Papers.

69. Christman, *Tin Horns and Calico,* 258–59, 263; Mayham, *Anti-Rent War,* 73; Gunn, *Decline of Authority,* 170–97; Matthew Griffin Diary, January 29, 1846.

70. Thomas Lawyer to Lyman Sanford, February 9, 1846, Box 3, Folder 7, Freeman Stanton to Lyman Sanford, January 22, 1846, Thomas Lawyer to Lyman Sanford, February 18, 1846, N. T. Rosseter to Lyman Sanford, February 23, 1846, Box 3, Folder 9, Thomas Lawyer to Lyman Sanford, February 13, 1845, Box 8, Folder 4, Sanford Papers.

71. N. T. Rosseter to Lyman Sanford, February 23, 1846, Box 3, Folder 9, ibid.

72. Minutes of the Blenheim Hill Anti-Rent Meeting, September 12, 1846, Box 1, Folder 1, Curtis Family Papers, SC19050, NYSL.

73. Thomas Lawyer to William C. Bouck, March 23, 1846, Box 1, Bouck Papers; Demosthenes Lawyer to William C. Bouck, April [?], 1846, Box 3, Folder 9, Demosthenes Lawyer to Lyman Sanford, April 12, 1846, Box 8, Folder 4, William C. Bouck to Lyman Sanford, March 21, 1846, Box 8, Folder 15, Sanford Papers.

74. I. W. Baird to William C. Bouck, April 14, 1846, Box 1, Bouck Papers.

75. I. W. Baird to William C. Bouck, April 23, 1846, Lyman Sanford to William C. Bouck, September 24, 1848, ibid.

76. C. W. Bouck to William C. Bouck, March 3, 1846, Elisha Brown to William C. Bouck, April 8, 1846, ibid.; William Lamont III to William C. Bouck, March 27, 1846, Box 8, Folder 12, Sanford Papers; Roscoe, *History of Schoharie County,* 159.

77. Albany *Freeholder,* April 29, May 6, 1846; Christman, *Tin Horns and Calico,* 268–70; Mayham, *Anti-Rent War,* 73; Gunn, *Decline of Authority,* 170–97; McCurdy, *Anti-Rent Era,* 260–63.

78. Albany *Freeholder,* April 29, May 6, 1846; Christman, *Tin Horns and Calico,* 268–70; Mayham, *Anti-Rent War,* 73; Gunn, *Decline of Authority,* 170–97; McCurdy, *Anti-Rent Era,* 260–63.

79. Albany *Freeholder,* August 12, 1846; Albany *Anti-Renter,* October 31, 1846; Christman, *Tin Horns and Calico,* 271–88.

80. Albany *Freeholder,* August 12, 1846; Ellis, *Landlords and Farmers,* 255–57; Christman, *Tin Horns and Calico,* 283.

81. Minutes of the Blenheim Hill Anti-Rent Meeting; Albany *Freeholder,* December 24, 1845.

82. Albany *Anti-Renter,* October 31, 1846.

83. Freeman Stanton to Lyman Sanford, January 25, 1847, Box 8, Folder 5, William C. Bouck to Lyman Sanford, May 22, 1847, Box 8, Folder 16, Sanford Papers; Albany *Freeholder,* February 17, September 15, October 13, 1847.

84. Albany *Freeholder,* September 15, October 6, October 20, 1847; N. T. Rosseter to Lyman Sanford, August 29, 1847, Box 8, Folder 5, Sanford Papers.

85. Quote from Albany *Freeholder,* October 20, 1847; Delaware *Gazette,* November 17, 1847.

86. Albany *Freeholder,* October 20, November 24, 1847.

87. Ibid., December 15, 1847.

88. Jacob Livingston to William H. Averell, October 8, 1846, Box 6, Averell Family Papers.

89. Ibid., January 17, 1847. Livingston felt that a dictatorship might be the only counter to the "anarchy" he saw arising in the state.

90. Ibid., April 14, September 22, 1846, February 25, 1848, September 24, 1850.

91. Summerhill, "Farmers' Republic," 233–42, 396–405.

92. Delaware *Gazette,* November 22, 1848; Cooperstown *Freeman's Journal,* November 25, 1848.

93. McCurdy, *Anti-Rent Era,* 276–81; Bloomville *Mirror,* December 29, 1851, November 8, 1853.

94. Thomas Summerhill, "Farming on Shares: Landlords, Tenants, and the Rise of the Hop and Dairy Economies in Central New York," *New York History* 76 (1995): 125–52; McCurdy, *Anti-Rent Era,* 295–336.

Chapter 4: New Crops, New Challenges

1. Silas Wellman to William Wellman, June 26, 1852, William Wellman Papers, #4-109, NYSHA.

2. Donald H. Parkerson, *The Agricultural Transition in New York State: Markets and Migration in Mid-Nineteenth-Century America* (Ames: Iowa State University Press, 1995), 79–124.

3. James Burnet to Lyman Sanford, December 14, 1847, Box 4, Folder 3, Lyman Sanford Papers, BY462, AIHA.

4. Manuscript United States Census, 1850, #432, Schedule I, Delaware County, Rolls 494, 495, Otsego County, Rolls 579, 580, Schoharie County, Rolls 595, 596, National Archives; George Clarke Account with Richard Cooper, 1850, MSS, George Clarke Rent Rolls, Cherry Valley and Long Patents, 1851, MSS, Box 45, George Hyde Clarke Papers, #2800, CUL; Albany *Freeholder,* April 17, April 24, June 19, 1850; Albert C. Mayham, *The Anti-Rent War on Blenheim Hill: An Episode of the 40's* (Jefferson, N.Y.: Frederick L. Frazee, 1906), 74–76; Henry Christman, *Tin Horns and Calico: An Episode in the Emergence of American Democracy* (New York: Henry Holt, 1945), 302–13; *New York State Census, 1855* (Albany: C. Van Benthuysen, 1857), 4–12. For a national comparison, see Jeremy Atack, "Tenants and Yeomen in the Nineteenth Century," *Agricultural History* 62 (Summer 1988): 6–32. I traced 109 of 114 Clarke tenants (96 percent). The aggregate numbers are from random samples of 402 households each in Otsego (9,087 total) and Schoharie (5,878) and 400 in Delaware (7,105). Atack found that only 8 percent of New York farmers were tenants in 1860. In 1855, Springfield, Clarke's home town, registered 57 percent, Butternuts 60 percent, and Otsego 64 percent. Anti-Rent towns showed mixed returns. In the Delaware County towns of Andes, Bovina, Delhi, Hamden, and Kortright, less than 20 percent of farm operators were tenants, while in Stamford (23 percent), Middletown (36 percent), and Roxbury (41 percent), tenancy was more widespread. Schoharie Anti-Rent towns likewise followed no discernible pattern, with Jefferson registering 78 percent, Blenheim and Fulton 76 percent, Conesville 74 percent, Summit 72 percent, Broome 63 percent, and Gilboa 61 percent freeholders.

5. Minutes of Journey to Otsego, Delaware, &c., July and August, 1850, Box 8, Folder 5, Goldsbrow Banyar Papers, DU10723, NYSL.

6. Albany *Freeholder,* March 13, 1850, February 19, March 19, May 7, 1851; *Supreme Court.*

The People of the State of New York Against George Clarke, February 21, 1850, MSS, Thomas Machin to George Clarke, March 2, 1850, Box 45, Gamaliel Bowdish to George Clarke, April 18, 1851, Box 1, Clarke Papers.

7. Thomas Machin to George Clarke, May 1, 1851, Box 45, Richard Cooper Journal, 1854–55, MSS, Box 100, Clarke Papers.

8. George C. Clyde to Richard Cooper, January 7, 1850, W. H. Coon to George Clarke, January 18, 1850, Box 45, ibid.

9. Bloomville *Mirror,* July 12, 1853.

10. Charles Harley Estate Inventories, 1839–49, Harley Family Papers, H Box, NYSHA.

11. Warrantee Deed: Hugh Innes to Moses Lyon, March 16, 1813, Warrantee Deed: Moses Lyon to Henry Lyon, May 29, 1815, Warrantee Deed: Henry and Rebecca Lyon to Moses F. Lyon, March 28, 1850, Mortgage Bond: Moses F. Lyon to Henry Lyon, March 28, 1850, Satisfaction of Mortgage: Seth Lyon to Moses F. Lyon, May 18, 1864, Miscellaneous Delaware County Papers, 1–88+, NYSHA; Seth Lyon to Henry Lyon, March 7, 1850, Lyon Family Papers, #66, NYSHA.

12. Edwin Williams, ed., *New York Annual Register, 1843* (New York: J. Disturnell, 1843), 59–64; *The Seventh Census of the United States, 1850* (Washington, D.C.: Robert Armstrong, 1853), 121–25; *Eighth Census of the United States, 1860: Agriculture* (Washington, D.C.: Government Printing Office, 1864), 100–101.

13. Williams, *New York Annual Register, 1843,* 59–64; *Seventh Census of the United States, 1850,* 121–25; *Eighth Census of the United States, 1860,* 100–101; David M. Ellis, *Landlords and Farmers in the Hudson-Mohawk Region, 1790–1850* (Ithaca: Cornell University Press, 1946), 102–11; Paul W. Gates, "Agricultural Change in New York State, 1850–1890," *New York State History* 50 (April 1969): 124. Anti-Rent leader Benjamin P. Curtis of Blenheim, Schoharie County, for example, grew oats and hay in the years following the uprising. Benjamin P. Curtis to William B. Curtis, August 22, 1847, Box 1, Folder 1, Curtis Family Papers, SC19050, NYSL.

14. Melvin W. Hill Diary, August 25, 1853, MSS, NYSHA.

15. "John W. Champlin Diary," *Delaware County History* 9 (Fall 1977): 6–14.

16. L. Ray Gunn, *The Decline of Authority: Public Economic Policy and Political Development in New York State, 1800–1860* (Ithaca: Cornell University Press, 1988), 198–245.

17. Elmer O. Frippen, *Rural New York* (New York: Macmillan, 1921), 195; George Rochefort Clarke to Alfred Clarke, July 24, 1838, Alfred Clarke to George Rochefort Clarke, July 25, 1838, T. H. Wheeler to Richard Cooper, January 20, 1841, Duncan C. Pell to George Clarke, February 14, 1846, George Clarke to Alfred Clarke, February 16, 1846, Duncan C. Pell to George Clarke, February 27, 1846, Box 1, T. O. Clarke to George Clarke, September 22, 1866, Box 3, George Clarke to Samuel Wilcox, January 14, 1847, Box 44, Thomas Machin to George Clarke, April 30, 1850, Box 45, S. S. Burnside to Richard Cooper, May 10, 1858, Box 47, Clarke Papers; Albany *Freeholder,* January 3, 1849, May 29, 1850. Clarke at first attempted to sell lots and, when he could not get the price he desired, offered annual cash leases. But tenants balked at what they considered high prices as well as annual leases, which they thought unduly exploitative.

18. *Transactions of the New York State Agricultural Society, 1856* (Albany: C. Van Benthuysen, 1856), 93–95; Herbert Myrick, *The Hop: Its Culture and Cure, Marketing and*

Manufacturing (New York: Orange Judd Co., 1914), 219–20; *Scientific American* 12 (January 21, 1865): 49; Cooperstown *Freeman's Journal,* February 2, 1855; Sharon Good, "The Hop Culture" (M.A. thesis, SUNY College at Oneonta, Cooperstown Graduate Program, 1968), 7–21, 70–75; Thomas Bridgemen, *The American Kitchen-Gardener* (New York: William Wood and Co., 1867), 83–89; Ezra Meeker, *Hop Culture in the United States* (Puyallup, Wash.: E. Meeker and Co., 1883), 3–4, 48–49; *Joseph Tunnicliff v. George R. Fowler,* MSS, n.d., Otsego County Court Records, #281, NYSHA; H. L. Routh and Sons to George Clarke, February 3, 1855, Box 46, George Clark to A. C. Smith, November 10, 1861, Box 49, Clarke Papers. For a recent study of hops culture in the United States, see Michael A. Tomlan, *Tinged with Gold: Hop Culture in the United States* (Athens: University of Georgia Press, 1992).

19. Brewster Conkling to George Clarke, December 16, 1853, Box 46, Hop Pole Contract: Joseph Carr and George Clarke, October 24, 1860, "Statement of Monies Borrowed of Bank of Cooperstown," April 1, 1861, Box 49, Clarke Papers.

20. George Clarke to Elijah Brown, March 24, 1864, Box 50, ibid.

21. Lease: George Clarke to Jacob M. Cooper, April 1, 1854, Box 44, ibid.

22. Receipt for 47,000 Hop Poles, S. and L. G. Wilkins to George Clarke, September 3, 1847, Box 44, George Clarke Fruit Account Books, 1854–59, MSS, Box 45, Richard Cooper Journal, 1854–55, MSS, George Clarke Journal, 1855–57, MSS, Box 100, "Statistics of the quantity of Hops raised by G. Clarke," 1858, Box 47, E. W. Wood to George Clarke, August 11, 1863, Box 50, ibid.

23. T. O. Clarke to George Clarke, November 2, 1861, March 15, March 18, March 29, April 5, April 8, August 23, 1862, June 12, 1866, T. O. Clarke to A. C. Smith, March 29, April 5, 1862, Box 3, Mortgage Agreement, Anna Pell to George Clarke, January 31, 1844, Box 44, ibid.

24. Clarke's hops operations and those of some of his competitors can be traced in Ralph Birdsall, *The Story of Cooperstown* (Cooperstown: Augur's Book Store, 1948), 239; H. L. Routh and Sons to George Clarke, February 3, 1855, Pier, Parker and Co. to George Clarke, October 8, 1856, Box 45, H. L. Routh and Sons to George Clarke, December 11, 1854, Cleveland and Co. to George Clarke, December 13, 1854, Box 46, Butler and Smith to George Clarke, March 23, April 16, 1859, April 30, June 15, June 28, 1859, Box 47, T. O. Clarke to George Clarke, March 15, April 5, May 3, July 16, 1862, D. Gregory to T. O. Clarke, August 13, August 21, 1867, Box 3, A. C. Smith to George Clarke, January 8, 1863, Box 50, George Clarke Journal, 1855–57, MSS, Box 100, Clarke Papers.

25. *Seventh Census of the United States, 1850,* 121–25; *Eighth Census of the United States, 1860,* 100–101; Cooperstown *Freeman's Journal,* August 4, 1854, February 2, 1855.

26. Cooperstown *Freeman's Journal,* February 2, 1855.

27. Eliakim R. Ford to Lyman Sanford, March 11, 1858, Box 5, Folder 4, Sanford Papers.

28. Jane Russell Averell to William Lawson Carter, July 3, 1856, Jane Russell Averell Letterbooks, Box 2, Clarke Papers.

29. T. O. Clarke to George Clarke, March 18, 1862, Box 3, ibid.

30. Myrick, *The Hop,* 233–34; Meeker, *Hop Culture,* 37–43, 47; Good, "Hop Culture," 84–94; H. L. Routh and Sons to George Clarke, February 3, 1855, "Statement of Hops Grown by George Clarke, 1856," Pier, Parker and Co. to George Clarke, October 8, 1856,

Box 46, Butler and Smith to George Clarke, March 23, April 16, April 30, June 28, 1859, Box 47, T. O. Clarke to George Clarke, May 3, 1862, D. Gregory to T. O. Clarke, August 13, August 21, 1867, Box 3, A. C. Smith to George Clarke, January 8, 1863, Box 50, Clarke Papers.

31. Eugene Milener, *Oneonta: The Development of a Railroad Town* (Deposit, N.Y.: Courier Printing, 1983), 473; New York, Vol. 516:147, R. G. Dun & Co. Collection, Baker Library, Harvard Business School, Harvard University; Willard V. Huntington, *Oneonta Memories* (San Francisco: Bancroft Co., 1891), 159.

32. Silas Dutcher Diaries, December 28, 1854, August 8, December 26, 1855, January 2, February 2, February 29, April 9, July 7, 1856, MSS, NYSHA.

33. Chattel Mortgage: George Clarke to John Chamberlain, June 25, 1857, Box 47, Chattel Mortgage: George Clarke to Jeremiah Aney, August 23, 1856, Chattel Mortgage: George Clarke to John Lambert, November 15, 1856, Box 46, Clarke Papers.

34. Jedediah Miller to Lyman Sanford, March 12, 1858, Box 5, Folder 4, Sanford Papers.

35. George and A. J. Wheeler and William H. Leeds, Articles of Agreement for Freighting and Forwarding Business, January 22, 1849, Box 2, Folder 1, George D. Wheeler Account Book, 1863–93, MSS, Box 3, George D. Wheeler Autobiography, MSS, Box 1, Folder 3, Wheeler Family Papers, #3878, CUL.

36. Schoharie *Patriot,* July 23, 1857; *Transactions of the New York State Agricultural Society, 1867* (Albany: C. Van Benthuysen and Sons, 1867), 2, 100–131; *Annual Descriptive Catalogue of Agricultural Implements, Horticultural Tools, and Field, Grain and Garden Seeds, for Sale at the Albany Agricultural Warehouse, and Seed Store* (Albany: C. Van Benthuysen, 1848); *Buffalo Mower and Reaper. Parkhurst's Patent, Improved Adjustable Mowing and Reaping Machine, With or Without Reaping Attachment. Miller, Bennett & Co., Manufacturers* (Buffalo, N.Y.: Murry, Rockwell and Co., 1858); *Pine's Patent Mowing Machines Manufactured by William P. Kellogg* (Lansingburgh, N.Y.: A. Kirkpatrick, 1862); "Kemp's & Burpee Manure Spreader," broadside (ca. 1870).

37. *Circular and Price List of Schenectady Agricultural Works, for 1877* (Schenectady, N.Y.: George Westinghouse, 1877); "Empire Agricultural Works Gold Medal Threshing Machine," broadside (ca. 1865), Box 21, Recall Notice for Hay Machinery, Levi Beardsley, July 25, 1861, Box 23, Warshaw Collection, #60, NMAH; *Transactions of the New York State Agricultural Society, 1867,* 122; Martin F. Noyes, *A History of Schoharie County* (Richmondville, N.Y.: Richmondville *Phoenix,* 1954), 66; Kenneth Fake et al., *Official History of the Town of Cobleskill, Schoharie County, New York* (Cobleskill, N.Y.: Cobleskill *Index,* 1937), 12; Huntington, *Oneonta Memories,* 158.

38. Cooperstown *Freeman's Journal,* February 2, 1855.

39. Rolla Tryon, *Household Manufactures in the United States, 1640–1860* (Chicago: University of Chicago Press, 1917), 7–10, 303–11; Sally McMurry, *Transforming Rural Life: Dairying Families and Agricultural Change, 1820–1885* (Baltimore: Johns Hopkins University Press, 1995), 62–122; Joan M. Jensen, *Loosening the Bonds: Mid-Atlantic Farm Women, 1750–1850* (New Haven: Yale University Press, 1986), xiii–xv, 36–91, 114–28; Nancy Grey Osterud, *Bonds of Community: The Lives of Farm Women in Nineteenth-Century New York* (Ithaca: Cornell University Press, 1991), 1–13, 26–27; Hal S. Barron, *Those Who Stayed Behind: Rural Society in Nineteenth-Century New England* (New York: Cambridge Uni-

versity Press, 1984), 94–99; Allan Kulikoff, *The Agrarian Origins of American Capitalism* (Charlottesville: University of Virginia Press, 1992), 34–59.

40. McMurry, *Transforming Rural Life*, 15–30; Osterud, *Bonds of Community*, 32–33; M. Francis Guenon, *A Treatise on Milch Cows: Principles and Practices of Butter Making* (New York: Greeley and McElrath, 1848). Osterud found that as late as the 1880s, farmers in Broome County on average milked just six cows.

41. Bloomville *Mirror*, December 29, 1851.

42. Ibid., November 8, 1853.

43. Ann E. Gould to Samuel Sherwood, May 27, 1855, Sherwood Family Papers, DCHA.

44. Matthew Griffin Diary, March 18, 1855, MSS, DCHA.

45. Anonymous to Lyman Sanford, July 5, 1853, Sanford Papers.

46. Cheese Contract: J. D. Shaul and R. Bomber, June 15, 1853, John D. Shaul Collection, #252, NYSHA. The document does not specify what "hundred" refers to.

47. Bloomville *Mirror*, November 1, 1859. For an excellent description of the barter system and the legal chaos it often created in central New York, see *Belknap & Hicks v. Benjamin Eckler*, 1859, MSS, Otsego County Court Records.

48. Jared Van Wagenen Jr., *Days of My Years: The Autobiography of a York State Farmer* (Cooperstown, N.Y.: NYSHA, 1962), x; Louise Nethaway Diary, November 8, 1865, MSS, NYSHA.

49. Chattel Mortgage: George Clarke to John Chamberlain, June 25, 1857, Box 47, Clarke Papers.

50. George Clarke to William Slocum, March 9, 1857, ibid.

51. Jon Bishop to Elihu Phinney, February 22, 1852, Phinney Family Collection, #235, NYSHA.

52. Jacob W. Cole to Jesse Sutliff, February 29, 1848, Sutliff Family Papers, #207, NYSHA. This was a very large operation by local standards, which suggests the level of production the new leases expected compared to average farms.

53. James Burnet to Lyman Sanford, June 18, 1867, Box 6, Folder 11, Sanford Papers.

54. Good, "Hop Culture," 93–94.

55. John Westover to Lyman Sanford, February 9, 1857, Box 3, Folder 18, Sanford Papers.

56. Eliakim R. Ford to Lyman Sanford, March 11, 1858, Box 5, Folder 4, ibid.

57. John Pindar to Lyman Sanford, October 3, 1856, Box 4, Folder 16, ibid.

58. George D. Taylor, *These Hills Are Not Barren: The Story of a Century Farm* (New York: Exposition Press, 1950), 50–52.

59. Benjamin Curtis to William B. Curtis, August 22, 1847, Curtis Family Papers; John N. Colburn Diaries, May 17, 1852, MSS, NYSHA; Taylor, *These Hills*, 44. Quote is from Taylor.

60. Seth Lyon to Henry Lyon, March 7, 1850, Lyon Family Papers.

61. Seth Lyon to Moses F. Lyon, March 13, 1861, ibid.

62. George P. Proctor Diary, March 11, March 13, 1853, MSS, NYSHA; John Robinson to Lyman Sanford, July 7, 1862, Box 5, Folder 17, Sanford Papers; Wayne Franklin, ed., *A Rural Carpenter's World: The Craft in a Nineteenth-Century New York Township* (Iowa City: University of Iowa Press, 1990), 4–5.

63. John N. Colburn Diaries, March 16, 1856, February 12, May 2, July 4, 1861, July 10, 1862.

64. Ibid., February 25, 1865, March 4, 1868. Colburn later took up hop farming in the 1870s.

65. Lucius Bushnell Diaries, May 8, 1857, November 15, 1859, June 4, December 8, 1877, MSS, NYSHA.

66. Catherine Bartoe to G. H. Edgerton, April 11, 1854, Edgerton Collection, DCHA.

67. Catherine Wood Diaries, April 24, September 20, September 23, December 31, 1857, January 30, 1884, MSS, NYSHA; Jensen, *Loosening the Bonds*, 57–76; Osterud, *Bonds of Community*, 123–36. For a view of the greater support networks available to married women, see Louise Nethaway Diary, March 29, May 30, June 14, December 24, 1865.

68. Rosetta Hammond Diary, March 15, 1857, MSS, NYSHA; Jensen, *Loosening the Bonds*, 91.

69. Good, "Hop Culture," 7–21, 70–75; Bloomville *Mirror*, October 15, 1860. Hops harvesting in central New York was organized in much the same way as in Kent in England, with women and children pickers predominating and considerable fears of pickers from outside being employed. See Alan Bignell, *Hopping Down in Kent* (London: Robert Hale, 1977), 40–67, and Gilda O'Neill, *Pull No More Bines: An Oral History of East London Women Hop Pickers* (London: The Women's Press, 1990), 6–11.

70. Thomas Clarke from Mrs. Clarke, September 9, [1840?], Emma Clarke to Thomas Clarke, September 14, 1863, Clarke Family Papers; Rosetta Hammond Diary, October 3, 1857; James Bryce to G. H. Edgerton, April 24, 1854, Edgerton Collection; Tryon, *Household Manufactures*, 7–10; Thomas Dublin, *Women at Work: The Transformation of Work and Community in Lowell, Massachusetts, 1826–1860* (New York: Columbia University Press, 1979), 1–144; Christine Stansell, *City of Women: Sex and Class in New York, 1789–1860* (Urbana: University of Illinois Press, 1987), 76–101.

71. Rosetta Hammond Diaries, September 6, September 7, October 3, 1857.

72. Ibid., September 9, September 12, September 15, 1857; Emma Clarke to Thomas Clarke, September 14, 1863, Clarke Family Papers; Bloomville *Mirror*, October 15, 1860.

73. Bloomville *Mirror*, October 4, 1859.

74. Clayton E. Risley, "Hop Picking Days," *New York Folklore Quarterly* 5 (Spring 1949): 18–24; Wheaton P. Webb, "Three Poems on New York State Folklore," *New York Folklore Quarterly* 11 (Summer 1955): 85–90; Bloomville *Mirror*, October 15, 1860; Good, "Hop Culture," 61–75.

75. Lucius Bushnell Diaries, September 6, 1859; Rosetta Hammond Diaries, February 9, September 18, October 9, 1859; Jensen, *Loosening the Bonds*, 114–28; Osterud, *Bonds of Community*, 89–108.

76. Bloomville *Mirror*, October 31, 1854.

77. Henry Conklin, *Through "Poverty's Vale": A Hardscrabble Boyhood in Upstate New York, 1832–1862* (Syracuse: Syracuse University Press, 1974), 199–200.

78. Martin Bruegel, *Farm, Shop, and Landing: The Rise of Market Society in the Hudson Valley, 1780–1860* (Durham, N.C.: Duke University Press, 2002), 186–89; Osterud, *Bonds of Community*, 109–14.

79. *Supreme Court. Edwin H. Cass v. Joseph Blanchard*, February 9, 1865, MSS, Otsego County Court Records. The succeeding paragraphs are based on this case.

80. Good, "Hop Culture," 38–60; Jensen, *Loosening the Bonds,* 79–128; McMurry, *Transforming Rural Life,* 100–147.

Chapter 5: Agrarianism Outflanked

1. Schoharie *Patriot,* November 25, 1858.
2. L. Ray Gunn, *The Decline of Authority: Public Economic Policy and Political Development in New York State, 1800–1860* (Ithaca: Cornell University Press, 1988), 181–245.
3. Ibid., 222–57.
4. Mark L. Berger, *The Revolution in the New York Party Systems, 1840–1860* (Port Washington, N.Y.: Kennikat Press, 1973), 2.
5. Jacob Livingston to William H. Averell, September 24, 1850, Box 6, Averell Family Papers, #82, NYSHA.
6. Albany *Freeholder,* September 4, 1850; Berger, *Revolution in the New York Party Systems,* 3; Daniel Walker Howe, *The Political Culture of American Whigs* (Chicago: University of Chicago Press, 1979), 8–9, 96–122.
7. Cooperstown *Freeman's Journal,* October 20, 1849.
8. Ibid., November 17, 1849, November 16, 1850.
9. Lyman Sanford to William C. Bouck, September 24, 1848, Thomas Lawyer to William C. Bouck, October 10, 1851, William C. Bouck Papers, #2206, CUL; A. Birdsall to Lyman Sanford, July 20, 1850, Charles Courter to Lyman Sanford, October 26, 1850, Box 4, Folder 7, Lyman Tremain to Lyman Sanford, September 5, 1851, Box 8, Folder 7, Lyman Sanford Papers, BY462, AIHA; Gunn, *Decline of Authority,* 211–12, 236–38; William E. Gienapp, *The Origins of the Republican Party, 1852–1856* (New York: Oxford University Press, 1987), 39–47.
10. Thomas Smith to William H. Averell, October 3, 1848, Box 6, Averell Family Papers; Albany *Freeholder,* October 23, 1850; Gienapp, *Origins of the Republican Party,* 40–47.
11. Delaware *Gazette,* October 17, 1849.
12. Ibid., November 28, 1849, November 20, 1850.
13. Francis W. Halsey, ed., *The Pioneers of Unadilla Village, 1784–1840 and Reminiscences of Village Life and of Panama and California from 1840 to 1850* (Unadilla, N.Y.: Vestry of St. Martin's Church, 1902), 220–99; Harvey Baker Reminiscences, Anne Manning Scrapbook, Huntington Memorial Library, Oneonta, N.Y., 2 (hereafter cited as Baker Reminiscences).
14. *The Centennial Celebration at Cherry Valley, Otsego County, N.Y., July 4th, 1840: The Addresses of William W. Campbell, Esq., and Gov. W. H. Seward, with Letters, Toasts, &c., &c.* (New York: Taylor and Clement, 1840), 33–40.
15. Erastus Crafts to William W. Campbell, March 25, 1851, Box 3, Campbell Family Papers, #313, NYSHA. For a sketch of Crafts's career, see Bernice Wardell, *The History of Laurens Township* (Laurens, N.Y.: Village Printer, 1975), 49, 61, 65, 74–75, 82–83.
16. Paul E. Johnson, *A Shopkeeper's Millennium: Society and Revivals in Rochester, New York, 1815–1837* (New York: Hill and Wang, 1978), 79–94, 116–41.
17. New York *Times,* September 1, 1859; Oneonta *Herald,* June 16, 1858.
18. Oneonta *Herald,* June 17, 1857.
19. Halsey, *Pioneers of Unadilla Village,* 167; New York, Vol. 516:148, 322, R. G. Dun &

Co. Collection, Baker Library, Harvard Business School, Harvard University; Cooperstown *Freeman's Journal,* April 12, 1851; Irene D. Neu, *Erastus Corning: Merchant and Financier* (Ithaca: Cornell University Press, 1960), 65–69, 73, 147; George R. Howell, ed., *History of Albany County, N.Y., from 1609 to 1886* (New York: W. W. Munsell, 1886), 33–34, 500, 531–32, 575.

20. Oneonta *Herald,* June 15, 1853, October 11, 1855, June 16, 1858; *Some Remarks on the Prospects of the Albany & Susquehanna Railroad as a Dividend-Paying Road* (Albany: J. Munsell, 1854), 26; Harry H. Pierce, *Railroads of New York: A Study of Government Aid, 1826–1875* (Cambridge: Harvard University Press, 1953), 60; Charles Francis Adams Jr., "An Erie Raid," in *High Finance in the Sixties: Chapters from the Early History of the Erie Railway,* ed. Frederick C. Hicks (New Haven: Yale University Press, 1929), 156–212; Cooperstown *Freeman's Journal,* April 5, 1851.

21. Schoharie *Patriot,* April 3, 1851; Cooperstown *Freeman's Journal,* April 5, 1851; *By-Laws of the Albany & Susquehanna Railroad Company* (Albany: J. Munsell, 1853); Minutes of the Board of Directors of the Albany & Susquehanna Railroad Company, April 19, 1851, Martha Phelps Collection, NYSL; William W. Snow to Robert H. Pruyn, April 1, 1851, Box 1, Folder 4, Robert H. Pruyn Papers, CH532, AIHA.

22. Articles of Association, Albany & Susquehanna Railroad Co. Minute Books, microfilm reel #6, D&H Collection, SUL; Manuscript United States Census, 1850, #432, Schedule I, Delaware County, Rolls 494, 495, Otsego County, Rolls 579, 580, Schoharie County, Rolls 595, 596, National Archives; *Seventh Census of the United States, 1850* (Washington, D.C.: Robert Armstrong, 1853), 112; Pierce, *Railroads of New York,* 49–53.

23. George Westinghouse to Robert H. Pruyn, November 18, 1852, Box 5, Folder 4, Pruyn Papers.

24. Charles L. Ryder, *The Man, the Railroad and the Bank: An Historical Narrative of the Life and Achievements of Charles Courter [1808–1879]* (Cobleskill, N.Y.: Cobleskill *Times-Journal,* 1960), 7, 17–19; New York, Vol. 566:38, R. G. Dun & Co. Collection, Baker Library, Harvard Business School, Harvard University; Charles Courter to Lyman Sanford, February 5, 1857, Box 4, Folder 18, Charles Courter to Lyman Sanford, September 28, 1865, Box 6, Folder 6, Charles Courter to Lyman Sanford, June 9, 1869, Box 7, Folder 1, Sanford Papers; H. T. Dana to W. L. M. Phelps, May 12, 1908, Martha Phelps Collection.

25. Cooperstown *Freeman's Journal,* March 3, 1854.

26. Gideon Hotchkiss to E. C. Delavan, April 24, 1851, Box 5, Folder 4, Pruyn Papers.

27. Bloomville *Mirror,* June 1, 1852; Pierce, *Railroads of New York,* 4, 49; *The Albany & Susquehanna Railroad, Its Probable Cost and Revenue* (Albany: J. Munsell, 1858), 13; Albany & Susquehanna Railroad Co. to William C. Bouck, July 4, 1851, Bouck Papers; E. Tompkins to E. C. Delavan, May 17, 1852, Box 5, Folder 4, Pruyn Papers; Cooperstown *Freeman's Journal,* May 31, August 9, 1851. Railroad promoters frequently made such claims. See Frederick A. Cleveland and Fred Wilber Powell, *Railroad Promotion and Capitalization in the United States* (New York: Longmans, Green and Co., 1909), 194.

28. Delaware *Gazette,* October 22, 1851.

29. Ibid., October 29, 1851.

30. Berger, *Revolution in the New York Party Systems,* 3; Delaware *Gazette,* November 19, 1851.

31. E. C. Delavan to R. H. Pruyn, April 6, 1851, Box 5, Folder 4, Pruyn Papers.

32. Unmarked newspaper clipping, Phelps Autobiography, Martha Phelps Collection.

33. *Some Considerations Respecting the Proposed Construction of the Albany & Susquehanna Railroad, January, 1852* (Albany: C. Van Benthuysen, 1852), 1–12; *Century of Progress: History of the Delaware and Hudson* (Albany: J. B. Lyon Co., 1925), 633.

34. William H. Averell to E. R. Ford, November 2, 1858, Box 6, Averell Family Papers.

35. Edward Tompkins to Robert H. Pruyn, April 12, 1852, Box 5, Folder 4, Pruyn Papers; Pierce, *Railroads of New York*, 53; Albany *Evening Journal*, February 12, 1858.

36. Oneonta *Herald*, January 25, 1854.

37. Edward Tompkins to E. C. Delavan, May 17, 1852, Box 5, Folder 4, Pruyn Papers.

38. *State of New York. Senate Document No. 88, March 17, 1858. Report of a Minority of the Committee on Railroads on Petitions for the Repeal of the Acts Extending the Time to Complete, and Authorizing Town Subscriptions to the Capital Stock of the Albany and Susquehanna Railroad* (1858), 3; W. L. M. Phelps Autobiography, MSS, Martha Phelps Collection; Oneonta *Herald*, June 15, 1853, October 11, 1855, June 16, 1858; *Century of Progress*, 631, 633; Unadilla *Times*, August 6, 1868; Edward Tompkins to Robert H. Pruyn, July 8, 1853, Box 5, Folder 5, Pruyn Papers; Albany *Evening Journal*, July 10, 1876; *An Act to Authorize the City of Albany to Make a Loan to the Albany & Susquehanna Railroad Company* (n.p., 1852).

39. *Act to Authorize the City of Albany;* Lee Benson, *Merchants, Farmers and Railroads: Railroad Regulation and New York Politics, 1850–1887* (Cambridge: Harvard University Press, 1955), 6–7.

40. George W. Chase to Robert H. Pruyn, July 2, 1852, Box 5, Folder 4, Pruyn Papers.

41. Schoharie *Patriot*, December 2, 1852; Delaware *Gazette*, November 10, November 17, 1852; Bloomville *Mirror*, October 19, November 16, 1852; Cooperstown *Freeman's Journal*, October 29, 1852. In Otsego County, Chase polled 5,311 votes while Gordon received 5,080, 560 less than Democrat Horatio Seymour tallied for governor. Chase secured most of the 500 votes cast for abolitionist gubernatorial candidate Minthorne Tompkins and gained 220 Democratic votes—probably Hards. Chase defeated Gordon more decisively in Delaware, 4,237 to 2,952, where Anti-Rent voters controlled a large portion of the vote.

42. Delaware *Gazette*, October 13, 1852.

43. Berger, *Revolution in the New York Party Systems*, 3–16.

44. Oneonta *Herald*, February 16, 1853.

45. *Report of a Minority*, 2.

46. Oneonta *Herald*, June 15, 1853.

47. Ibid., August 3, 1853; *Report of a Minority*, 2–3.

48. E. R. Ford to Robert H. Pruyn, January 8, 1852, Edward Tompkins to Robert H. Pruyn, May 17, 1852, Box 5, Folder 4, William W. Snow to Robert H. Pruyn, January 23, 1853, Box 5, Folder 5, Charles Courter to Robert H. Pruyn, September 11, 1853, Box 5, Folder 6, Pruyn Papers; *Century of Progress*, 634; Cooperstown *Freeman's Journal*, November 25, 1853; Oneonta *Herald*, September 7, September 14, September 21, 1853.

49. Charles Courter to Robert H. Pruyn, September 11, 1853, Box 5, Folder 6, Pruyn Papers.

50. Delaware *Gazette*, November 2, November 23, 1853; Cooperstown *Freeman's Journal*,

November 4, November 11, 1853; Schoharie *Patriot,* December 12, 1853; Oneonta *Herald,* January 18, 1854; Gienapp, *Origins of the Republican Party,* 39–47, 66–67.

51. Oneonta *Herald,* September 7, September 14, September 21, 1853, February 22, 1854; New York *Times,* July 20, July 28, 1854.

52. Stockholder Petition to the Board of Directors of the Albany & Susquehanna Railroad Co., April 15, 1854, vol. 4, W. L. M. Phelps Scrapbooks, NYSL; Cleveland and Powell, *Railroad Promotion,* 190.

53. Minutes of the Board of Directors of the Albany & Susquehanna Railroad Co., May 23, 1854, microfilm reel #6, D&H Collection.

54. Ibid., June 7, July 19, 1854, June 5, 1855; Cooperstown *Freeman's Journal,* June 2, 1854; New York *Times,* July 20, July 28, 1854; Baker Reminiscences, 179; Unadilla *Times,* July 30, 1868; George W. Chase to Robert H. Pruyn, July 2, 1852, Box 5, Folder 4, Pruyn Papers; *Century of Progress,* 634–35.

55. Baker Reminiscences, 179. See also Cooperstown *Freeman's Journal,* March 17, April 28, June 2, 1854, and Pierce, *Railroads of New York,* 48, 69.

56. *Some Remarks on the Prospects of the Albany & Susquehanna Railroad as a Dividend-Paying Road,* 30–31.

57. Notices to the stockholders of the Albany & Susquehanna Railroad Co., vol. 5, Phelps Scrapbooks; Minutes of the Board of Directors of the Albany & Susquehanna Railroad Co., December 7, 1853, microfilm reel #6, D&H Collection; Eugene Milener, *Oneonta: The Development of a Railroad Town* (Deposit, N.Y.: Courier Printing, 1983), 62.

58. Oneonta *Herald,* January 18, 1854.

59. *Opinion of Judge Willard in the Cases of the Albany & Susquehanna Railroad Company Against Abraham A. Stanton and others, May 23, 1854* (Albany: H. H. Van Dyck, 1854), 13.

60. Ibid., 13.

61. Ibid., 13; Gunn, *Decline of Authority,* 216–19; Morton J. Horwitz, *The Transformation of American Law, 1780–1860* (Cambridge: Harvard University Press, 1977), 253–66.

62. *In the Supreme Court of the State of New York. The Albany & Susquehanna Railroad Company agst. Edward E. Kendrick* (ca. 1855).

63. Oneonta *Herald,* June 7, 1854.

64. Berger, *Revolution in the New York Party Systems,* 59–69.

65. Bloomville *Mirror,* October 10, 1854. See also Eric Foner, *Free Soil, Free Labor, Free Men: The Ideology of the Republican Party before the Civil War* (New York: Oxford University Press, 1970), 169–70, and Gienapp, *Origins of the Republican Party,* 69–87.

66. Delaware *Gazette,* November 1, 1854.

67. Cooperstown *Freeman's Journal,* November 17, 1854; Gienapp, *Origins of the Republican Party,* 157–60.

68. Delaware *Gazette,* November 22, 1854. The tally was as follows: Seymour, 2,828; Clark, 2,772; Know-Nothing Daniel Ullman, 558; and Hard Greene Bronson, 326. In the congressional race, the vote was 3,150 for the Whigs and 2,231 for the Softs.

69. Schoharie *Patriot,* November 2, November 23, 1854. Clark polled 1,833 votes, Seymour 1,712, Bronson 1,481, and Ullman 1,138. Mayham lost, 1,612 to 1,297.

70. Minutes of the Board of Directors of the Albany & Susquehanna Railroad Co., ca. 1854, vol. 5, Phelps Scrapbooks; New York *Times,* August 16, 1854; Albany *Evening Journal,*

April 5, 1855; Pierce, *Railroads of New York*, iv, 21; *Century of Progress*, 635; Benson, *Merchants, Farmers, and Railroads*, 1–9; Gunn, *Decline of Authority*, 210–16.

71. Delaware *Gazette*, October 17, 1855; Berger, *Revolution in the New York Party Systems*, 96. Quote is from the *Gazette*.

72. John Blakely to Lyman Sanford, October 26, 1855, T. Durand to Lyman Sanford, November 3, 1855, Charles Courter to ——— Erskine, November 5, 1855, Box 8, Folder 7, Sanford Papers; Schoharie *Patriot*, November 22, 1855. The race pitted Democrats, Republicans, and Know-Nothings against each other. Ramsey ran behind the Democrat in Schoharie and the Know-Nothing in Delaware. Because of the bargain, Ramsey received 347 more votes than the Republican candidate for secretary of state in Schoharie, which was key to victory in the close contest. Sanford, Courter, and Ramsey must have had a very accurate estimate of the Delaware County vote ahead of time in order to pull off this coup, but no evidence remains to confirm it.

73. Cooperstown *Freeman's Journal*, November 2, 1855.

74. Ibid., November 16, 1855; Gienapp, *Origins of the Republican Party*, 223–34. Republican Preston King received 2,872 votes; Soft Israel T. Hatch, 2,540; American Joel T. Headley, 1,958; and Hard Aaron Ward, 533.

75. Albany *Argus*, January 22, 1856.

76. Quote from Albany *Atlas*, February 14, 1856; Minutes of the Board of Directors of the Albany & Susquehanna Railroad Co., March [?], 1856, microfilm reel #6, D&H Collection.

77. Charles Courter to Lyman Sanford, April 1, 1856, Box 8, Folder 8, Sanford Papers; Town Bond Account Books of the Albany & Susquehanna Railroad Co., MSS, Phelps Collection; Pierce, *Railroads of New York*, 27; Cleveland and Powell, *Railroad Promotion*, 206.

78. Edward Tompkins to E. P. Prentice, April 9, 1856, quoted in Pierce, *Railroads of New York*, 49.

79. *The Albany & Susquehanna Railroad, Its Probably Cost and Revenue*, 14–15; *Report of the Majority of the Committee on Railroads on Petitions for the Repeal of the Acts Authorizing Town Subscriptions to the Capital Stock of the Albany and Susquehanna Railroad* (Albany: C. Van Benthuysen, 1858), 12; *Address by the Directors of the Albany & Susquehanna R.R. Company, Together with the Acts Authorizing Town Subscriptions* (Albany: J. Munsell, 1856), 3–7; Baker Reminiscences, 180; Minutes of the Board of Directors of the Albany & Susquehanna Railroad Co., March 1856, August 13, 1856, microfilm reel #6, D&H Collection. The board set October 15 as the final day to secure the approval of the towns or shut down the company. They felt that the towns of Colesville, Bainbridge, Unadilla, Maryland, Worcester, Decatur, Westford, Richmondville, Cobleskill, Esperance, Carlisle, Seward, and Summit would support the measure.

80. Cooperstown *Freeman's Journal*, May 23, 1856.

81. Schoharie *Patriot*, June 19, 1856.

82. Ibid., June 26, 1856.

83. Cooperstown *Freeman's Journal*, May 23, 1856; Baker Reminiscences, 185.

84. T. D. Bailey to Ezra P. Prentice, April 23, 1856, Letterbooks of the Albany & Susquehanna Railroad Co., NYSL.

85. *In the Supreme Court of the State of New York, Richard J. Grant agst. Charles Courter* (Albany: Atlas and Argus Printing, 1857), 1–9. See also *Grant v. Courter*, 24 Barb. 232;

In the Supreme Court of the State of New York. Richard J. Grant agst. Charles Courter. Points on the Part of the Plaintiff (ca. 1856), 1–2; C. L. Ryder, *The Man, the Railroad and the Bank,* 2; and Pierce, *Railroads of New York,* 26–28.

86. *In the Supreme Court of the State of New York, Richard J. Grant agst. Charles Courter,* 1–9.

87. Ibid.

88. Ibid.

89. Jane Russell Averell to William Lawson Carter, June 5, 1856, Jane Russell Averell Carter Letterbooks, Box 2, George Hyde Clarke Papers, #2800, CUL.

90. Levi C. Turner to William W. Campbell, June 22, 1856, Frank M. Rotch to William W. Campbell, August 25, 1856, Box 3, Levi Beardsley to William W. Campbell, December 10, 1855, Box 5, Campbell Family Papers.

91. Oneonta *Herald,* October 1, 1856.

92. Egbert Olcott to Cleveland J. Campbell, October 23, 1856, Box 7, Campbell Family Papers.

93. Jane Russell Averell to William Lawson Carter, October 19, 1856, Carter Letterbooks, Box 2, Clarke Papers.

94. Egbert Olcott to Cleveland J. Campbell, October 23, 1856, Box 7, Campbell Family Papers.

95. Delaware *Gazette,* November 19, 1856; Cooperstown *Freeman's Journal,* November 14, 1856. The gubernatorial vote was split the following way in Delaware: Republican John A. King, 4,088; Soft Democrat A. J. Parker, 2,386; Know-Nothing Erastus Brooks, 1,981. In Otsego, the numbers were equally decisive: King, 6,213; Parker, 3,683; Brooks, 1,310. There is some indication that either Hards were voting Know-Nothing or Know-Nothings had Democratic sympathies, for Democratic votes went up when the nativists did not field candidates.

96. Schoharie *Patriot,* October 16, October 30, November 11, November 20, 1856; Charles Courter to Lyman Sanford, February 19, March 3, April 1, 1856, H. Schoolcraft to Lyman Sanford, August 29, 1856, Charles Courter to Lyman Sanford, November 3, 1856, Box 8, Folder 8, Sanford Papers. The gubernatorial tally was Democrat A. J. Parker, 2,958; Republican John A. King, 2,178; Know-Nothing Erastus Brooks, 1,700.

97. John H. Angle to Lyman Sanford, July 7, 1857, Box 5, Folder 2, Sanford Papers.

98. Charles Courter to Lyman Sanford, February 18, 1857, Box 8, Folder 8, ibid.

99. Ibid., August 3, 1857, Box 5, Folder 2; Schoharie *Patriot,* October 16, October 23, October 30, November 19, 1857.

100. Ransom Balcom to William W. Campbell, October 19, [ca. 1857], Box 6, Zebulon E. Goodrich to William W. Campbell, November 4, 1857, Box 1, Campbell Family Papers, #306, NYSHA; Bloomville *Mirror,* January 6, 1857; Delaware *Gazette,* October 14, October 21, October 28, November 19, 1857.

101. Oneonta *Herald,* November 18, 1857; Cooperstown *Freeman's Journal,* October 30, November 13, 1857; George D. Beers to William W. Campbell, October 28, 1857, Box 5, Campbell Family Papers, #313.

102. Oneonta *Herald,* March 17, 1858. See also Albany *Evening Journal,* February 12, 1858; Benson, *Merchants, Farmers, and Railroads,* 11–12; and Pierce, *Railroads of New York,* 27–28.

103. Cooperstown *Freeman's Journal,* January 15, 1858.

104. Schoharie *Patriot,* February 4, 1858.

105. J. H. Salisbury to Lyman Sanford, February 15, 1858, Box 5, Folder 3, Sanford Papers.

106. Oneonta *Herald,* March 17, 1858.

107. Albany *Evening Journal,* February 12, 1858.

108. *Railroad Influences. Speech of Hon. A. H. Laflin, of the Twentieth District, (Otsego and Herkimer,) Delivered in the Senate, Tuesday, March 2d, 1858* (n.p., 1858).

109. Baker Reminiscences, 181–82.

110. Ibid., *Report of a Minority;* Cooperstown *Freeman's Journal,* April 2, 1858.

111. Baker Reminiscences, 195.

112. Cooperstown *Freeman's Journal,* July 23, August 20, 1858; Minutes of the Board of Directors of the Albany & Susquehanna Railroad Co., August 25, 1858, microfilm reel #6, D&H Collection; Joseph H. Ramsey to William W. Campbell, March 1, 1859, Box 5, Campbell Family Papers.

113. Schoharie *Patriot,* November 25, 1858.

114. Cooperstown *Freeman's Journal,* October 5, October 19, November 2, 1860; Delaware *Gazette,* October 17, 1860; Gunn, *Decline of Authority,* 220–21.

115. Cooperstown *Freeman's Journal,* October 12, 1860.

116. Ibid., October 5, 1860.

117. Ibid., October 19, November 2, 1860.

118. Edgar Ryder, "Civil War Days in Cobleskill," *Schoharie County Historical Review* 25 (1961): 7–10.

119. This is elaborated in Thomas Summerhill, "The Farmers' Republic: Agrarian Protest and the Capitalist Transformation of Upstate New York, 1840–1900" (Ph.D. diss., University of California, San Diego, 1993), 377–88.

120. *In Assembly, March 28, 1859. The Bill Granting Aid by the State to the Albany and Susquehanna Railroad,* (n.p., 1859), 3; Albany *Argus,* March 14, 1894; Baker Reminiscences, 185; Oneonta *Herald,* April 15, 1863; Pierce, *Railroads of New York,* 55; *Century of Progress,* 636–39.

121. Pierce, *Railroads of New York,* 21. The A&S claimed in 1865 that it had received subscriptions for $1,479,400 from individuals and $950,000 from townships. Of the $2,429,400 subscribed, the company had collected $788,892.57 from individuals and $558,300 from towns, or $1,347,192.57.

Chapter 6: The Maturation of the Market Economy

1. Lewis M. Hall, *Gleanings from a Country Journal: Life of the Southern Tier of New York State in 1870* (Utica, N.Y.: Brodock Press, 1975), 31–32; Samuel H. Grant to John Waterman, May 18, 1865, Misc. Manuscripts, Box G, NYSHA; Paul W. Gates, *Agriculture and the Civil War* (New York: Alfred A. Knopf, 1965), 139–40; Jackson Lears, "Packaging the Folk: Tradition and Amnesia in American Advertising, 1880–1940," in *Folk Roots, New Roots: Folklore in American Life,* ed. Jane Becker and Barbara Franco (Lexington, Mass.: Museum of Our National Heritage, 1988), 107; Hal S. Barron, *Mixed Harvest: The Second Great Transformation in the Rural North, 1870–1930* (Chapel Hill: University of North Carolina Press, 1997), 7–16.

2. Solon J. Buck, *The Granger Movement: A Study of Agricultural Organization and Its*

Political, Economic, and Social Manifestations 1870–1880 (Cambridge: Harvard University Press, 1933), 3; Lee Benson, *Merchants, Farmers, and Railroads: Railroad Regulation and New York Politics, 1850–1887* (Cambridge: Harvard University Press, 1955), 80–114; Jeffrey G. Williamson, *Late Nineteenth-Century American Development: A General Equilibrium History* (New York: Cambridge University Press, 1974), 46–163.

3. *Eighth Census of the United States, 1860: Agriculture* (Washington, D.C.: Government Printing Office, 1864), 121–25; *Ninth Census of the United States, 1870: Agriculture* (Washington, D.C.: Government Printing Office, 1872), 210–13, 314–17; *Tenth Census of the United States, 1880: Agriculture* (Washington, D.C.: Government Printing Office, 1883), 103–301; *Eleventh Census of the United States, 1890: Agriculture* (Washington, D.C.: Government Printing Office, 1896), 340–483; *Twelfth Census of the United States, 1900: Agriculture* (Washington, D.C.: United States Census Office, 1901), 462–63, 612–13; Hall, *Gleanings from a Country Journal*, 57; Paul W. Gates, "Agricultural Change in New York State, 1850–1890," *New York History* 50 (April 1969): 115–41; Paula Baker, *The Moral Frameworks of Public Life: Gender, Politics, and the State in Rural New York, 1870–1930* (New York: Oxford University Press, 1991), 18–19.

4. *Ninth Census of the United States, 1870: Agriculture*, 210–13, 314–17; *Tenth Census of the United States, 1880: Agriculture*, 103–301; *Eleventh Census of the United States, 1890: Agriculture*, 340–483; *Twelfth Census of the United States, 1900: Agriculture*, 462–63, 612–13; George Hyde Clarke to Mary G. Carter, March 18, 1884, Box 5, George Hyde Clarke Papers, #2800, CUL; Ezra Meeker, *Hop Culture in the United States* (Puyallup, Wash.: E. Meeker and Co., 1883), 48; Herbert Myrick, *The Hop: Its Culture and Cure, Marketing and Manufacturing* (New York: Orange Judd Co., 1914), 243–47; Elmer O. Frippen, *Rural New York* (New York: Macmillan, 1921), 195; *Emmet Wells' Weekly Hop Circular*, August 7, 1888. New York State yields averaged 559 pounds per acre in 1879. Oregon farmers expected consistently more than 1,000 pounds per acre, California growers attained over 1,400 pounds per acre, and Washington farmers produced 1,600 pounds per acre in the 1880s.

5. *Ninth Census of the United States, 1870: Agriculture*, 210–13, 314–17; *Tenth Census of the United States, 1880: Agriculture*, 103–301; *Eleventh Census of the United States, 1890: Agriculture*, 340–483; *Twelfth Census of the United States, 1900: Agriculture*, 462–63, 612–13; Willard V. Huntington, *Oneonta Memories* (San Francisco: Bancroft Co., 1891), 198–200; Baker, *Moral Frameworks of Public Life*, 5; Martin F. Noyes, *A History of Schoharie County* (Richmondville, N.Y.: Richmondville *Phoenix*, 1954), 91; William F. Helmer, *O&W: The Long Life and Slow Death of the New York, Ontario & Western Railway* (Berkeley: Howell-North, 1959), 18, 59, 62, 67. The shift to fresh milk production was pronounced between 1870 and 1900; Delaware County produced 198,326 gallons in 1870 and 20,058,381 in 1900. In that same time period, production rose from 4,808,627 to 15,587,968 in Otsego and from 217,346 to 3,240,568 in Schoharie.

6. Jeremy Atack, "Tenants and Yeomen in the Nineteenth Century," *Agricultural History* 62 (Summer 1988): 6–32. The figures gleaned from the federal census probably underestimate tenancy in New York. Other sources suggest that a large number of farmers engaged in some tenancy arrangements and that tenants often shared parcels of property. The enumerator likely asked a farmer if he owned the farm he lived on. But men with some property often farmed other land on cash or share terms, especially hops yards and pastures for cattle or sheep. As freehold farm sizes decreased, the need to enter such agreements would increase.

7. *Ninth Census of the United States, 1870: Agriculture,* 210, 314–17; *Tenth Census of the United States, 1880: Agriculture,* 74–77; *Eleventh Census of the United States, 1890: Agriculture,* 220–21; *Twelfth Census of the United States, 1900: Agriculture,* 106–9.

8. Quote is from Sharon Good, "The Hop Culture" (M.A. thesis, SUNY College at Oneonta, Cooperstown Graduate Program, 1968), 93. See also Hall, *Gleanings from a Country Journal,* 52, and Frippen, *Rural New York,* 195.

9. *Tenth Census of the United States, 1880: Agriculture,* 74–77; *Eleventh Census of the United States, 1890: Agriculture,* 220–21; *Twelfth Census of the United States, 1900: Agriculture,* 106–9.

10. Huntington, *Oneonta Memories,* 134–35, 164; *Tenth Census of the United States, 1880: Agriculture,* 74–77; *Eleventh Census of the United States, 1890: Agriculture,* 220–21; *Twelfth Census of the United States, 1900: Agriculture,* 106–9.

11. Lears, "Packaging the Folk," 107. A sense of the dimensions of these changes can be gleaned from the following: Charles Broughton Diaries, July 5, 1884, MSS, NYSHA; M. W. Frisbee Diaries, December 31, 1870, December 31, 1882, MSS, NYSHA; Lucius Bushnell Diaries, January 1, 1880, February 12, 1881, April 27, June 1, 1892, October 16, 1893, MSS, NYSHA.

12. Jared Van Wagenen Jr., *Days of My Years: The Autobiography of a York State Farmer* (Cooperstown: NYSHA, 1962), xii, 59.

13. Hop Growth Account, 1881, MSS, Box 63, Clarke Papers; Huntington, *Oneonta Memories,* 69.

14. George Hyde Clarke to Mary G. Carter, March 7, 1885, Box 5, Clarke Papers.

15. Ibid., March 18, 1884; Paul W. Gates, *Landlords and Tenants on the Prairie Frontier* (Ithaca: Cornell University Press, 1973), 165, 256.

16. George Hyde Clarke to Mary G. Carter, May 10, 1885, Box 5, Clarke Papers.

17. George Hyde Clarke to George Clarke, April 29, 1881, George Hyde Clarke to Mary G. Carter, March 17, 1885, Box 5, George Hyde Clarke to George Clarke, August 3, 1885, Box 7, ibid.

18. George Clarke to Marie Clarke, July 25, 1875, Box 3, ibid.

19. George Hyde Clarke to Mary G. Carter, March 7, 1885, Box 5, ibid.

20. Ira Sherman, "Goin' Hop Pickin': Hop-Growing Was Once a Big Industry in Central New York," *Courier Magazine* 4, no. 7 (July 1955): 28–31; Good, "Hop Culture," 95–98; George Hyde Clarke to Mary G. Carter, May 12, 1885, Box 5, Clarke Papers; Myrick, *The Hop,* 243–47; Hall, *Gleanings from a Country Journal,* 83. Production previously considered large paled next to Clark's. In 1870, Hall noted, the Morris *Chronicle* had considered Schuyler Cummings's 50,000 pounds of hops on his Middlefield farm, his forty-eight pickers, and $3,300 in picking and curing costs (above board) extremely large.

21. *"Grip's" Valley Gazette* 5 (1895): 20, 26; Zenas France to ——— Robinson, ca. 1871, telegram, Zenas France to Q. K. McKendrick, September 13, 1877, Box 2, Clifford France Timebook, 1884–91, Vol. 6, MSS, Clifford France Papers, #160, NYSHA.

22. John D. Shaul to G. W. Wilkins, February 19, 1877, Box 1, John D. Shaul Collection, #252, NYSHA.

23. John Marsh Account Book, 1895–97, MSS, Marsh Family Papers, #194, NYSHA; "To Whom It May Concern," political broadside, David Wilber, 1888, MSS, Political Ephemera, NYSHA; Good, "Hop Culture," 61–75.

24. Ford Welles to Clifford France, March 12, 1888, Clifford France to Charles Ehlerman, March 31, 1888, Clifford France Letterbooks, Vol. 14, Hop Contract Form, May 2, 1888, MSS, Box 1, France Papers.

25. Quote from *Delaware Dairyman,* March 19, 1886; Myrick, *The Hop,* 251; Meeker, *Hop Culture,* 63–65; Cornelius Shaul to J. D. Shaul, November 25, 1883, Box 1, Shaul Collection; J. R. Sutliff Daybooks, 1892–95, MSS, #A1S98, NYSHA.

26. George Hyde Clarke to Mary G. Carter, September 17, 1885, Box 7, Clarke Papers. The 1882 prices were extremely high. In 1870, the price was between twelve and fourteen cents per pound. Hall, *Gleanings from a Country Journal,* 79.

27. George Clarke to ———, March 10, 1882, Box 64, George Barnard to Alexander Yates, January 4, 1888, Report of Assignee of George Clarke, George Barnard, January 31, 1888, Report of Assignee of George Clarke, George Barnard, July 27, 1888, Box 71, Allison Butts to George Clarke, February 3, 1889, Box 73, Clarke Papers. Clarke owned 40,900 acres in 1882 in Otsego, Montgomery, Oneida, Dutchess, Saratoga, Fulton, and Greene Counties. Creditors held (by his count) mortgages amounting to $205,000 against the estate, leaving a net value of over $1,385,000. His 10,000 Otsego acres were worth $300,000. In 1887, $183,438 of Clarke's real estate was sold to cover debts, with $227,451 more sold from January to July 1888. His estate was gutted in Montgomery, Otsego, and Oneida Counties.

28. Sherman, "Goin' Hop Pickin'," 30–31; Myrick, *The Hop,* 253; Good, "Hop Culture," 94–100; Frippen, *Rural New York,* 195; broadside, George Hyde Clarke for State Assembly, October 15, 1906, Box 152, Clarke Papers.

29. George Hyde Clarke to Mary G. Carter, March 7, 1885, Box 5, Clarke Papers; Emma Clarke to Anne Clarke, May 7, 1872, Clarke Family Papers, #115, NYSHA; Share Lease: John D. Shaul to John Boocock, April 1, 1876, John D. Shaul's Accounts for John Boocock, 1873–80, MSS, Box 1, Shaul Collection; Share Lease: A. H. Becker to S. E. Preston and Abram W. Hyney, April 1, 1874, Share Lease: A. H. Becker to Edith Becker, April 9, 1875, Lease: Rachel Becker to Edith Becker, April 1, 1881, Lease: Rachel Becker to Edith Becker, January 20, 1882, Lease: Harmon Howland to Laura Preston, February 4, 1885, Lease: Daniel S. Hoyt to Daniel H. and Mary Conrad, August 20, 1889, Lease: Sarah J. and Delphine Antisdel to Isaac Crippen, January 29, 1895, Share Lease: Edith Becker to Herman Falmsbee and Marvin Stringer, December 16, 1901, Box 1, Becker Family Papers, #297, NYSHA; *Delaware Dairyman,* February 5, 1886. For similar changes out West, see Gates, *Landlords and Tenants,* 309.

30. Quote from Good, "Hop Culture," 94; Isaac Peaslee to Clifford France, March 18, 1884, Clifford France Letterbooks, Vol. 14, France Papers; H. F. Phinney Real Estate Account Book, MSS, Vol. 6, Phinney Family Collection, #235, NYSHA; Caleb Clarke Real Estate Account Book, MSS, Box 1, Caleb Clarke Papers, #108, NYSHA.

31. Sherman, "Goin' Hop Pickin'," 28–31; Myrick, *The Hop,* 251–52; Clayton E. Risley, "Hop Picking Days," *New York Folklore Quarterly* 5 (Spring 1949): 18–24.

32. *Transactions of the New York State Agricultural Society, 1867* (Albany: C. Van Benthuysen and Sons, 1867), 185; LaWanda Cox, "The American Agricultural Wage Earner, 1865–1900: The Emergence of a Modern Labor Problem," *Agricultural History* 22 (April 1948): 95–114.

33. Helmer, *O&W,* 25–35.

34. Cooperstown *Freeman's Journal,* August 21, 1886. The increase in nativism during the late nineteenth century cannot be attributed to a rise in the immigrant population. In 1900, all three counties had fewer immigrants than in 1870. Delaware County decreased from 3,469 (8 percent of the total population of 42,972) to 2,227 (5 percent of the total 46,413), Otsego fell from 2,733 (6 percent of the total 48,967) to 2,294 (5 percent of the total 48,939), and Schoharie dropped from 901 (3 percent of the total 33,340) to 659 (2 percent of the total 26,854). During that period, however, Italians and Poles began to supplant the Irish, English, and Scots. The new immigrants dressed much differently and did not share a common language with farmers, which likely contributed to much of the popular reaction against them. They probably made up the bulk of the migrant workers brought into the area at harvest as well. *Ninth Census of the United States, 1870: Population,* 365–66; *Twelfth Census of the United States, 1900: Population,* 645–47.

35. *Transactions of the New York State Agricultural Society, 1867,* 184–89; Cox, "American Agricultural Wage Earner," 99–100; George Hyde Clarke to Mary G. Carter, March 14, 1885, Box 5, Clarke Papers; Labor contract: Amos Brayman and Clifford and Maria France, April 1, 1883, Box 1, France Papers; Emma Clarke to Tommie Clarke, November 1, 1864, Box 1, Clarke Family Papers; Van Wagenen, *Days of My Years,* xii. Brayman's contract with the Frances typified those of most "hired men" on upper-class farms. Clifford France paid him $300 per year, prorated for sick days, and required Brayman and his wife to milk the cows, produce the butter, and provide additional labor for butter-making if necessary.

36. Sherman, "Goin' Hop Pickin'," 28–31; Risley, "Hop Picking Days," 18–24; John D. Shaul Accounts for John Boocock, 1873–80, MSS, Box 1, Shaul Collection; Good, "Hop Culture," 76–82.

37. Hall, *Gleanings from a Country Journal,* 64.

38. Ibid., 71. See also Cooperstown *Freeman's Journal,* September 19, 1878.

39. *Delaware Dairyman,* March 19, 1886.

40. Risley, "Hop Picking Days," 18–24; Sherman, "Goin' Hop Pickin'," 28–31; Good, "Hop Culture," 76–82; Cox, "American Agricultural Wage Earner," 102, 105–6.

41. Good, "Hop Culture," 38–60.

42. Cooperstown *Freeman's Journal,* November 21, 1878.

43. Ibid., September 19, 1878; Good, "Hop Culture," 38–60; Sherman, "Goin' Hop Pickin'," 28–31; Risley, "Hop Picking Days," 18–24.

44. Sherman, "Goin' Hop Pickin'," 28–31; Risley, "Hop Picking Days," 18–24; Huntington, *Oneonta Memories,* 195; Good, "Hop Culture," 78–82.

45. Albert C. Mayham, *The Anti-Rent War on Blenheim Hill: An Episode of the 40's* (Jefferson, N.Y.: Frederick L. Frazee, 1906), 87; Emma Jarvis to Will Clarke, December 8, 1887, Box 1, Clarke Family Papers; Elbridge Hunter Diaries, 1875, 1886, 1889, passim, MSS, DCHA; *Delaware Dairyman,* December 13, 1878.

46. Emma Jarvis to Will Clarke, March 7, 1888, Box 1, Clarke Family Papers; George D. Taylor, *These Hills Are Not Barren: The Story of a Century Farm* (New York: Exposition Press, 1950), 24–27, 33–35, 54; Van Wagenen, *Days of My Years,* xii; Walter E. Bard Diary, April 11, April 22, September 6, 1876, MSS, NYSHA.

47. Nancy Grey Osterud, *Bonds of Community: The Lives of Farm Women in Nineteenth-Century New York* (Ithaca: Cornell University Press, 1991), 139–227; Baker, *Moral Frame-*

works of Public Life, xv–xviii; Taylor, *These Hills,* 27, 35; *Delaware Dairyman,* February 19, May 7, 1886.

48. Taylor, *These Hills,* 35.

49. M. W. Frisbee Diary, September 9, 1882.

50. Charles Broughton Diaries, October 1, 1889, December 4, 1890; Walter E. Bard Diary, December 22, 1876.

51. Letter placed within Charles Broughton Diary, Addison J. Blumberg to Charles Broughton, May 21, 1876.

52. Charles Broughton Diaries, October 27, 1876.

53. Lucius Bushnell Diaries, January 9, November 22, 1859, February 5, 1878.

54. Ibid., October 4, 1879, February 23, 1881, February 27, 1890.

55. Charles Broughton Diaries, October 20, 1900, January 22, June 15, 1901.

56. *Transactions of the New York State Agricultural Society, 1867,* 184; Lucius Bushnell Diaries, December 30, 1886; Cheese Factory Returns, George Clarke Estate, 1885, Box 68, Clarke Papers; John D. Shaul to E. Burrell and Co., March 21, 1878, Box 1, Shaul Collection; Gates, "Agricultural Change," 118, 125; Hall, *Gleanings from a Country Journal,* 38; Sally McMurry, *Transforming Rural Life: Dairying Families and Agricultural Change, 1820–1885* (Baltimore: Johns Hopkins University Press, 1995), 148–93.

57. Taylor, *These Hills,* 33–34.

58. Charles Broughton Diaries, July 17, 1896; A. G. Beardslee Milk Receipts, 1881–92, MSS, Beardslee Family Papers, #88, NYSHA.

59. Cooperstown *Freeman's Journal,* March 4, 1875.

60. Taylor, *These Hills,* 54; "In the Interest of Commerce and Fair Dealing," Pro-Oleomargarine Petition, Members of the New York Product Exchange to the United States Congress, 1881, Dairy, Box 1, Folder 21, *Fraudulent Butter, and Its Ruinous Effects on American Dairying: Address of Joseph H. Reall, President of the American Agricultural and Dairy Association, Before the Committee on Agriculture of the House of Representatives, April 3, 1886* (1886), *Butter vs. Oleomargarine, A Speech by the Hon. Warner Miller, of New York, Chairman of the Committee on Agriculture and Forestry, in the United States Senate, July 17, 1886* (1886), Dairy, Box 1, Folder 15, Warshaw Collection, #60, NMAH.

61. Helmer, *O&W,* 62; Noyes, *History of Schoharie County,* 66.

62. Taylor, *These Hills,* 41.

63. Ibid., 40–41, 74.

64. Van Wagenen, *Days of My Years,* x; *Patent Round Silos* (Cobleskill, N.Y.: Harder Manufacturing Co., 1902); *Why It Pays: The Smalley Ensilage and Fodder Cutters* (Buffalo: Gies and Co., 1889).

65. Charles Broughton Diaries, November 23, 1895.

66. Taylor, *These Hills,* 74.

67. Robert H. Wiebe, *The Search for Order, 1877–1920* (New York: Hill and Wang, 1967), 44.

Chapter 7: Tenant Unrest and Elite Cooperation

1. Solon J. Buck, *The Granger Movement: A Study of Agricultural Organization and Its Political, Economic, and Social Manifestations 1870–1880* (Cambridge: Harvard University

Press, 1933); Lee Benson, *Merchants, Farmers, and Railroads: Railroad Regulation and New York Politics, 1850–1887* (Cambridge: Harvard University Press, 1955); Robert H. Wiebe, *The Search for Order, 1877–1920* (New York: Hill and Wang, 1967); Lawrence Goodwyn, *Democratic Promise: The Populist Moment in America* (New York: Oxford University Press, 1976); Paula Baker, *The Moral Frameworks of Public Life: Gender, Politics, and the State in Rural New York, 1870–1930* (New York: Oxford University Press, 1991). These works are part of a much longer list of studies that touch on the themes discussed in this chapter but also offer a range of arguments about the process of political change in the late nineteenth century.

2. Richard L. McCormick, *From Realignment to Reform: Political Change in New York State, 1893–1910* (Ithaca: Cornell University Press, 1981), 31; Benson, *Merchants, Farmers, and Railroads,* vii.

3. Baker, *Moral Frameworks of Public Life,* 3–154, recognizes central New York farmers' activism on their own behalf, particularly from within the Grange, but does not fully explore the relationship between the multiple farm organizations.

4. Barron found that such divisions, if expressed in less political terms, developed concurrently in Vermont. Hal S. Barron, *Those Who Stayed Behind: Rural Society in Nineteenth-Century New England* (New York: Cambridge University Press, 1984), 14–15, 31–36.

5. George Clarke to Marie Clarke, July 25, 1875, Box 3, George Hyde Clarke Papers, #2800, CUL.

6. George Hyde Clarke to Mary G. Carter, March 3, 1885, Box 5, ibid.

7. Ibid., March 21, 1885.

8. George Hyde Clarke to George Clarke, October 25, 1883, ibid.

9. Henry L. Stanton to George Clarke, July 28, 1868, George W. Hubbell to George Clarke, August 4, 1868, Chauncey Andrews to George Clarke, August 19, 1868, Box 54, George Clarke Journal, 1874, MSS, Box 102, ibid.

10. *Supreme Court. George Clarke Against Nicholas Fero, Jr.* (Fonda, N.Y.: Mohawk Valley Democrat Steam Print, 1876).

11. Ibid.; *Joseph A. Sherman v. George Clarke,* December 17, 1880, MSS, Box 62, Clarke Papers. The Sherman case illustrated the complexity of Clarke's arrangements with tenants. Sherman's father had taken a three-life lease from Clarke in 1838. He then rented the land to other tenants. When he died, his wife sold her interest to Clarke for seventy dollars per year for the duration of the contract (her life). She and her son sued Clarke for back rents on the property.

12. Albany *Freeholder,* March 13, December 18, 1850, March 26, 1851; *In the Court of Appeals. The People of the State of New York v. George Clarke. Opinion of the Court* (Albany: Munsell, 1854); Richard Cooper Journal, 1854–55, MSS, Box 100, "Supreme Court of the State of New York. George Clarke v. Charles Montanye. Complaint," August 31, 1878, MSS, Box 59, Clarke Papers.

13. Gamaliel Bowdish to Duncan C. Pell, April 18, 1851, Box 1, T. O. Clarke to George Clarke, September 22, 1866, George Clarke to Marie Clarke, July 25, 1875, Box 3, George Clarke to T. C. Smith, November 21, 1871, Box 55, George Clarke Journal, 1874, MSS, Box 102, George Clarke to Charleston Tenants' Committee, February 2, 1874, Minutes of the Montgomery County Mutual Tenants' Association, April 28, 1874, MSS, Box 57, T. C. Smith

to George Clarke, January 24, 1876, Box 58, *George Clarke v. Charles Montanye,* Box 59, *Sherman v. George Clarke,* Box 62, Clarke Papers.

14. George Clarke to Home Fire Insurance Company, March 14, 1871, George Clarke to T. C. Smith, November 21, December 17, 1871, Box 55, "Lenox" to the editor of the Montgomery County *Republican,* undated, 1874, George Clarke to the editor of the Utica *Morning Herald,* May 31, 1875, Box 57, George Clarke to Marie Clarke, July 25, 1875, Box 3, *George Clarke v. Charles Montanye,* Box 59, ibid.; Paul W. Gates, *Landlords and Tenants on the Prairie Frontier* (Ithaca: Cornell University Press, 1973), 281–97. As Gates noted, the period was marked by tenant unrest throughout the North that centered on the very issues that animated the Montgomery and Otsego troubles: reimbursement for improvements made to the property, taxation, and the right of landlords to reenter lease lands.

15. *George Clarke v. Charles Montanye,* Box 59, Minutes of the Montgomery County Mutual Tenants' Association, April 28, 1874, MSS, Box 57, Clarke Papers.

16. "Well Wisher" to George Clarke, April 3, 1877, Box 58, ibid.

17. George Clarke to the editor of the Utica *Morning Herald,* May 31, 1875, Box 57, ibid.

18. George Clarke to the Corresponding Committee of the Montgomery County Mutual Tenants' Association, February 25, 1874, George Clarke to A. F. Wellmarth, March 31, 1875, ibid.

19. Montgomery County *Republican,* January 27, 1874, quoted in *George Clarke v. Charles Montanye,* Box 59, ibid.

20. *George Clarke v. Charles Montanye,* Box 59, ibid.

21. George Clarke to the editor of the Utica *Morning Herald,* May 31, 1875, Box 57, ibid. It is worth noting that "playing Indian" was common among local boys, which likely made the choice of Captain Jack and the Modocs fairly natural. See Willard V. Huntington, *Oneonta Memories* (San Francisco: Bancroft Co., 1891), 127–28, and Philip J. Deloria, *Playing Indian* (New Haven: Yale University Press, 1998), 38–70. For a discussion of western tenant associations' use of similar republican organizational forms, see Gates, *Landlords and Tenants,* 285–86.

22. *George Clarke v. Charles Montanye,* Box 59, Clarke Papers.

23. Ibid.; George Clarke to the editor of the Utica *Morning Herald,* May 31, 1875, Box 57, Clarke Papers.

24. George Clarke Journal, 1874, MSS, Box 102, George Clarke to Aetna Insurance Company, April 17, 1876, Box 58, *George Clarke v. Charles Montanye,* Box 59, Clarke Papers.

25. Montgomery County *Republican,* January 27, 1874, quoted in *George Clarke v. Charles Montanye,* Box 59, ibid.

26. See Eric J. Hobsbawm and George Rudé, *Captain Swing: A Social History of the Great English Agricultural Uprising of 1830* (New York: W. W. Norton, 1968), 195–250, for similarities between the Captain Jack insurgents and English land rioters in the midst of a commercial transition in the 1840s.

27. T. C. Smith to George Clarke, January 24, 1876, February 20, 1877, Box 58, Clarke Papers.

28. *Ninth Census of the United States, 1870: Agriculture* (Washington, D.C.: Government Printing Office, 1872), 210, 314–17; *Tenth Census of the United States, 1880: Agriculture* (Washington, D.C.: Government Printing Office, 1883), 74–77; Cooperstown *Freeman's*

Journal, October 3, 1872; Wiebe, *Search for Order* (New York: Hill and Wang, 1967), 16–17.

29. T. O. Clarke to George Clarke, May 6, June 11, 1868, George Clarke to T. O. Clarke, June 12, 1868, Box 3, J. W. Alexander to Henry Day, December 31, 1870, Box 55, Clarke Papers.

30. R. B. Gillies to George Clarke, July 10, 1871, A. M. McLean to T. C. Smith, July 31, 1871, George Clarke to T. C. Smith, September 12, October 1, October 27, November 19, November 20, November 21, December 17, 1871, Box 5, ibid.

31. George Clarke to T. C. Smith, December 17, 1871, ibid.

32. George Clarke to T. C. Smith, March 20, 1872, Box 56, T. C. Smith to George Clarke, January 7, 1876, February 20, 1877, George Clarke to T. C. Smith, March 8, 1877, A. Bowman to George Clarke, March 30, 1878, D. K. Colburn to George Clarke, August 14, 1878, Box 58, Bill, North American Detective Police Agency to George Clarke, September 19, 1878, Box 59, T. C. Smith to George Clarke, December 21, 1880, Box 61, T. C. Smith to George Clarke, May 10, 1883, Box 65, James Young to George Clarke, March 13, 1884, Box 66, ibid.

33. George Clarke to T. C. Smith, February 16, 1872, Box 56, ibid.; *Emmet Wells' Weekly Hop Circular,* August 7, 1888; Herbert Myrick, *The Hop: Its Culture and Cure, Marketing and Manufacturing* (New York: Orange Judd Co., 1914), 243–47.

34. T. C. Smith to George Clarke, October 19, 1875, Box 57, Chattel Mortgage: George Clarke to Stephen Kilmartin, July 17, 1879, Box 60, Chattel Mortgage: George Clarke to George Steger, August 12, 1880, Box 62, Alva Mayne to T. C. Smith, April 9, 1885, Box 68, Clarke Papers.

35. Cooperstown *Freeman's Journal,* October 12, 1855, October 23, 1886.

36. T. O. Clarke to George Clarke, July 20, 1865, Box 3, Clarke Papers; H. F. Phinney Diary, May 30, 1872, Vol. 6, Box 1, Phinney Family Collection, #235, NYSHA.

37. T. O. Clarke to George Clarke, July 20, 1865, Box 3, Clarke Papers.

38. Cooperstown *Freeman's Journal,* October 24, 1878, October 23, 1886; Myrick, *The Hop,* 243–47; Silas Dutcher Diaries, February 29, 1856, June 27, 1864, MSS, NYSHA.

39. *Century of Progress: History of the Delaware and Hudson Company* (Albany: J. B. Lyon Co., 1925), 631–52; Benson, *Merchants, Farmers, and Railroads,* 21–26; Cooperstown *Freeman's Journal,* March 12, March 26, April 16, 1874; James C. Mohr, *The Radical Republicans and Reform in New York during Reconstruction* (Ithaca: Cornell University Press, 1973), xiii, 1–20; Charles Francis Adams Jr., "An Erie Raid," 156–212, Albert Stinckney, "The Lawyer and His Clients," 213–45, George Ticknor Curtis, "An Inquiry into the Albany & Susquehanna Railroad Litigation of 1869 and Mr. David Dudley Fields' Connection Therewith," 246–350, and Jeremiah H. Black, "A Great Lawsuit and a Field Fight," 351–86, all in *High Finance in the Sixties: Chapters from the Early History of the Erie Railway,* ed. Frederick C. Hicks (New Haven: Yale University Press, 1929); John Eddy to Augustus R. Elwood, October 25, 1869, Augustus R. Elwood Papers, #96, NYSHA.

40. Cooperstown *Freeman's Journal,* October 23, 1886; John Kerwin to Clifford France, March 2, 1887, Box 1, Clifford France Papers, #160, NYSHA.

41. David Wilber, "To Whom It May Concern," political broadside, 1888, Political Ephemera, NYSHA.

42. Cooperstown *Freeman's Journal,* October 24, 1878; Benson, *Merchants, Farmers, and*

Railroads, 102–4; David Montgomery, *Beyond Equality: Labor and the Radical Republicans, 1862–1872* (New York: Alfred A. Knopf, 1967), 340–56, 425–47.

43. George Clarke to Marie Clarke, July 25, 1875, Box 3, Clarke Papers.

44. George Clarke to Thomas Patten, undated, 1878, Box 58, ibid.

45. Cooperstown *Freeman's Journal,* September 19, October 17, 1878.

46. Franklin *Register,* October 11, 1878.

47. Cooperstown *Freeman's Journal,* November 21, 1878; Benson, *Merchants, Farmers, and Railroads,* 102–4; Baker, *Moral Frameworks of Public Life,* 45–48.

48. Henry Nichols to George Clarke, January 16, 1882, Box 64, Clarke Papers.

49. George Clarke to Thomas Patten, undated, 1878, Box 59, Henry Nichols to George Clarke, January 16, 1882, Box 64, ibid.; Benson, *Merchants, Farmers, and Railroads,* 151–73.

50. E. M. Harris to Marie Clarke, December 24, 1889, Box 73, Clarke Papers.

51. Cooperstown *Freeman's Journal,* October 17, 1878; Smith Edick to George Clarke, March 19, 1887, Box 70, George Hyde Clarke to George Clarke, January 5, 1889, Box 9, Clarke Papers; Sharon Good, "The Hop Culture" (M.A. thesis, SUNY College at Oneonta, Cooperstown Graduate Program, 1968), 95–98.

52. H. V. Pindar to George Hyde Clarke, April 11, 1892, Box 74, "George Hyde Clarke for Assembly," political broadside, 1906, Box 152, Clarke Papers; Cooperstown *Freeman's Journal,* October 23, 1886; Good, "Hop Culture," 93, 95–98; Ralph Birdsall, *The Story of Cooperstown* (Cooperstown: Augur's Book Store, 1948), 239.

53. Cooperstown *Freeman's Journal,* September 8, November 3, 1892; George Hyde Clarke, "Reasons for Establishing a Hop Growers' Exchange, and What It Might Accomplish," broadside, ca. 1895, Box 152, Clarke Papers.

54. Clarke, "Reasons for Establishing a Hop Growers' Exchange."

55. Ibid.; Myrick, *The Hop,* 243–47; Ira Sherman, "Goin' Hop Pickin': Hop-Growing Was Once a Big Industry in Central New York," *Courier Magazine* 4, no. 7 (July 1955): 30.

56. George D. Wheeler Autobiography, MSS, Box 1, Folder 3, Minute Book of the Deposit Farmer's Club, MSS, Box 3, Wheeler Family Papers, #3878, CUL.

57. Cooperstown *Freeman's Journal,* January 29, 1874, August 14, 1886; Franklin *Register,* December 13, 1878; *Delaware Dairyman,* February 5, 1886; Elmer O. Frippen, *Rural New York* (New York: Macmillan, 1921), 263; Patent Rights Circular, New York State Dairymen's Protective Association, 1881, Box 1, Folder 21, "In the Interest of Commerce and Fair Dealing," Pro-Oleomargarine Petition, Members of the New York Product Exchange to the United States Congress, 1881, Dairy, Box 1, Folder 21, *Fraudulent Butter, and Its Ruinous Effects on American Dairying: Address of Joseph H. Reall, President of the American Agricultural and Dairy Association, Before the Committee on Agriculture of the House of Representatives, April 3, 1886* (1886), *Butter vs. Oleomargarine, A Speech by the Hon. Warner Miller, of New York, Chairman of the Committee on Agriculture and Forestry, in the United States Senate, July 17, 1886* (1886), Dairy, Box 1, Folder 15, Warshaw Collection, #60, NMAH; Minute Book of the Farmer's Club of Deposit, January 2, 1886, MSS, Box 3, Wheeler Family Papers; Benson, *Merchants, Farmers, and Railroads,* 80–114.

58. George D. Wheeler Account Book, 1863–93, MSS, Minute Book of the Farmer's Club of Deposit, June 21, 1888, MSS, Box 3, George D. Wheeler Autobiography, MSS, Box 1, Folder 3, Wheeler Family Papers.

59. Minute Book of the Farmer's Club of Deposit, December 3, 1887, MSS, Box 3, ibid.

60. Ibid., April 19, 1884; Baker, *Moral Frameworks of Public Life,* 3–23.

61. *Delaware Dairyman,* April 2, 1886.

62. Minute Books of the Farmer's Club of Deposit, November 28, 1885, January 2, 1886, MSS, Box 3, Wheeler Family Papers; Cooperstown *Freeman's Journal,* October 24, August 14, 1878; *Delaware Dairyman,* February 5, March 5, April 23, May 7, 1886; John W. McArthur, *New Developments: Including the Grange, Anti-Monopoly Farmers' Alliance, Co-operative Fire Insurance, and the Economic Barn, to which is Added an Account of Artificial Butter* (Oneonta, N.Y.: Oneonta Press, 1886), 143–48.

63. Minute Books of the Farmer's Club of Deposit, January 2, February 26, 1887, MSS, Box 3, Wheeler Family Papers.

64. Ibid., March 12, 1887.

65. Ibid.

66. Ibid., March 3, 1888.

67. Ibid.; E. R. Wattles to George D. Wheeler, March 1, 1889, Box 3, Wheeler Family Papers; Baker, *Moral Frameworks of Public Life,* 3–23.

68. Minutes of the Deposit Farmer's Club, March 15, March 19, December 3, 1887, March 3, 1888, MSS, Box 3, Wheeler Family Papers; *Delaware Dairyman,* February 12, 1886.

69. Minutes of the Deposit Farmer's Club, October 8, 1887, MSS, Box 3, Wheeler Family Papers; R. L. McCormick, *From Realignment to Reform,* 26; Frippen, *Rural New York,* 257–300; Jared Van Wagenen Jr., *Days of My Years: The Autobiography of a York State Farmer* (Cooperstown, N.Y.: NYSHA, 1962), 121–27; Mohr, *Radical Republicans,* 153; Jeffrey W. Moss and Cynthia B. Lass, "A History of Farmer's Institutes," *Agricultural History* 62 (Spring 1988): 150–63; Alan I. Marcus, "The Ivory Silo: Farmer-Agricultural College Tensions in the 1870s and 1880s," *Agricultural History* 60 (Spring 1986): 22–36.

70. Charles Broughton Diaries, January 12, 1889, MSS, NYSHA.

71. Van Wagenen, *Days of My Years,* 56, 60–141; John R. McMahon, ed., *How These Farmers Succeeded* (New York: Henry Holt, 1919), 129–30.

72. McMahon, *How These Farmers Succeeded,* 127.

73. Ibid., 126–46.

Chapter 8: The Grange Movement, 1874–1900

1. Oliver H. Kelley, *Origin and Progress of the Order of the Patrons of Husbandry in the United States: A History from 1866 to 1873* (Philadelphia: J. A. Wagenseller, 1875), 13–20; Leonard L. Allen, *History of the New York State Grange* (Watertown, N.Y.: Hungerford-Holbrook Co., 1934), 39–40, 189–93; John W. McArthur, *New Developments: Including the Grange, Anti-Monopoly Farmers' Alliance, Co-operative Fire Insurance, and the Economic Barn, to which is Added an Account of Artificial Butter* (Oneonta, N.Y.: Oneonta Press, 1886), 1–10; Solon J. Buck, *The Granger Movement: A Study of Agricultural Organization and Its Political, Economic, and Social Manifestations 1870–1880* (Cambridge: Harvard University Press, 1933), 40–79; Richard Hofstadter, *The Age of Reform* (New York: Vintage Books, 1955), 36–63; Paula Baker, *The Moral Frameworks of Public Life: Gender, Politics, and the State in Rural New York, 1870–1930* (New York: Oxford University Press, 1991),

3–154; Thomas A. Woods, *Knights of the Plow: Oliver H. Kelley and the Origins of the Grange in Republican Ideology* (Ames: Iowa State University Press, 1991), xv–xxii; Donald B. Marti, *Women of the Grange: Mutuality and Sisterhood in Rural America, 1866–1920* (New York: Greenwood Press, 1991), 1–13.

2. Buck, *Granger Movement,* 61–62; Lee Benson, *Merchants, Farmers, and Railroads: Railroad Regulation and New York Politics, 1850–1887* (Cambridge: Harvard University Press, 1955), 80–93; D. Sven Nordin, *Rich Harvest: A History of the Grange, 1867–1900* (Jackson: University Press of Mississippi, 1974), 30–44; Morton Rothstein, "Farmer Movements and Organizations: Numbers, Gains, Losses," *Agricultural History* 62 (Summer 1988): 161–81; Nancy Grey Osterud, *Bonds of Community: The Lives of Farm Women in Nineteenth-Century New York* (Ithaca: Cornell University Press, 1991), 254–62.

3. Cooperstown *Freeman's Journal,* February 2, February 19, March 19, 1874; Gracchus Americanus [Thomas S. Goodwin], *The Grange: A Study in the Science of Society* (New York: G. P. Putnam's Sons, 1874), 22–25; McArthur, *New Developments,* 9–10; Buck, *Granger Movement,* 3; Baker, *Moral Frameworks of Public Life,* xv–xviii, 24–55; Rothstein, "Farmer Movements," 161–81. My contention that the Grange, despite its social conservatism, can still be considered "radical" in its anticapitalist ideas is partly derived from Robert D. Johnston, *The Radical Middle Class: Populist Democracy and the Question of Capitalism in Progressive Era Portland, Oregon* (Princeton: Princeton University Press, 2003).

4. Baker, *Moral Frameworks of Public Life,* xv–xviii, 8–14. Jeffrey Ostler discovered that in states where farmers held the balance between the two major parties, they tended to avoid third-party insurgencies. During most of the late nineteenth century, this applied to New York. See Jeffrey Ostler, *Prairie Populism: The Fate of Agrarian Radicalism in Kansas, Nebraska, and Iowa, 1880–1892* (Lawrence: University Press of Kansas, 1993).

5. Manuscript Membership Rolls of the New York State Grange, Seward, Schoharie County, Subordinate Grange #344, Huntersland, Schoharie County, Subordinate Grange #375, Rock Valley, Delaware County, Subordinate Grange #470, China, Delaware County, Subordinate Grange #475, Elk Creek, Otsego County, Subordinate Grange #506, Westville, Otsego County, Subordinate Grange #540, in Office of the State Master of the New York State Grange, Cortland, New York; Manuscript United States Census, 1880, #T9.1, Schedules 1 and 2, Albany County, Roll 808, Broome County, Rolls 810, 811, Delaware County, Rolls 822, 823, Otsego County, Rolls 915, 916, Schoharie County, Rolls 930, 931, Sullivan County, Roll 936, National Archives; Osterud, *Bonds of Community,* 257–58.

6. Manuscript United States Census, 1880, Schedule 2, Schoharie County, Rolls 930, 931.

7. Ibid., Otsego County, Rolls 915, 916.

8. Ibid., Delaware County, Rolls 822, 823, Broome County, Rolls 810, 811, Sullivan County, Roll 936.

9. Ibid., Schedules 1 and 2, Delaware County, Rolls 822, 823, Broome County, Rolls 810, 811; Minutes of the Deposit Farmer's Club, April 10, 1884, MSS, Box 3, Wheeler Family Papers, #3837, CUL; Rothstein, "Farmer Movements," 161–81.

10. Allen, *History of the New York State Grange,* 190–217; Nordin, *Rich Harvest,* 30–44; Membership Rolls of Rock Valley Subordinate Grange #470, 1883–94, MSS, Box 130, New York State Grange Papers, CUL. Few Grange records have survived from the nineteenth century, but one can estimate membership roughly with extant sources. The Rock Valley

Grange had 26 charter members, and another 122 Grangers joined between 1883 and 1894. Assuming that other Granges enjoyed similar growth, the seventy-eight chapters in Delaware, Otsego, and Schoharie Counties (with over 1,600 charter members) would have drawn nearly 8,000 members into the organization by 1894.

11. Baker, *Moral Frameworks of Public Life,* 8–14, 20–55; Americanus, *The Grange,* 83–97; Benson, *Merchants, Farmers, and Railroads,* 80–93.

12. Baker, *Moral Frameworks of Public Life,* 3–154; Benson, *Merchants, Farmers, and Railroads,* 97–98; *Delaware Dairyman,* March 19, 1886; Cooperstown *Freeman's Journal,* September 19, November 21, 1878; Sharon Good, "The Hop Culture" (M.A. thesis, SUNY College at Oneonta, Cooperstown Graduate Program, 1968), 38–60, 78–82.

13. Quote from McArthur, *New Developments,* 52; Baker, *Moral Frameworks of Public Life,* 60–65; Buck, *Granger Movement,* 281; Osterud, *Bonds of Community,* 254–62; Donald B. Marti, "Sisters of the Grange: Rural Feminism in the Late Nineteenth Century," *Agricultural History* 58 (April 1984): 247–61; Donald B. Marti, "Woman's Work in the Grange: Mary Ann Mayo of Michigan, 1882–1903," *Agricultural History* 56 (April 1982): 439–52.

14. *Transactions of the New York State Agricultural Society, 1867* (Albany: C. Van Benthuysen and Sons, 1867), 185; LaWanda Cox, "The American Agricultural Wage Earner, 1865–1900: The Emergence of a Modern Labor Problem," *Agricultural History* 22 (April 1948): 95–114.

15. McArthur, *New Developments,* 52–54; Osterud, *Bonds of Community,* 1–15, 139–227; Baker, *Moral Frameworks of Public Life,* 90–118; Aileen S. Kraditor, *The Ideas of the Woman Suffrage Movement, 1890–1920* (New York: Columbia University Press, 1965), 43–74, 123–62.

16. McArthur, *New Developments,* 83–84.

17. Americanus, *The Grange,* 106–47; Rothstein, "Farmer Movements," 161–81.

18. *Patrons of Husbandry Declaration of Purposes, Constitution and By-Laws of the National Grange, Constitution and By-Laws of the New York State Grange* (1908), 3–4 (pamphlet).

19. Ibid., 4; McArthur, *New Developments,* 38–47; Rothstein, "Farmer Movements," 161–81; Baker, *Moral Frameworks of Public Life,* 1–24.

20. Baker, *Moral Frameworks of Public Life,* 8–14, 20–23; McArthur, *New Developments,* 47–50.

21. McArthur, *New Developments,* 32.

22. Ibid., 32–33; Mark C. Carnes, *Secret Ritual and Manhood in Victorian America* (New Haven: Yale University Press, 1989), 8, 22.

23. *Manual of Subordinate Granges of the Patrons of Husbandry* (Philadelphia: J. A. Wagenseller, 1873), 7.

24. Ibid.; Baker, *Moral Frameworks of Public Life,* 14–19; Carnes, *Secret Ritual,* 125–27.

25. *Manual of Subordinate Granges,* 7–13.

26. Ibid., 19–25; Baker, *Moral Frameworks of Public Life,* 8–14, 20–23.

27. *Manual of Subordinate Granges,* 40.

28. James L. Orr, ed., *Grange Melodies* (Philadelphia: George S. Ferguson, 1923), 34–35.

29. Ibid., 60–61.

30. Ibid., 4.

31. Michael E. McGerr, *The Decline of Popular Politics: The American North, 1865–1928* (New York: Oxford University Press, 1986), 13.

32. Baker, *Moral Frameworks of Public Life,* 24–55.

33. *Manual of Subordinate Granges,* 15–16.

34. Ibid., 26–28.

35. Ibid., 46.

36. Orr, *Grange Melodies,* 109; Osterud, *Bonds of Community,* 249–88; Minute Books of Rock Valley, Delaware County, Subordinate Grange #470, May 17, October 4, 1890, August 20, 1892, MSS, Box 130, New York State Grange Papers; Kraditor, *Ideas of the Woman Suffrage Movement,* 96–122.

37. Baker, *Moral Frameworks of Public Life,* 55–89. See also Michael L. Goldberg, *An Army of Women: Gender and Politics in Gilded Age Kansas* (Baltimore: Johns Hopkins University Press, 1997), 39–40, for a short discussion of the conventional gender ideas of the Grange.

38. McArthur, *New Developments,* 73; Robert H. Wiebe, *The Search for Order, 1877–1920* (New York: Hill and Wang, 1967), 17; Baker, *Moral Frameworks of Public Life,* 8–14; Jackson Lears, "Packaging the Folk: Tradition and Amnesia in American Advertising, 1880–1940," in *Folk Roots, New Roots: Folklore in American Life,* ed. Jane Becker and Barbara Franco (Lexington, Mass.: Museum of Our National Heritage, 1988), 103–40.

39. McArthur, *New Developments,* 76.

40. Ibid., 47–50, 72–76.

41. Americanus, *The Grange,* 180–216; Cooperstown *Freeman's Journal,* February 5, March 12, March 26, 1874; *Delaware Dairyman,* February 5, 1886; McArthur, *New Developments,* 91–99; Minute Book of Sidney, Delaware County, Subordinate Grange #729, February 3, 1897, MSS, Box 58, New York State Grange Papers; Benson, *Merchants, Farmers, and Railroads,* 80–114; Buck, *Granger Movement,* 9–11; William F. Helmer, *O&W: The Long Life and Slow Death of the New York, Ontario & Western Railway* (Berkeley, Calif.: Howell-North, 1959), 55; Gabriel Kolko, *Railroads and Regulation, 1877–1916* (Princeton: Princeton University Press, 1965), 20–21, 232.

42. McArthur, *New Developments,* 37.

43. Ibid., 91–112.

44. Ibid.; Cooperstown *Freeman's Journal,* March 26, December 31, 1874; Wiebe, *Search for Order,* 5–7.

45. Kolko, *Railroads and Regulation;* Wiebe, *Search for Order,* 8–9; Stephen Skowronek, *Building a New American State: The Expansion of National Administrative Capacities, 1877–1920* (Cambridge: Cambridge University Press, 1982), 125–30.

46. Americanus, *The Grange,* 22–39; Baker, *Moral Frameworks of Public Life,* 24–55.

47. Benson, *Merchants, Farmers, and Railroads,* 94–114; Baker, *Moral Frameworks of Public Life,* 6–8, 20–23.

48. Buck, *Granger Movement,* 80–122; Benson, *Merchants, Farmers, and Railroads,* 80–114.

49. Baker, *Moral Frameworks of Public Life,* 3–154.

50. Ibid.; Marti, "Sisters of the Grange," 247–61; Marti, "Woman's Work," 439–52; Osterud, *Bonds of Community,* 249–74; Kraditor, *Ideas of the Woman Suffrage Movement,* 1–13, 43–74; Richard L. McCormick, *From Realignment to Reform: Political Change in New*

York State, 1893–1910 (Ithaca: Cornell University Press, 1981), 40–68, 94–98; Allen, *History of the New York State Grange,* 52, 60; Cooperstown *Freeman's Journal,* February 19, 1874; Minute Book of Rock Valley, Delaware County, Subordinate Grange #470, February 21, 1891, MSS, Box 130, New York State Grange Papers.

51. R. L. McCormick, *From Realignment to Reform,* 33–34, 37, 63, 94–95; Franklin *Register,* October 11, December 20, 1878, February 19, 1886; *Address and Resolutions at the Fourth Annual Meeting of the North American Hop Growers' Association, July 11, 1882* (1882), George Hyde Clarke Papers, #2800, CUL; Clayton E. Risley, "Hop Picking Days," *New York Folklore Quarterly* 5 (Spring 1949): 18–21; George D. Taylor, *These Hills Are Not Barren: The Story of a Century Farm* (New York: Exposition Press, 1950), 46, 112; Jared Van Wagenen Jr., *Days of My Years: The Autobiography of a York State Farmer* (Cooperstown, N.Y.: NYSHA, 1962), 137–39.

52. Cooperstown *Freeman's Journal,* November 20, 1872, November 20, 1886, November 26, 1896; Delaware *Gazette,* November 20, 1872, November 24, 1886, November 11, 1896.

53. Allen, *History of the New York State Grange,* 47, 51–52, 61; Buck, *Granger Movement,* 279–301; Baker, *Moral Frameworks of Public Life,* 24–154; Kraditor, *Ideas of the Woman Suffrage Movement,* 43–95; Osterud, *Bonds of Community,* 258–59; Marti, "Sisters of the Grange," 247–61.

54. McArthur, *New Developments,* 53; Osterud, *Bonds of Community,* 260.

55. Minute Books of Rock Valley, Delaware County, Subordinate Grange #470, May 17, October 4, December 20, 1890, April 30, August 20, September 13, December 3, 1892, December 2, 1893, July 31, August 4, December 15, 1894, MSS, Box 130, Minute Books of Sidney, Delaware County, Subordinate Grange #729, October 14, November 21, December 5, 1891, December 17, 1892, March 7, December 19, 1893, January 16, December 17, 1894, May 21, December 4, 1895, October 21, December 2, 1896, December 1, 1897, December 7, 1898, December 13, 1899, December 12, 1900, MSS, Box 58, New York State Grange Papers; Baker, *Moral Frameworks of Public Life,* 68–75; Osterud, *Bonds of Community,* 260–62, 275–88; Marti, "Sisters of the Grange," 247–61.

56. McArthur, *New Developments,* 35; Buck, *Granger Movement,* 64.

57. Baker, *Moral Frameworks of Public Life,* 45–55; Allen, *History of the New York State Grange,* 39–146; Buck, *Granger Movement,* 71–73, 80–122; Benson, *Merchants, Farmers, and Railroads,* 80–93.

58. McArthur, *New Developments,* 5.

59. Osterud, *Bonds of Community,* 53–85, 231–74; Wiebe, *Search for Order,* 9.

60. McArthur, *New Developments,* 109–12; Americanus, *The Grange,* 180–216.

61. Benson, *Merchants, Farmers, and Railroads,* 108–9, 115–32, 172–73; R. L. McCormick, *From Realignment to Reform,* 26.

62. Allen, *History of the New York State Grange,* 60; Secretary's Book, West Exeter, Otsego County, Subordinate Grange #824, March 8, 1902, MSS, Box 134, Minute Books of Rock Valley, Delaware County, Subordinate Grange #470, September 20, 1890, August 18, 1894, MSS, Box 130, New York State Grange Papers.

63. Quote in Allen, *History of the New York State Grange,* 50–51; Baker, *Moral Frameworks of Public Life,* 20–23; Elmer O. Frippen, *Rural New York* (New York: Macmillan, 1921), 263, 267; Osterud, *Bonds of Community,* 260.

64. Allen, *History of the New York State Grange,* 46, 51, 55, 120–24; Buck, *Granger Movement,* 238–78.

65. Minute Books of Rock Valley, Delaware County, Subordinate Grange #470, May 2, 1891, March 5, December 17, 1892, January 7, 1893, January 6, May 19, June 30, November 17, December 1, 1894, February 15, 1895, February 20, 1897, MSS, Box 130, Minute Books of Sidney, Delaware County, Subordinate Grange #729, March 31, June 3, June 17, September 2, November 21, December 12, 1891, April 18, 1893, March 18, June 17, October 21, 1896, September 21, 1898, April 26, June 28, October 11, 1899, MSS, Box 58, New York State Grange Papers; Baker, *Moral Frameworks of Public Life,* 20–23; Osterud, *Bonds of Community,* 260.

66. McArthur, *New Developments,* 118–26.

67. Allen, *History of the New York State Grange,* 47–48, 53, 55, 60, 133–37; Benson, *Merchants, Farmers, and Railroads,* 80–114; Cooperstown *Freeman's Journal,* March 19, April 2, 1874, December 20, 1878; Minute Book of Sidney, Delaware County, Subordinate Grange #729, December 1, 1897, MSS, Box 58, New York State Grange Papers.

68. Allen, *History of the New York State Grange,* 47–48, 53, 55, 60, 133–37; Benson, *Merchants, Farmers, and Railroads,* 80–114.

69. Secretary's Book, West Exeter, Otsego County, Subordinate Grange #824, August 27, 1902, May 13, June 24, November 25, 1903, July 27, 1904, MSS, Box 134, Minute Books of Sidney, Delaware County, Subordinate Grange #729, November 7, 1893, June 19, 1894, November 3, 1897, July 6, December 21, 1898, June 14, 1899, MSS, Box 58, Minute Books of Rock Valley, Delaware County, Subordinate Grange #470, February 15, 1890, June 20, 1891, MSS, Box 130, New York State Grange Papers; Buck, *Granger Movement,* 279–301.

70. Minute Book of Sidney, Delaware County, Subordinate Grange #729, January 16, 1892, MSS, Box 58, Minute Book of Rock Valley, Delaware County, Subordinate Grange #470, April 1, 1893, MSS, Box 158, New York State Grange Papers; *Delaware Dairyman,* March 5, 1886; Baker, *Moral Frameworks of Public Life,* 3–154.

71. Secretary's Book, West Exeter, Otsego County, Subordinate Grange #824, October 23, November 27, December 28, 1901, November 26, 1902, MSS, Box 134, Minute Books of Sidney, Delaware County, Subordinate Grange #729, May 15, 1891, October 1, 1892, MSS, Box 58, Minute Books of Rock Valley, Delaware County, Subordinate Grange #470, August 19, 1890, September 17, 1892, January 30, April 5, 1897, MSS, Box 130, New York State Grange Papers; Baker, *Moral Frameworks of Public Life,* 20–23; Jeffrey W. Moss and Cynthia B. Lass, "A History of Farmer's Institutes," *Agricultural History* 62 (Spring 1988): 150–63.

72. Baker, *Moral Frameworks of Public Life,* 119–54; Allen, *History of the New York State Grange,* 56, 76, 138–46; Secretary's Book, West Exeter, Otsego County, Subordinate Grange #824, February 25, 1903, July 27, 1904, MSS, Box 134, Minute Book of Sidney, Delaware County, Subordinate Grange #729, April 14, 1904, MSS, Box 58, New York State Grange Papers; James C. Mohr, *The Radical Republicans and Reform in New York during Reconstruction* (Ithaca: Cornell University Press, 1973), 153; Alan I. Marcus, "The Ivory Silo: Farmer-Agricultural College Tensions in the 1870s and 1880s," *Agricultural History* 60 (Spring 1986): 22–36.

Conclusion

1. Solon J. Buck, *The Granger Movement: A Study of Agricultural Organization and Its Political, Economic, and Social Manifestations 1870–1880* (Cambridge: Harvard University Press, 1933), 302–12; Lee Benson, *Merchants, Farmers, and Railroads: Railroad Regulation and New York Politics, 1850–1887* (Cambridge: Harvard University Press, 1955), 150–73.

2. Thomas A. Woods, *Knights of the Plow: Oliver H. Kelley and the Origins of the Grange in Republican Ideology* (Ames: Iowa State University Press, 1991), 3–21.

3. Jabez D. Hammond, *The History of the Political Parties of the State of New York* (Syracuse: Hall, Mills, and Co., 1852), 1:488–89.

4. *The Softs: The True Democracy of the State of New York* (New York: n.p., 1856), 33.

5. Jared Van Wagenen Jr., *Days of My Years: The Autobiography of a York State Farmer* (Cooperstown, N.Y.: NYSHA, 1962), 60–141.

6. Elizabeth Sanders, *The Roots of Reform: Farmers, Workers, and the American State, 1877–1917* (Chicago: University of Chicago Press, 1999), 409–19; Hal S. Barron, *Mixed Harvest: The Second Great Transformation in the Rural North, 1870–1930* (Chapel Hill: University of North Carolina Press, 1997), 7–16; Linda G. Ford, "Another Double Burden: Farm Women and Agrarian Activism in Depression Era New York State," *New York History* 75 (1994): 373–96; Emelyn E. Gardner, *Folklore from the Schoharie Hills, New York* (Ann Arbor: University of Michigan Press, 1937), 47–84; Fred Lape, *A Farm and Village Boyhood* (Syracuse: Syracuse University Press, 1980), 1–15, 69–82, 113–32; Jeremy Atack and Fred Bateman, *To Their Own Soil: Agriculture in the Antebellum North* (Ames: Iowa State University Press, 1987), 10–14.

Index

THOMAS SUMMERHILL is an associate professor of history at Michigan State University. He is the coeditor with James C. Scott of *Transatlantic Rebels: Agrarian Radicalism in Comparative Context.* He has held fellowships at the Yale University Program in Agrarian Studies and at the National Museum of American History, Smithsonian Institution. In addition to authoring other publications in agrarian history, he has served as a consultant for museums and government agencies engaged in historic preservation.

The University of Illinois Press
is a founding member of the
Association of American University Presses.

Composed in 10.5/13 Adobe Minion
at the University of Illinois Press
Manufactured by Thomson-Shore, Inc.

University of Illinois Press
1325 South Oak Street
Champaign, IL 61820-6903
www.press.uillinois.edu